阿科米星
Atelier Archmixing 2009–2019

阿科米星 著
Atelier Archmixing

同济大学出版社
Tongji University Press

改变是一个常识

Change is a Common Sense

目录
Contents

综述
Review

阿科米星：朝向日常与变化的实践

Atelier Archmixing: Practice Toward Everydayness and Change

李翔宁

LI Xiangning

当代中国建筑经历了狂飙式的 40 年发展，创造出了全球瞩目的成就，其成果既体现在数量上，也体现在质量上。在我看来，中国众多创造出优秀作品的建筑师们，正在努力把当代中国建筑提升到一个更高的层面，找到中国建筑特有的道路。在这些努力中，或许存在两条路径或者说两种实践方式：一是在建筑设计最硬核的设计语言上持续不懈地探索，二是在中国社会与城市文化的当代性命题中发掘突破固有建筑学学科内涵的养料。

事实上，这两条路径也是 20 世纪以来的建筑师们为自己的职业生涯确定方向时必须直面的选择。众多被中国建筑师们当作职业生涯参照系的建筑先贤们，如可以归入前者的路易·康、斯卡帕、卒姆托，或是可以归入后者的柯布西耶、文丘里、库哈斯，都成为在这两条道路上前行的重要灯塔。

当然，在当代中国建筑的语境中，这两条看起来给建筑师们期许与允诺的道路上同样也布满着荆棘和危险。前者容易使人堕入形式主义的巢臼，或者过度个人风格化的泥沼；而后者则存在和消费文化同流合污，或者脱离建筑学的范畴而成为政治正确的社会运动的鼓吹者。

在 21 世纪第三个十年的开端回望，或许会发现过去的 40 年间在这两条道路上真正深耕并有独特建树的中国建筑师并不多。我们既缺少像西扎或 RCR 那样数十年如一日坚持发展一种建筑形式语言的探索，又缺少像 FAT 或者犬吠工作室这样植根当代文化并给予批判性呈现的建筑实验。而庄慎和阿科米星在过去的十年间显然走在了后一条道路上，用他们自己的话来说，“从纷繁复杂的日常城市和建筑中发现建筑学的新经验”，而正是这种“在中国社会不确定的语境中探索建筑学还能做什么”，也正“意味着开放的边界和更多的可能”。[1]

记得多年前一次出差前往阿那亚的途中，同行的一名上海建筑师让我一个个点评当代中国的明星建筑师。虽然没有办法根据建筑师们的实践水准排个座次，但当被问到哪些建筑师独辟蹊径开创了有价值的新方法时，我第一个想到的就是庄慎。今天想来我当时给出的理由是非常直接和感性的，“因为庄慎的设计中，有着当代中国建筑师群体很少有的特殊状态——轻”。这不是指结构意义上的轻盈，而是指一种态度——他既不像一些建筑师孜孜以求醉心于创造完美的形式，也不像另一些为“理论化焦虑”所扰的建筑师穷经皓首为自己

With forty-year rapid development, contemporary Chinese architecture has made remarkable achievements that have received global attention and reflected well both on quantity and quality. In my opinion, brilliant architects in China are working hard to push contemporary Chinese architecture to a new height with their outstanding buildings, and paving a unique path of development for Chinese architecture. There may be two paths or practice ways in this endeavor: a sustained exploration on core architectural design languages, and an exploration on resources that may overturn conventional architecture discipline in the context of contemporary Chinese society and urban culture.

In fact, since the 20th century, architects would inevitably pick one of these two paths in their career planning. Architects like Louis I. Kahn, Carlo Scarpa, Peter Zumthor picked the former path, while the others like Le Corbusier, Robert Venturi, Rem Koolhaas picked the latter. All these prominent predecessors are guides and lighthouses for Chinese architects.

However, in the context of contemporary Chinese architecture, great difficulty and danger are also found on these two promising career paths. On the first path, architects may easily fall into formalism or indulge in excessive personal style, while on the second, they tempt to go with the stream of consumer culture, or rather lose touch with architectural discipline and become a politically correct advocate for social movement.

When entering the third decade of the 21st century and looking back upon the past 40 years, it would be hard to find enough Chinese architects on either path who have stayed focused and scored unique achievements. We are starved of Alvaro Siza's or RCR's (Rafael Aranda, Carme Pigem & Ramon Vilalta Arquitectes) persistence on developing a formal language in architecture, as well as FAT's (Future Architecture Thinking) or Atelier Bow-Wow's architectural experiments which are rooted in contemporary culture and critically presented. Zhuang Shen and Atelier Archmixing obviously picked the second path during the past decade. In their own words: “To extract new architectural experience from complex everyday city and buildings.” The question “what else can architecture achieve in this uncertain context of Chinese society” exactly “means a broader range of exploration and more possibilities”.[1]

On a business trip to Aranya many years ago, I was asked by an architect from Shanghai to comment on contemporary Chinese starchitects. I failed to rank them by their practices, but when he asked who has invented a new way of creating valuable working methods, the first name that

的作品寻找理论阐释和附加意义。他不拘泥于形式的纯净统一，甚至用不同材料的拼贴和并置，来完成每一次场所营造的实验，因地制宜的形态和举重若轻的设计策略，让他的建筑具有一种舒适宜人的在地性。

这种印象应该直接来自 2008 年我在成都青城山策划的建筑师集群项目，八位建筑师每人设计八座联排院宅中的一座。在那个项目中，庄慎提交了一个诗意的设计——“一院一世界”[图1]：非常节制、有限的形式手段，经过巧妙控制的、压低了的视线，让青城山完全被收入建筑的主院落，而狭长的水院又将青城山景倒映其中。这好像是我第一次直接和庄慎共事的建筑项目。

1 一院一世界
A Court, A World

仔细想来，庄慎和阿科米星的作品在当代中国建筑实践中占据的独特位置，在直观感受层面的“轻”，只是一种感觉可以进入的表象，而在这种举重若轻的建筑态度背后，或许可以解读出一种对价值的思考。正如卡尔维诺在《新千年文学备忘录》中指出的，“轻”相对于试图赋予语言力量、密度，以及事物、形体和感觉的具体性的方法而言，是“一种试图把语言变成无重量的元素，它像一朵云那样飘浮在事物的上空”，但这种轻“是与精确和坚定为伍，而不是与含糊和随意为伍”。[2] 或许轻松和随性的天性，使他不愿意将建筑师视为一个居高临下、曲高和寡的职业，这也使得在明星建筑师们成为大众传媒的宠儿而自觉不自觉地以生活方式和潮流引领者自居的今天，庄慎的轻装简行甚至有点不修边幅的形象，和他对于建筑实践的态度，对于“作为个人的建筑师及其设计作品与大众真实生活之间的关系日渐疏离”[3] 的认识有着准确的对位关系，从而发展出一套基于日常生活场景与消费文化逻辑的设计策略。

或许也正是出于这样的原因，在本文开头提到的两种实践路径或策略之间，庄慎选择了后者。今天的中国建筑师们，在经历了快速建造也快速被消费的几十年之后，意识到中国需要严肃认真的建筑学。同样出于对中国建筑学从未真正经历过现代主义启蒙这一现实的痛悟，许多真正认真思考建筑学命题的建筑师们重新回到现代主义的广袤田野深耕细作，发掘着诸如柯布西耶或路易·康的资源，不自觉地走向第一条道路，重新回到建筑学学科的核心议题如空间、结构、材料，乃至弗兰姆普敦为反对后现代和消费文化所祭起的建构文化大

came to my mind was Zhuang Shen. The reason I gave seems quite straight-forward and emotional today. “Because Zhuang Shen’s designs contain a state, lightness, which is rarely found in contemporary Chinese architects.” This doesn’t mean that he applied light structures in his design, but refers to his working attitudes. Zhuang Shen has never been lost in assiduously creating perfect forms, neither has he been eager to seek theoretical interpretation and additional value for his own work like those architects affected by theoretical anxiety. All his buildings have been comfortably localized by never confining the work to a unified form, miraculously completing architectural experiments with the collage and juxtaposition of varied materials, applying architectural forms fit for local conditions and developing effective while easy design strategies.

This impression is derived from the group design project I initiated in Mount Qingcheng, Chengdu (2008). Each of the eight architects was supposed to design one of the eight townhouses with courtyards. Zhuang Shen submitted a poetic design, “A Court, A World” [Fig. 1]. With quite temperate and limited approaches, when people’s sights were cleverly lowered, Mount Qingcheng was entirely absorbed in the main court and reflected in the narrow water court. This might be the first time I cooperated directly with Zhuang Shen in a building project.

Both Zhuang Shen’s and Atelier Archmixing’s work dominated a unique position in contemporary Chinese architectural practice. Considering that “lightness” contained in their work in an intuitive sense may be only a perceptual representation, behind this effective and easy architectural attitude is a reflection on values. As Italo Calvino wrote in *Lezioni Americane*, “In comparison with the tendency that tries to give language the weight, density, and concreteness of things, bodies, and sensations, ‘lightness’ tries to make language into a weightless element that hovers above things like a cloud.”[2] But such “lightness” “go with precision and determination, not with vagueness and the haphazard”. Possibly because of his relaxing and causal style, he is unwilling to be a condescending architect who is too talented to be appreciated. In this day and age, as mass media darlings, starchitects voluntarily or involuntarily regard themselves as leaders of lifestyle and trend. Zhuang Shen’s casual and slovenly image has been well matched with his attitudes toward architectural practice and his awareness of “the increasing dissociation of individual architects and their designs from the public and their real lives”.[3] Therefore, he successfully developed a series of design strategies based on everyday scenes and logic of consumer culture.

Perhaps it is for this reason that Zhuang Shen picked the second of these two practical paths

旗，或多或少都带着向现代主义回归的意味。正如前文所述，第一条路径的危险是脱离了当下中国社会现实的极端挑战，而回到西方近一个世纪前的文化语境下所创造的“革命性”的现代语言，一种在柯布身后可能是以形式的语言固化下来的现代性的残影，并试图用这种纯净的现代主义语言对抗后现代与消费文化的肆虐。

然而，正如马歇尔·伯曼为他讨论现代性体验的名著《一切坚固的东西都烟消云散了》的新版撰写了序言《宽广开放的理解方式》，并认为“宽广开放的理解方式仅仅是许多可能的理解方式之一，但它确具有一些优点……这种理解方式为在过去、现在与将来之间展开对话创造了条件。它跨越了物理的和社会的空间，揭示出伟大的艺术家与普通老百姓之间的统一，它拓宽了我们对自己经验的理解，向我们表明我们的生活中存在着比我们原有的设想更多的东西，赋予我们的日常生活一种新的广度与深度”。他始终认为，“任何一种现代主义的模式都不可能是最终的、不可变更的。对于排斥或敌视变化的现代主义而言，更确切地说，对于寻求一个伟大的变化而后不再寻求任何变化的现代主义而言，这个问题尤为尖锐”[4]。

2　同济大学中德学院
Sino-German College, Tongji University

相对于这种回到曲高和寡的现代主义正统的方式（事实上庄慎早年的作品同济大学中德学院[图2]就已经在柯布式现代主义的语汇方面有上乘的表现），在过去的十年，庄慎似乎更愿意从一些商业性的项目入手，更好地理解日常生活和消费文化的逻辑。面对消费文化，既不能与其同流合污，又不能转身装作视而不见，与其被动地被消费文化所主宰，不如创造性地发掘其中的力量。这和罗兰·巴特对大众文化的态度或许有所共鸣。

作为对都市现实和消费文化的回应，近十年来，阿科米星完成了一系列城市类型的项目，这或许被庄慎认为是“商业项目促使我们真正认真地开始关注当今社会以及消费文化的逻辑、大众文化的现象与规律，并认识到关乎建筑师社会理想的现实对象与内容”[5]。庄慎非常敏锐地感受到，当代中国社会的发展造就了一个在政府项目之外的社会，或者说是商业开发的业主群落，他们正视商业价值创造的诉求并尊重社会和人文价值的体现，阿科米星很审慎地挑选甲方和项目，可以更多地在项目策划的前期就提供自己的建议，这些对于现实的考量最后常常体现在项目独特的任务书中，这也是庄慎所追求的“混合”，拓展而非自我封闭建

or strategies mentioned at the beginning of this article. Having gone through decades of rapid construction and being quickly consumed, Chinese architects in this age have realized that China needs a serious architecture. Also, out of the fact that Chinese architecture has never been enlightened by modernism, many architects who really think seriously about architecture started fresh and hard work on modernism, exploring Le Corbusier's or Louis I. Kahn's resources. They unwittingly set out on the first path, focusing on key issues of architecture discipline: space, structure, material and even Kenneth Frampton's banner of tectonic culture against postmodernism and consumer culture. All this has implied more or less the return of modernism. As previously mentioned, the danger on the first path lies in ignoring extreme challenges of contemporary social realities in China, going back to the "revolutionary" modern language created in the cultural context a century ago, as a sort of modern ghost shadow probably fixed by the formal language after Le Corbusier, and trying to combat the raging postmodernism and consumer culture with the pure language of modernism.

However, exactly as in the preface "The Broad and Open Way" written by Marshall Berman for the new edition of his masterpiece *All That Is Solid Melts into Air*, a book discussing the experience of modernity, "the broad and open way is the only one of many possible ways, but it has advantages... It creates conditions for dialogue among the past, the present and the future. It cuts across physical and social space, and reveals solidarities between great artists and ordinary people. It enlarges our vision of our own experience, shows us that there are more to our lives than we thought, and gives our days a new resonance and depth." He always holds that "no mode of modernism can ever be definitive. This problem is especially acute for a modernism that forecloses or is hostile to change, or rather, a modernism that seeks one great change, and then no more".

3　上海文化信息产业园一期B4/B5地块
B4/B5 Blocks of Shanghai Culture & Information Industrial Park, Phase I

In comparison with returning to the highbrow orthodox modernism (in fact, with his early work Tongji University Sino-German College Building [Fig. 2], Zhuang Shen did quite well in developing the Corbu-style Modernism), he seemed to be more than willing to better understand everyday life and the logic of consumer culture through commercial projects during the past ten years. He can neither go along with consumer culture, nor can he turn a blind eye to it. So he would explore the power it contains in a creative way rather than being dominated. Perhaps, it is a resonance with Roland Barthes's attitude toward mass culture.[4]

In recent ten years, Atelier Archmixing has responded to urban reality and consumer cul-

筑师能量的边界，用组织者的角色替代传统的设计师的角色，从而真正找到一条打通个人化的设计思维与大众生活现实的管道。从庄慎和阿科米星的项目中可以看到突破传统的建筑组织方式和类型、切实呼应建筑介入城市的期待。

在上海文化信息产业园一期 B4/B5 地块的项目中[图3]，建筑师希望创造不同高度、有私密感的庭院，采用了“悬挂的庭院”这一核心概念。通过将“庭院”悬挑在矩形办公单元楼不同高度和不同的位置这一简单的动作，就为园区带来了开放的地面空间和丰富的空间体验。在细部构造上，建筑非常适宜地选择了简单高效的策略。悬挑庭院的包裹材料经过长时间的推敲，选用了类似江南园林木格窗的比例，使人在很远的地方也能感受到一种半透明的效果。为了追求建筑体量的简洁轻盈，建筑师将建筑的窗做成外平，这种做法一般需要较为精密的特殊构件才能实现。为了配合标准建筑构件和传统安装门窗的做法，施工时将一个混凝土垛头浇筑在窗洞外，通过立面保温材料将厚度处理平整。这样一个简单的土建处理取得了非常好的视觉效果，同时可以保证不因为外平的窗发生渗水，这种高效和具有针对性的策略同样出现于阿科米星后来的项目中。

张江集电港三期华鑫慧天地[图4]是阿科米星新近完成的一个从城市设计到建筑单体的复杂建筑群落。混合多样的建筑类型使得整个园区具有了特殊的空间活力。而这些大小不同的建筑体，都围绕着一个基本的网格，从平面布局到建筑立面的构成，都呈现一种像素化的特征。建筑体量通过底层退让和大跨度的悬挑，形成了丰富多变的小型广场，使得建筑群具有了某种微型城市的场所气质。主体建筑立面上严格的网格和随机布置的凸窗之间形成了有趣的张力，同样小建筑群立面上的视觉张力则来自三种不同尺度的开窗形式的变换。

在另一个领域，阿科米星用实践证明了另一种建筑介入社会的方式。轻巧的、快速的、易于操作且效果显著的设计涌现出来。位于衡山路 890 弄（衡山和集）的发光砖墙[图5]、徐汇龙华街道敬老院外立面[图6]与棉仓城市客厅[图7]是这一类城市改造项目的代表作品。即便只是立面改造，阿科米星的作品也早已超越了“换个皮”的层级，并在使用、空间和技术等层面上给出丰富可能性。

4　张江集电港三期华鑫慧天地
Zhangjiang IC Harbor Phase III China Fortune Wisdom Mark

ture with a range of urban projects. According to Zhuang Shen, “it was commercial projects that prompted us to seriously pay attention to the logic of today’s society and consumer culture, to the phenomena and principle of popular culture, and to find out real objects and contents relating to architects’ social ideals.”[5] Zhuang Shen keenly sensed that the development of contemporary Chinese society had made a group of clients who were developing social or commercial projects instead of governmental ones. They acknowledge the pursuit of commercial value and respect the reflection of social and humanistic values. Atelier Archmixing picks its customers and projects with extra prudence, seeking the possibility to provide suggestions in the early stage of planning. These considerations on reality can usually be found in the special program. Zhuang Shen has been seeking a “mixing” strategy, to extend instead of self-limit architect’s working boundary, to move from the role of a conventional designer to an organizer, and explore a connection between personalized design thinking and public life reality essentially. In projects designed by Zhuang Shen and Atelier Archmixing, we can find a breakthrough in the architectural organization mode and types, which echoes the expectation for intervening urban issues through architectural practices.

In the project B4/B5 Blocks of Shanghai Culture & Information Industrial Park, Phase I [Fig. 3], the architect applied “Hanging Courtyard” as a key concept to create courtyards of different heights with a sense of privacy. The park was provided with open ground space and rich spatial experience by hanging these “courtyards” at different levels and positions of the rectangular office unit. Simple and effective strategies were also applied in detail. After a long discussion on the coating material of the “Hanging Courtyard”, the architect took reference of wooden lattice windows in traditional garden buildings in regions south of the Yangtze River and applied a similar hole scale to achieve a sense of semi-transparency from a distance. By making a neat and light volume, windows were flush with the surface. Instead of applying fine special joints to create this effect, the architect followed the traditional construction technique and standard window hardware components, only adding a concrete buttress head outside the window opening, then filling up the surface with facade insulation material after installed the windows. Such a plain construction method not only achieved satisfying visual effect but also avoided water leakage of the window. Similar efficient and targeted strategies have also been evident in Atelier Archmixing’s later projects.

5　衡山路890弄（衡山和集）8号楼外立面改造
Facade renovation for No. 8 Building, Lane 890, Hengshan Road

Zhangjiang IC Harbour Phase III China Fortune Wisdom Mark [Fig. 4] is an architectural complex

6 徐汇区龙华街道敬老院立面改造
Facade Renovation for Longhua Street Elder Care Center

商业空间与建筑艺术本不是一对矛盾的词语，但在当代中国建筑实践中，经常会被对立地看待。商业项目似乎被视作炫目浮夸的代名词，艺术项目则和现实保持着冷漠的距离。衡山路 890 弄（衡山和集）8 号楼外立面改造很好地给出了可以平衡两者的设计，项目所在地衡山坊历史街区紧邻衡复历史风貌区的边界，置身一条时尚的商业步行街。虽然整体保持了原有的花园洋房特征，但对建筑立面的控制相对宽松。这让立面的变化成为可能，建筑师正是抓住了这一可能性，设计了符合商业气质的会发光的砖墙，在同质性的砖墙中植入了完全异质的元素：同样尺寸的砖，传统的青砖被建筑师以特殊定制的发光砖取代。白天不起眼的砖墙到了夜间透出斑斓的光彩，创造了令人赞叹的特殊场所气氛。为此阿科米星研发了新的发光砖构造，表面的拓彩岩石板在白天与青砖无异，夜晚则会透光，形成有质感的发光体。昼夜更替，建筑也具有了两种不同的表情。变化的建筑面貌在满足建筑商业氛围的同时，也达到了装置艺术的效果，并营造了衡山坊区域的标识性。如果说庄慎对于城市消费文化的关注和思考还能在他为数不多的商业建筑中体现的话，这栋建筑一定是具有代表性的。

徐汇龙华街道敬老院外立面则由一个简单的外立面改造项目，变为具有实质意义的空间改造项目。原本单独的阳台被打通连接，文化信息产业园“悬浮的庭院”被“拿来”变成多组公共活动空间。在收拾建筑立面的同时，建筑师零敲碎打地选取每个适合的位置进行空间的置入，就使原本做个表皮的工作变得有厚度和有实用意义。

在棉仓城市客厅项目中，阿科米星对既有工业建筑的改造提出了“屋中屋”的概念。为了给服装公司提供一个成衣和餐饮品牌的线下体验店，建筑师将两个由钢结构和钢木结构组成的坡屋顶玻璃体量置入工厂内，通过将建筑、结构和设备进行整合，棉仓城市客厅可以在大跨度的工厂内提供稳定的室内环境。我们又看到了突出建筑体量的、提示性的“悬挂庭院”和立面的色彩划分，对于基本手段的反复使用被看作十分重要的建筑设计方法。

棉仓城市客厅的组成方式其实来自阿科米星在上海新天地设计的临时读书空间［图 8］，相同的架构既可以用于城市开放空间，也可以放入高大的工厂空间，建筑师能够控制构架的预算和把握建造的速度，甚至有一定产品化的倾向。现代主义萌芽期德意志制造联盟的工作，

7 棉仓城市客厅
Cotton Lab Urban Lounge

8 新天地临时读书空间
Temporary Reading Pavilion in Xintiandi

newly completed by Atelier Archmixing, from urban design to a single building proposal. A mix of building types invigorated the place with spatial vitality. However, these buildings of different sizes were designed based on a basic grid. Both the architectural layout and the building elevation presented a feature of pixelation. Small multi-functional plazas were formed between individual buildings through setting back the ground floor and applying long-span cantilever. The whole complex looks like a miniature city. For those major buildings, there is an interesting tension between the strict grid on the main elevations and the randomly designed bay windows, while the visual tension on the small building group comes from three different ways in which windows are opened.

In another field, Atelier Archmixing has proved with many practices a different architectural intervention into the society. These practices are light, rapid, simple to operate and dramatically effective, among which typical urban renovation cases include the illuminated brick wall located on Lane 890, Hengshan Road [Fig. 5], the renovation of the Longhua Street Elder Care Center [Fig. 6] and the Cotton Lab Urban Lounge [Fig. 7]. Even for facade renovation, Atelier Archmixing has done more than changing a “skin”, that rich possibilities in use, space and technology have been explored.

There is no inherited conflict between commercial space and art of architecture, but they are often regarded as two opposite words in contemporary Chinese architectural practice. On the one side, the commercial project seems like a synonym of vertigo and exaggeration; on the other side, the art project keeps indifferent distance with reality. In the design of the Facade Renovation for No.8 Building, Lane 890, Hengshan Road, the architect has found a great balance between them. The project belongs to a boutique commercial area, Hengshanfang, a historical renovation plot adjacent to Hengshan Road-Fuxing Road Historical and Cultural Area. Although the garden villas of the district must be preserved, there were few restrictions for the facade renovation. The architect took advantage of this possibility of surface change to design a commercially valuable luminous brick wall, integrating the homogeneous wall with heterogeneous elements. In addition to conventional black bricks, a customized luminous brick of the same size was introduced. A normal brick wall in daytime starts flashing and creates a luring atmosphere when night falls. In order to make this possible, Atelier Archmixing has developed a new luminous brick. The surface photosensitive mineral resin sheet looks just like black bricks during the day time, but at night, it turns to be a luminous object with rich textures. Days and nights, the building obtains two different facial expressions. This changing architectural appearance has not only succeeded in an attractive com-

以及他们将工业化生产的体系和建筑设计的变革紧密联系起来的倡议在今天的中国依然成立，一个工业生产的大国和建筑建造的大国如何打破两者的壁垒，这和庄慎所关注的建筑作为科技的使用端被技术化的可能有着密切的因果联系。[6]

如果说当代建筑中商业文化和城市开发所要求的视觉主导的经验可以作为建筑现实的一个面向，那么与之相生相克的另一个面向则是日常性的实践。这种源自日复一日的生活实践的最普通的行为，却在文丘里那里和一种奇观化的都市现实相勾连，使我们能够理解像拉斯维加斯那样疯狂的城市背后却遵循日常生活基本而简单的理性逻辑。对庄慎而言，对抗商业和消费侵蚀的一种力量或许也来自身边日常性的都市实践，以细微的甚至带有现象学意味的敏感来珍视这种流水般静默的空间实践。

在第 16 届威尼斯国际建筑双年展中国国家馆中，“我们的乡村”主题展展出了阿科米星设计的位于南京桦墅周冲村的乡村工作室［图 9］，与其他项目一起，作为当代中国建筑乡村实践文化类建筑的代表，引起广泛关注。桦墅乡村工作室将两座旧房改建为乡村书院和活动室，如果熟悉庄慎前期作品的话，可以再次找到“悬挂的庭院”的身影。一个架高的观景台和一个弯弯的露台创造了两个物质化的庭院，用以重新梳理建筑室内外的关系。如果说桦墅乡村工作室的操作是一种加建和对比的处理，那么双栖斋和黎里则构筑出了独特的整体性乡村建筑艺术。空间、光线、材质、装饰的潜质都被建筑师敏锐地挖掘出来。

9　桦墅乡村工作室
Huashu Rural Studio

在双栖斋中［图 10］，可以明确的一件事是，在乡村地区追求建筑精致品质的疲惫感在对空间氛围的精准把控中消失殆尽。建筑由一个 56 平方米的猪圈改造而来，围绕场地中的树木，使用拆除下来的旧砖做墙。屋顶略带坡度，用来支撑反梁的结构柱与树木一起，并置了人工和自然的景致。很难去定义它到底是一栋房子还是一个庭院，压低的屋顶给树木更为透气的空间，也使内部的明暗变化愈发明显。同样向屋顶争取资源的还有黎里，建筑原本是一处被民宅包围的小厂房，在对多稿新建方案并不满意和控制造价的要求下，建筑师选择直接对老建筑进行改造。通过对入口流线的调整，建筑将原本南侧的卫生间和主要的大空间进行了分区处理，人们先由昏暗的备弄进入一个内凹的门斗，紧接着是一个具有舞台感的双坡顶空

mercial atmosphere, but also an art effect of the installation. An iconic identity is thus achieved for Hengshanfang. This project is a vivid representative of Zhuang Shen's introduction of his concern and consideration of urban consumer culture into the few commercial projects he has proposed.

In the project Facade Renovation for Longhua Street Elder Care Center, the architect successfully changed the simple facade renovation into a significant spatial renovation. All isolated balconies were cross connected, and the idea of "Hanging Courtyard" used in the project Shanghai Cultural and Information Industry Park was copied to create various sets of public activity space. While resetting the facade the architect utilized every adequate place to design a new space. The project was more than a renovation of facade, and it turned out to be a piece of meaningful and practical work.

10　双栖斋
Twin Trees Pavilion

In the project of Cotton Lab Urban Lounge, Atelier Archmixing put forward the concept of "House in House" in the renovation of this existing industrial building. To provide the clothing company with an offline brand experience store for clothing and catering, the architect designed two glass volumes in the factory, one with a steel structured sloping roof and the other with a timber-steel structured sloping roof. With the integration of building, structure and equipment, Cotton Lab Urban Lounge furnished the factory with a stable indoor environment. In this project, once more we found that the suggestive "Hanging Courtyard" and color scheme on the facade played a part in highlighting the building volume. The repeated use of basic approaches is seen as an extremely important design method in architecture.

The composition of Cotton Lab Urban Lounge was drawn from Atelier Archmixing's other project Temporary Reading Pavilion, Xintiandi, Shanghai [Fig. 8]. The same structure could be used in an urban open space as well as a tall and big plant space. Architects could control the construction budget and speed, even gravitated toward productization. In today's China, the work of Deutscher Werkbund in the infancy of modernism and its initiative, which advocated combining the industrial production system with the architectural design revolution, still makes sense. The thinking of breaking down barriers between industrial power and a building giant has directly provoked Zhuang Shen's concern about the technologization of architecture as the user's end of technology.[6]

If the visual-oriented experience required by commercial culture and urban development constitutes one side of architectural reality, the practice on everydayness must be the other side of the coin. However, Robert Venturi has linked such perfectly ordinary behavior originated from

间。室内被定义为次暗，氛围的营造就全靠点状的天窗，大小不一的 40 个亮子将自然光引入室内。而屋顶的乐趣可以通过民宿部分体验——庄慎在一次对谈中提到了儿时对于外婆家江南小镇的记忆，人与并不常接触的瓦屋面产生了近距离的关系。平台上的符号与提亮的窗口组成了一个巨大的屋顶装置，这是黎里最为动人的瞬间，庄慎似乎总能在富含中国传统文化韵味的空间中敏锐而自由地找到属于建筑的当代性。与桦墅乡村工作室一样，人们也都顺利地被他从地面带到了屋顶，仿佛都在等待登高后透出的第一口气。

另一个有趣的案例是阿科米星为莫干山庾村文化市集蚕种场所做的改造[图 11]，在临时性的要求下，改造趋向了一种展览的思考方式。由于蚕种场地面空间较为零碎，加之场地存在高差，如何有效地组织室外空间成为了真正的命题。设计最终使用竹棚搭建的方式，通过明确的矩形中心和不规则的外边界创造了几组重复的室外空间节点，集市等室外活动便可以通过几个中心“庭院”组织起来。放弃坚实的覆盖而选择竹棚不仅是出于性价比的考虑，也同时考虑通过现场工匠的手工调整来获得最佳的效果。蚕种场的竹棚完全可以被看作一次空间装置的实验，在理性层面，它是建筑空间创造的一部分，在感性层面，它可以成为一件独立于建筑空间之外的装置作品。富春俱舍走马楼中的“光锥”也具有同样的效果。

11　莫干山庾村文化市集蚕种场改造
Silkworm Hatchery Renovation, Yucun Culture Market, Mogan Mountain

而在城市环境中，相似的介入手法也被应用到都市空间的更新中。永嘉路口袋广场、宝山陈化成纪念馆移建改造[图 12]都被设计建造成城市中可被体验的空间装置。陈化成纪念馆移建改造项目位于一座传统风格的老建筑旁边，阿科米星的策略是保持老建筑的本体不做任何改动，但通过添加一座由钢、木材共同建造的回廊向老建筑致意，通过回廊围合的院落空间在老建筑和新的场所之间建立了一道过渡性的屏障，保证了观看老建筑时所需的视觉距离，维护了老建筑相对私密和安静的空间氛围，这也暗示着两处空间在建造时间上的距离。而回廊的节点设计在某种程度上延续着传统木结构建筑榫卯的特质，回廊的细部干净明了，创造了有韵律的光影，毫不拖泥带水。

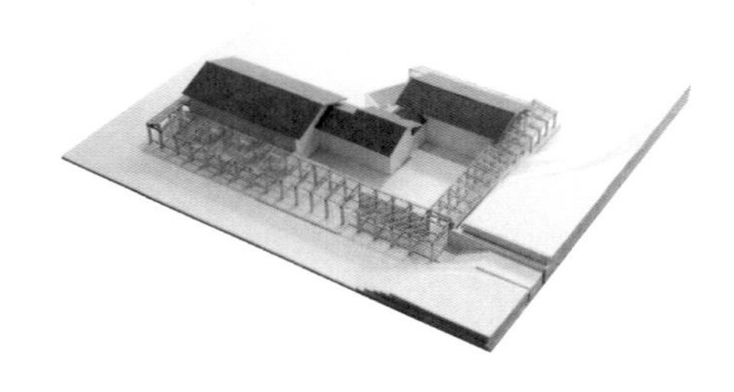

12　宝山陈化成纪念馆移建改造
Removal Renovation of Chen Huacheng Memorial

这几个小项目为我们呈现的是阿科米星对于“日常”的诗意再造，他们相信眼前的城市和乡村有与众不同之处，并且积极地转化来自其中的启示。庄慎在其作品中提及的日常也并

routine life to the spectacular urban reality. Through his study, we can understand that the crazy city of Las Vegas follows the basic and simple rational logic of everyday life. For Zhuang Shen, the urban practice on everydayness could be one source for resisting the erosion of commercialization and consumption. That's why he cherishes these silent urban spatial practices so much and has examined them with phenomenological sensitiveness.

One of Atelier Archmixing's projects, Huashu Rural Studio at Zhoucong Village, Nanjing [Fig. 9], was exhibited in the themed Exhibition *Building a Future Countryside* in the China Pavilion at the 2018 Venice Architectural Biennale. This project, together with other representative Chinese contemporary rural cultural practices exhibited, has aroused wide public concern. In Huashu Rural Workshop, two old houses were respectively renovated into a rural academy and a function room. If you have already known Zhuang Shen's early works, you will definitely track down here again his "Hanging Courtyard". An elevated observation deck and a long curved terrace as two materialized "courtyards" were designed to reorganize the inside-outside relationship. If the design of Huashu Rural Studio is seen as a treatment of addition and contrast, then Twin Trees Pavilion and LILI in Lili, Suzhou, have built an integral rural architectural art. The architect has tapped the potentials of space, light, material and decoration with his keen insight.

In the project of Twin Trees Pavilion [Fig. 10], tiresome caused by the pursuit of fine architectural quality in rural areas was swept off by the architect's accurate control of the spatial atmosphere. The pavilion was renovated from a 56m^2 pigpen. The bricks dismantled were recycled to build the new walls around the trees. The roof with a gentle slope, a structural column supporting the upstand beam, together with two trees, created a dialogue between the artificial and natural sceneries. It is hard to define whether this pavilion is a house or a courtyard. Through its low roof design, trees were provided with plenty of space for fresh growth, and sharper contrast between light and shade could be felt inside. LILI is another case in which the architect made the best use of the roof. The building used to be a workshop surrounded by sloping roof dwellings. Since the client was not satisfied with various reconstruction plans and there was a budget control, the architect decided to renovate the old building. The entrance circulation was adjusted to block the main space from the southern public toilet. A dim back alley at the west leads the visitors to a concaved glass foyer, and a stage-liked space covered with double slope top is just ahead. Indoor space is slightly brighter, and sunlight through 40 scattered roof windows of varying sizes drives out the gloomi-

非是我们所认知的普遍意义的日常，他的工作更像是照亮荧光粉的紫外光射灯，将日常中的某些价值“高亮”了出来。

如果说阿科米星的实践与理论思考是基于他们对大众文化和消费社会之现实的观察，并将其与日常性的生活和空间实践联系起来，尝试在实践中批判性地重建日常生活场景及其秩序，那么有一点需要特别强调，即在阿科米星看来，日常生活的力量还在于一种潜移默化的、改变的过程：自发性的或者有导向性的自我调整的过程，在时间的维度上保持一种不断变革的力量。因为对他而言，改变在当代中国已经成为了一种“日常”。当然，这既可以体现在建筑师设计策略上的改变，更可以体现在建筑师职业身份的改变，也即以社区参与者的身份促成这种变革。庄慎采取了一种非常特别的实践方式：阿科米星事务所在近五年内每年一搬迁。我把这视作一种都市游牧。

自 2014 年始，他们开始了每年换一个工作室的计划，至今已在上海更换了 6 个工作地点。建筑师通常希望工作的空间尽量稳定，可以使自己逐渐进入一个舒适区，而这样的感觉在搬家计划下变得荡然无存。较早的长顺路工作室是令人印象深刻的一处工作室，由社区托儿所改造而成，庭院被重新整理，长满了花草。工作室内除了办公空间还有可以用于展览和聚会的区域，几年来举办过很多高品质的讲座和沙龙。2014 年庄慎搬离了这里，结束了相对安逸的状态。[图 13]

13　阿科米星长顺路工作室（2009–2014）
Atelier Archmixing's Changshun Road Office

不同于其他工作室的搬迁，阿科米星不是因为租金和场地等原因被动开始游牧，而是他们主动地搬离了熟悉的街区。用庄慎自己的话讲，搬家可以开启完全不一样的城市建筑的观察角度。“工作空间少了以往固定的氛围，简易的环境更像是一种匆匆的撤离，建筑师要面对的心情与很多城市当代人一样了。设计工作者被剥离掉以往氛围暗示的意义，正如建筑被剥离掉各种人为赋予的意义一样，不是一下子能够适应和接受的，也很难判断它的好坏。”这种主动的游牧或许是在自我暗示一种不断的变化、更新和迭代。

观察 20 世纪最后十年以来的中国城市变迁，这种不用深思熟虑就可以迅速建造，再通过实践中的效果反馈迅速调整与变化的模式，成为中国城市极速迭代演进的基本逻辑。这种

ness. The joy of the roof was the privilege of B&B visitors. Zhuang Shen once recalled in a conversation his childhood memory of Grandmother's home in a small town in the south of Yangtze River, where people built an intimacy with the usually unfamiliar tile roof. The graffiti on the roof platform and the bright windows constituted a huge roof installation art; this is the most touching portrayal of LILI. Zhuang Shen seems talented in exposing the architectural contemporaneity with keen insight and ease in those spaces with Chinese traditional culture. Just as in Huashu Rural Workshop, the visitors once again followed the architect from the ground to the roof, as if they were attracted toward the first fresh breath on the top.

Another interesting project is Silkworm Hatchery Renovation, Yucun Culture Market, Mogan Mountain [Fig. 11]. At the request of being a temporary market, the architect decided to put a concept of exhibition into the renovation. Since the space in the Silkworm Hatchery was fragmentized and the ground floor was uneven, how to organize the outdoor space effectively became a real mission. The design ended up using "Bamboo Shelters" to duplicate several outdoor space joints, all with the shape of a distinct rectangular center and irregular edges. Outdoor activity spaces including the market could thus be organized through these central "courtyards". Using bamboo shelter instead of solid covering resulted not only from the economical restriction but also the potential flexibility of onsite manual adjustment. The "Bamboo Shelters" could be regarded as an experiment on spatial installation. Rationally, it became part of the architectural space creation; emotionally, it could be seen as an installation independent of the architectural space. The"Light Cone"in the project of Fuchun Kosa Zoumalou has achieved the same effect.

A similar design method has also been applied to the renovation of urban spaces. Two Shanghai projects, Pocket Plaza, Yongjia Road and Removal Renovation of Chen Huancheng Memorial in Baoshan district [Fig. 12] were both designed and built into accessible urban spatial installations. Chen Huacheng Memorial is located next to a traditional-style Confucian Temple. In Atelier Archmixing's design, the existing building was full kept and a timber-steel structured gallery was added to show respect to the original context and enclose courtyard between the old building and the new construction, which served as a transitional space. In this way, the memorial hall obtained a comfortable view distance as well as a relatively private and tranquil spatial atmosphere. This also implied that these two buildings were built at different times. The joint design of the gallery implicates the feature of a mortise-and-tenon joint in traditional Chinese buildings of the wooden structure. With

超越了中世纪以来城市发展节奏的极速迭代模式，或者更适应 21 世纪数码时代的精神：我们已经熟悉了电子产品不必等到所有用户满意就可以推陈出新，迅速推出更高版本，而这种迭代使得改变本身不是一种途径，或许改变已经成为了一种驱动力，推动着电子产品、工业系统，乃至建筑、城市和整个社会跃迁式的前进。庄慎自觉不自觉地感受到这种持续变化的动力给一个建筑师事务所带来的可能性，一种更开放、更多元、更异质、更混合的未来路径。

这是一次拥抱未知和改变的身体力行，是一次对建筑师身份的自我迭代。庄慎和他的同事们正在进行一种社会设计实践，他们不仅作为建筑师介入建设项目和城乡空间，同时也将改造用于自身，以获得一种来自社会的反作用。和传统的建筑师通过设计介入社会不一样，阿科米星事务所通过游牧，一边观察着城市，一边将自己作为一个社会结构的参与者，尝试不同类型的建筑被改造和用作建筑师事务所的可能性。同样，建筑师事务所的结构和行为模式也在这种变迁中被不断重构。

14 阿科米星桂林路小白楼（2018–）
Archmixing's Little White House on Guilin Road

也许若干年后，这一游牧计划会被看作一次大型且旷日持久的艺术计划，甚至被美术馆展出。我们不完全确信阿科米星这颗持续游牧、不断变化迭代的星球会按着何种轨迹发展，至少我们可以看到最新版本的桂林路小白楼工作室已经开始试验个人移动办公的模式，员工通过网络协同工作，不是必须出现在工作室上班了[图 14]。这对每天打卡，计算加班费的传统工作方式提出了新的挑战。所有人都知道建筑师是个古老的职业，可以走出来的从来都是革新者和不惧怕改变的人。

回溯阿科米星成立以来的十年历程，我们看到了一条在都市消费文化的现实逻辑和日常性的痕迹之间穿行的道路，从城市的变迁中发掘着变革的力量，并将这种思考的“重”转化为一种行动的“轻”，这种轻，既是事务所抛弃辎重，便捷游牧的物理状态，更是一种放弃建筑师职业的铠甲，寻求马歇尔·伯曼所说的一种更宽广开放的理解方式。挟着都市游牧和自我迭代所分泌的能量，在这种不断朝向日常和变化的实践道路上“精确和坚定”地前行，或许是阿科米星永葆青春的方法。

concise details, the gallery shapes rhythmic light and shadow without any irrelevance.

These are the small projects that Atelier Archmixing completed as its poetic reinvention of “everydayness”. The architects are firmly convinced that the city where they live is different from those rural areas, and they have been actively turning this recognition into practice. The everydayness contained in Zhuang Shen's work is not actually the everydayness we have generally known. His work feels more like an ultraviolet spotlight revealing the fluorescent trace, which “highlights” some values in the everydayness.

If Atelier Archmixing's practice and theoretical reflections are based on their observation on the reality of mass culture and consumer society, connecting such observation with everyday life and spatial practice and attempting to critically reconstruct everyday living scenes and order, it should be highly stressed that in the view of Atelier Archmixing, the strength of everyday life also lies in the imperceptible process of change. This is a process of spontaneous or oriented self-adjustment, which in the temporal dimension remains a power of constant evolutions. For Atelier Archmixing, in contemporary China, change has become an “everydayness”. Certainly, such change can be considered in various aspects: change of architect's design strategy or professional identity, and the latter means the architect promotes change as a community participant. Zhuang Shen has conducted a very special practice: Atelier Archmixing annually relocated its office in the past five years. I see this relocation as urban nomadism.

Since 2014, Atelier Archmixing has initiated a “New Year, New Office” program, and the studio so far has successively been relocated in six workplaces. Due to this plan, it is impossible for architects to work in a relatively stable space and stay in the comfort zone as usual. The studio on Changshun Road in the early years was hugely impressive. It was renovated from a community nursery. The courtyard was rearranged and covered by flowers and plants. There was office space and extra space for exhibition and gathering where quality lectures and seminars were held. However, Atelier Archmixing ended such an easy and comfortable state and left this office in 2014. [Fig. 13]

Different from other office moves, Atelier Archmixing didn't start its urban nomadism because of rent and place shortage. They moved away from familiar blocks on purpose. According to Zhuang Shen, through annual moving, they can obtain an entirely different perspective on urban architecture. “There is no regular atmosphere in the workspace and it feels more like a hasty

注释

1 庄慎、华霞虹：《改变即日常：阿科米星的实践综述》，《建筑师》2014 年第 2 期，第 131 页。

2 伊塔洛·卡尔维诺：《新千年文学备忘录》，译林出版社，2015，第 14–16 页。

3 庄慎、华霞虹：《选择在个人与大众之间》，《建筑师》2012 年第 6 期，第 43 页。

4 马歇尔·伯曼：《一切坚固的东西都烟消云散了：现代性体验》，徐大建、张辑译，商务印书馆，2003，第 1–4 页。

5 庄慎、华霞虹：《选择在个人与大众之间》，《建筑师》2012 年第 6 期，第 45 页。

6 庄慎：《看不见的改变：论使用端空间技术化的可能》，《时代建筑》2018 年第 3 期，第 32–35 页。

evacuation from a simple work environment. The architects will have the same mood as today's urbanites. Exactly as a building being stripped of various man-made meanings, when the designers have been kept away from the significance embodied in the past atmosphere, it is not easy to get used to and accept this situation in a short time, and it is even harder to tell whether such practice is positive or negative. This intentional urban nomadism perhaps is a self-suggestion of constant change, renovation and update.

Observing China's urban change from the last decade of the 20th century, it can be found that China's rapid urban development is driven by fast construction without careful consideration and fast adjustment and change according to the effect feedback in practice. This unprecedented urban iteration mode, faster than any other urban development since the Middle Ages, probably is more adaptable to the Digital Age Spirit in the 21st century. It has been a widespread common sense that sustaining innovations and quick updates of electronics are usually put forth without measuring all users' satisfaction degree. The innovation or update has made changes no longer a path but probably a driving force which pushes forward the rapid development of electronics, industrial system as well as architecture, city and the whole society. Zhuang Shen intentionally or unintentionally sensed the possibility brought by this constant changing force to a design studio, a future path which is more open, diverse, heterogeneous and mixed.

It is Atelier Archmixing's earnest practice to embrace uncertainty and change, and also, an update of the architect's own identity. Zhuang Shen and his colleagues are implementing a kind of social design practice, not only intervening into projects and urban and rural spaces as architects, but also reinventing themselves to get a retroaction from the society. Different from the traditional architects' intervention into the society, through urban nomadism, Atelier Archmixing observes the city and sees itself as a participant of social structure, trying all kinds of renovation and the possibility of working in these spaces of diverse types. The structure and working pattern of the studio are constantly reconstructed in this nomadism.

Several years later, this urban nomadism may be regarded as a major prolonged art project and even be exhibited in the art gallery. We can't fully predict what track Atelier Archmixing, this constantly moving and changing planet will follow. At the very least, we have seen that they have started a mobile office mode since they moved in the latest studio, the "Little White House" on Guilin Road [Fig. 14]. The staffs can work together via Internet instead of attendance in the office every day. This working mode has brought up a new challenge against traditional everyday punch in and overtime remuneration. As is known to all, the architect is a traditional profession, but those prevail have always been innovative and never feared change.

Reviewing the first decade for Atelier Archmixing, we found that the studio has found a path by carrying out its practice between the realistic logic of consumer culture and everydayness, exploring the strength of changes in urban development and turning its "weighted" reflection into "light" practice. Such "light" practice lies in not only a nomadic physical condition but also a pursuit of Marshall Berman's broader and more open way when removing architects' professional armor. I believe, concisely and firmly marching on the path of practice toward everydayness and change with the energy derived from urban nomadism and self-update, Atelier Archmixing will always retain the vitality.

Notes

1 Zhuang Shen and Hua Xiahong, "Change is More: Atelier Archmixing's Practice" (in Chinese), in *The Architect,* no.2 (2014): 131.

2 Italo Calvino, *Six Memos for the Next Millennium* (Houghton Mifflin Harcourt, 2016), 18–19.

3 Zhuang Shen and Hua Xiahong, "Standing Between the Individual and Public" (in Chinese), in *The Architect*, no.6 (2012): 43.

4 Marshall Berman, *All That is Solid Melts Into Air: The Experience of Modernity* (Verso, 1983), 5–7.

5 Zhuang Shen and Hua Xiahong, "Standing Between the Individual and Public" (in Chinese), in *The Architect*, no.6 (2012): 45.

6 Zhuang Shen, "Invisible Change: Technologizing Space at the User's End" (in Chinese), in *Time+Architecture*, no.3 (2018): 32–35.

李翔宁

同济大学建筑与城市规划学院教授、博士生导师、副院长，长江学者青年学者，知名建筑理论家、评论家和策展人，哈佛大学客座教授。

任中国建筑学会建筑评论委员会副理事长兼秘书长，国际建筑评论家委员会委员，国际建筑杂志《Architecture China》主编。曾在达姆施塔特工业大学，东京工业大学、UCLA 等大学任教。他担任《THE PLAN》《Le Visiteur》等国际刊物编委。担任密斯凡德罗奖欧盟建筑奖、CICA 建筑写作奖、PLAN 建筑奖、西班牙国际建筑奖等国际奖项评委。他还曾担任米兰三年展中国建筑师展、哈佛大学中国建筑展、深圳双年展、西岸双年展、上海城市空间艺术季等重大展览的策展人。2017 年釜山建筑文化节艺术总监和 2018 威尼斯双年展中国国家馆策展人。他的近期著作包括《上海制造》《上海城市更新 · 五种策略》《走向批判的实用主义——当代中国建筑》等。

LI Xiangning

Deputy dean and full professor in history, theory and criticism at Tongji University College of Architecture and Urban Planning, member of CICA (Comité International des Critiques d'Architecture), and Secretary General of China Architectural Society Architectural Criticism Committee.

He is Editor-in-Chief of the international magazine *Architecture China*. He serves as an advisory board member of Confluence Institute for Innovation and Creative Strategies in Architecture, and editorial board member of magazines including *THE PLAN* and *Le Visiteur* (Journal of Société Française des Architectes).

He has been a jury member of many international awards and competitions, including Spanish International Architectural Award and Mies van der Rohe Award the European Union Prize for Contemporary Architecture. He taught and lectured in universities and institutes, including Princeton University, TU Darmstadt, UCLA, USC, Chalmers University, IUAV, IAAC, Canadian Center for Architecture. He was a Visiting Professor in Architecture at Harvard GSD in 2016.

He is co-curator of 2011 Chendu Biennale, 2011 Shenzhen Biennale, academic director and curator of 2013 Shenzhen/Hong Kong Bi-city Biennale, curator of Shanghai Westbund 2013 Biennale, and curator for 2015 and 2017 Shanghai Urban Space Art Season. He serves as a director for the 2017 Busan Architectural Culture Festival (Korea) and recently curated the Chinese Pavilion at 2018 Venice Biennale.

His recent books include *The Real and the Imagined: A Study of Value in Contemporary Urban Theory* (2009), *Updating China: Projects for a Sustainable Future* (2010), *Made in Shanghai* (2014), *Shanghai Regeneration* (2017) and *Towards a Critical Pragmatism: Contemporary Chinese Architecture* (2018).

对谈
Dialogues

李兴钢 × 庄慎：本土实践中的有效性与普遍性

LI Xinggang × ZHUANG Shen: The Effectiveness and Universality in Local Practice

1　安龙森林公园东部码头小镇商业建筑
Commercial Building at East Dock Town in Anlong Forest Park

时间：2019 年 8 月 19 日
地点：北京李兴钢建筑工作室

庄慎（后简称“庄”）：我们两个人都是一直在国内工作的。我是 1997 年参加工作的。1990 年代那会儿，国家的建设量大得惊人，年轻建筑师几乎是被抛进那个时代的，我们算是被大量项目砸出来的建筑师。

李兴钢（后简称“李”）：对，我们这种成长方式跟国外建筑师完全不一样。刚毕业参加工作的时候，我发现学校里学的东西基本都“失效”了，面对真实的项目、甲方，我们做的都太“学生气”了，一开始真的很受挫。因此刚开始在设计院工作的时候，那种职业化的工作方式我还是很认可的，但是对于设计的方式和结果，我觉得好像还不能仅仅是这样。所以毕业后相当于从头学习和再思考，一边学，一边想，一边做。

庄：嗯。在面对实践机会时认知与方法的准备非常重要。我国的城市化现象是独特的。我一直在想，在我们碰到的问题当中，有没有通识性的方法或者通识性的认知？

李：也许就是做设计本身的东西，那些本能的、具体的手法。不论现在有什么样的新认知和方向，原来那些设计的本能，包括受教育时候受到的强烈影响，都会变成你血液的一部分，一定会起作用的。

庄：一定会是这样。

李：你们的工作案例里，我特别喜欢千岛湖安龙森林公园那个项目［图 1］。没建吗？第一眼看到这栋房子，会觉得很感动，有一种诗意。我觉得就是刚才说到的你骨子里内在的东西在起作用。
这个项目很有意思，下面的房子看似是永久性的，却是允许改变的、临时性的，而上面的木构平台看似很临时，然而它却是不可改变的。这具有某种思辨性，类似于临时和永久，改变和不变，日常和仪式感的那种对应，这里面还隐藏有生活和诗意。如果这是一个改造项目，你在上面做新的介入，那是相对顺理成章的，因为它已经有一个前提，你可以借力、跟它对话。然而这是个新建项目，实际上是你自己在制造基础，然后再去介入。

2　家琨建筑：成都西村大院
Jiakun Architects: West Court, Chengdu

Time: Aug. 19, 2019
Venue: Atelier Li Xinggang, Beijing

ZHUANG Shen (ZS): Both of us have been working domestically all along. I started working in 1997. The country started construction on a surprisingly large-scale in the 1990s. Young architects of our generation were almost tossed into that trend. You can say we were “forced” into architects by the large amounts of projects.

LI Xinggang (LX): Indeed. The way we became an architect was totally different from that of foreign architects. When we first entered the workplace upon graduation, what we learned at school was basically “useless”. When we faced real projects and clients, we felt our work was too “childish”. It was really frustrating at the beginning. So I did approve of the professional style of working at the first start in an architectural design institute. But as for design methods and results, I feel there’s something more than that, thinking is necessary. So basically, when you graduate, you are thrown into a new start and have to keep rethinking. You learn, think and carry out.

ZS: I agree. It is crucial to be prepared in thoughts and methods when faced with chances of practice. The process of urbanization is unique in our country. I’ve thought whether there is a general method or cognition about the problems we encounter?

LX: In another aspect, even if you just focus on the design itself, the instinctive and specific techniques are necessary. No matter how novel the study is, those that are instinctive, including the strong influence from our education, will become part of you and have its effects.

ZS: Can’t agree more.

LX: Among your cases, I especially enjoy the project of the Commercial Building at East Dock Town in Anlong Forest Park [Fig. 1]. Is it an unbuilt case? Upon first glance at the houses, you’d be moved and have a poetic feeling. I think this is due to what we just mentioned before, the thing in your bone.
This project is quite interesting. The houses below seem permanent yet can be changed, whereas the above wooded platform looks temporary but is hard to change. There are elements of thinking and analyzing in it. Let’s say here, it is the comparison between temporary and permanent, changing and consistent, everyday and ritual, where embody daily life and poetic feelings as well. If this case is given as a renovation project, and you step in to make changes, it’s quite natural, because there is already a foundation, and you can step on that to initiate interactions. But this case is a

庄：这个项目的设计由两个因素决定：一是来自传统的形式逻辑的手法——依附，轻质的木构依附在大的实体上面。第二，跟商业使用、消费逻辑有关。这是一组商业建筑，交付之后一定会面临更换门面、挂上招牌等不可能去控制的事情，所以我们采用了一个策略：把不大会被改变的木平台设计为控制性的突出要素。

李：这样反而是有中国的特色，和家琨的西村大院［图2］有点像。这也让我想起，2016 年我在南京做的"瞬时桃花源"［图3］，当时的设计也探讨了瞬时和永恒，那个"瞬时"体现出来的思想和诗意可能是很多年大家都在追求的"永恒"。类似的思辨，实际上是内心的一种"你的感受"在起作用。

庄：也可以说，形式背后是有通识的意义并可传达的。

李：你对城市里那种临时性搭建的领悟，肯定是受到感动，内心产生了共鸣，你才会捕捉到它，甚至把它变成你选择实践方向的某种策略和切入点。这种临时性所透露出来的深层次的生活诗意，我觉得在你这里是一直存在的。

庄：你说到了建筑的一个重要层面。如果说从使用建造的有效性去看待城市的日常生活是一种冷静无温度的视角的话，那么你说的这个层面使它变得更有温度。

形成千岛湖那个设计的第三个方面，跟改造里"调整"的方式有关系。我们经过千岛湖，观察到当地都愿意到屋外吃饭看风景，这个使用习惯是对房屋空间的一种调整使用，所以木平台的作用除了形式控制，还可以看作是一种使用改造与调整。我们故意把新建看成是一个改造过程。

李：所以这个项目又实现了一种普适性，它也完全可以转化成另外一个设计。

庄：是的，这种工作里可能产生的普适性是一种工作动力：在日常工作里有机会去做一点跟促进建筑学科有关系的事情。

李：对，"普适性"这个词挺好的，它跟社会性有关系。我是非常认同这类通过普适性作品的力量达到某种社会性的建筑师的，他是靠专业的思考和实践去达到，不是附加到一种社会关系里。

brand-new one. In fact, you build your own foundation and then make changes.

ZS: The project could be decided by several factors. The first is techniques from traditional formal logic. As you know, the wooden structure is light in texture and could attach to a bigger volume. The second has to do with commercial use and consumption logic. This is a group of commercial buildings. When it is finished, the facades are sure to be changed. We can't control it. So we turned to control the seemingly unchangeable wooden platform. This is a prominent element and our strategy.

LX: This is indeed of Chinese characteristics and similar to West Village designed by Liu Jiakun [Fig. 2]. This also reminds me of "Instant Garden" [Fig. 3] I did in Nanjing in 2016, which revolved around "momentary" and "permanent" as well. The thinking and poetic feeling embodied at the moment is probably the "permanence" what people are after most of the years. Such similar thinking is in fact a reflection of "your feelings".

ZS: You can also say that there is a universal meaning and it could be communicated via form.

LX: Like what you grasp when you see those temporary constructions in the cities, there must be a moment when you feel connected and touched, so you capture it and even turn this moment into some strategies and starting point for your architectural practice. Such deep liveliness and poetic meaning revealed through this temporariness are always present in your place.

ZS: You just mentioned an important layer in architecture. If we see urban everyday life from the effectiveness of construction and usage, it is calm-minded and cold, while your perspective infuses warmness into it.

The third aspect contributing to the Thousand Islands Lake project is related to the way of "adaptation" in renovation. We treated this construction as a renovation process on purpose. As we passed by the lakes, we discovered that local people prefer to dine outside to enjoy the scenery. Such a habit is an adapted usage of residential space. So the wooden platform is both a form in control and a renovation and adaptation in use.

LX: Thus this project is made universalized; it could be turned into another design as well.

ZS: Yes. Such universality created during work is a motivation for us to seize chance in everyday work to do something that can improve architectural discipline.

LX: "Universality" is a proper word here. It is of social significance. If someone's work has reached a certain level of universality, the so-called wide social recognition is what I approve of. These archi-

3 李兴钢建筑工作室：瞬时桃花源
Atelier Li Xinggang: Instant Garden

关于普适性我有两个层面的理解，一种是我们通常所说的，以个性的力量达到对社会更广泛的影响。这点是需要追求的，不管是独立事务所，还是大院的建筑师，因为你动用很多社会资源，就有这种社会责任，我觉得这是应该的。第二个层面的普适性就是从专业的角度来讲，只要是各种城市现象或人们的各种生活空间，你的思考和实践都需要能够去适应、去覆盖不同的建筑类型。不管是理论研究还是设计实践，都需要去追求这个方向，比如说我自己，我选择项目的原则是我跟甲方能不能达成某种关键性的共识，但是对项目的类型我是不挑的。

如果你的思想真正有价值、有力量，我相信它应该能够获得这种普适性，而只有这种普适性才会对这个行业、这个专业，甚至对这个社会，有更大的影响。

庄：在这个方面实际做事时往往会碰到不少需要克服的限制，独立事务所力量微小，有它的问题。大设计院下的工作室这方面的条件是否更强一些？

李：这要看从哪个方面说。一方面，设计院有很强大的生产和商业特性，但是在某些局部、特殊的情况下，又可以对此进行适当的抵抗甚至"批判"；另一方面，也还是得警惕建筑的社会性被过度和片面强调。这里存在两个层面的事情，一个层面是建筑设计本身，它有属于上层建筑、形而上的社会责任，需要表达自身的一种立场，另一层面它同时又是一个服务性的行业，确实需要资本支持，会由经济基础决定，这两个方面和两个层面都需要达到一定的平衡才是合适的。之前我们谈到"自然"这个词，我对它的另一种理解是"自然而然"，这更接近中国人对"自然"的含义的说法。自然而然地去做这件事情，然后批判性有了，社会性也有了，同时它的社会性和批判性是通过建筑自身的专业性来呈现的，我觉得这样的一种状态是最好的。

庄：我们的工作内容与方式可能有点区别，相比于大机构的重大项目，我们的项目更日常化；相比于大机构完整的内部组织生态，我们的对外合作更多样。这两者我觉得挺有意思的，像大石块之间会有小细缝，我们小沙子就灌进去。

李：我们是有一些做大项目的机会，但也有限制，因为你想抵抗，想批判，想不做成常见的

tects reach there through professional thinking and practice rather than simply attaching their work to the social relationship.

For universality, I have two understandings. One is the commonly-mentioned: to exert wider influence on society via the power of distinction. This is worth pursuing no matter you are running an independent studio or working for big state-owned institutions. Since you are unitizing various social resources, you have such social responsibility. I think this is mandatory. The second is the universality stems from the professional perspective. It covers all types of buildings. Your thinking and practice should be able to adapt to and cover all kinds of urban situations and living spaces. This is worth pursuing both in theoretical research and design practice. Take myself as an example, my principle in picking a project is whether I could reach consensus with the other party in some key aspects, but I'm open to projects of all types.

If your thoughts are truly valuable and powerful, I believe it would acquire such universality and only the universality in this kind would bring dramatic influence to the industry, to this profession, and even to the society.

ZS: In practice, there inevitably would be many restrictions to overcome. Independent studios are limited in their power and have their own issues. Would studios under big institutions be better-conditioned?

LX: This depends on which aspect you are talking about. In one aspect, design institutes are powerful in production and quite commercialized. But under certain conditions and in some special domains, these situations need resistance and critical reflection. In the other aspect, we should be alerted to avoid over-and-partial emphasis on the social nature of buildings. Here I'm talking about two dimensions. One is about the architecture design itself, which has metaphysical social duty unique to top-structure. It needs to take a stance. But on the other hand, it belongs to service industry as well, which requires capitals and would be decided by economic base. These two aspects and dimensions need to reach a certain proper balance. We mentioned "nature" before. I understand this from a different perspective: "to come naturally". This is closer to how Chinese interpret "natural". You carry on naturally, and then you acquire the ability to be critical and of social responsibility. Meanwhile, such critical and social status is presented through the building's own expertise. This is the best state in my opinion.

ZS: We might differ in what we work with and how we work. Compared with major projects

设计院的那种设计，那你就可能会面临失败、碰壁，所以其实这么多年来，我的大项目作品还是寥寥无几。

庄：你是说没做到你理想状态的作品？

但是这些已经非常好了，很难得，个性非常强烈，而且通过阅读你的著作，可以了解到你对于这方面有连续性的思考。

李：我觉得这也算是作为大院建筑师的一种责任吧。

庄："责任"的说法令人兴奋。我觉得针对学科的或者是对于一些现实问题的解决态度，我们会有某种一致性的追求。

李：现在确实是这种生态越来越好，越来越多的人会认同这一点。

庄：大院、独立事务所，实际上只是在不同的工作条件下工作，面临的很多问题是共通的。虽然有很多问题，但也不是说全是坏事情。好处是我们承接了那么多项目，我们很多的感性经验、肌肉经验，远远多过我们能够讲述出来的，这其中现实的智慧非常多，虽然可能还不能形成理论，严密整合后系统地分享出来。

李：从横向的比较来讲，我们跟外来者、海归人士对中国现实的理解和适应，实际上是不一样的。这也是一件好事，从国外带着工具、手段甚至思想进来的时候，反而会跟中国的现实在不同的层面上有隔膜，会有一种总是碰不到"实质"的感觉。这种感觉会从很多非常具体的地方呈现出来。

庄：其实我觉得未来中国的一些学科的进步，也还是需要用实践砸出来的。

李：对，硬核的实践非常重要。我们现在的问题是碰到的挑战太大了，很多既有的思想和实践其实是失效的。

庄：我记得你刚才也提到一个失效的时间段，是在刚毕业的时候。到现在也还是失效的。这个问题我觉得也正是工作的动力。

李：当然这是另一个层次上的"失效"了。所以我感觉你们抓到的这个方向和切入点很有现实意义，很"有效"。你们把失效的东西通过一种对使用的关注、对改变的关注，对临时性

at big institutes, our projects are closer to everyday life. Compared with a rather complete organization in big institutes, we have more various cooperations with other studios. It's like whenever there is a little gap among big rocks, we would fill in like sands.

LX: We indeed have the opportunity for big projects but there are also limits. Because when you want to resist and criticize or go beyond the conventional design from regular institutes, you may as well face failure and rejection. So in fact, I have had few huge projects in recent years.

ZS: Do you mean you haven't had an ideal work as you desired yet?

But I think these are already excellent and extraordinary. They are quite unique. Also, through your writing, we learn that you are continuously thinking about this.

LX: This might also be considered as a duty for architects in big state-owned institutes.

ZS: "Duty" sounds stimulating. I think we may share one pursuit, be it the solution to a discipline or some real problems.

LX: Such an ecological cycle is improving. There will be more people who agree with us.

ZS: Whether it is big state-owned institutes or independent studios, we are in fact only working under different conditions, and the problems we face are common. Despite many problems, they are not all bad. It is positive that with these projects, we have obtained rich experiences and practice, far more than we can tell. Among them, most are sudden flashes of inspiration from reality, not ready to be theoretically organized and systemized for sharing.

LX: Compared with foreign architects or overseas returnees, how we understand and adapt to the reality in China is different. This sounds good too. Applying foreign materials, methods or even thoughts to deal with local reality would lead to misunderstanding of Chinese reality on different levels. You would always feel you can't approach the "reality". Such feelings would ooze out of many specific situations.

ZS: Actually, I think if we expect improvements in certain disciplines in our country, we need to be patient. It takes time.

LX: Yes. Real practice is crucial. Now, the challenge is overwhelming and the existing thoughts and practices are useless.

ZS: I remember you just mentioned there is a useless period after graduation. I think it has been continuing till now and it is our career motivation.

LX: This is certainly another "useless". So I think the direction and starting point you seize is realis-

和永久性的这种思辨，把那个“效”再找回来。

庄：我看你特别喜欢中国传统文化，你的办公室、书房有很多这方面的书籍。

李：我的确对历史感兴趣，因为我觉得历史和传统里面有很多好的东西，但是现在它被隐藏起来了，被忽视，或者被误读。这些东西往大里说，其实就是解决人的问题，解决人的生存问题，解决跟建筑有关系的所有事情，它是一个非常“有效”的系统。我们现在所碰到的很多问题，其实以前人家都碰到过，并且解决了。

记得有一次在天津大学开会，我回答刘东洋老师的提问，我说自己对历史的兴趣不是纯粹历史学科性质的，我还是在用当代的眼光，用一个当代建筑师的眼光去看历史上发生的那些事情，当时的建筑师他们怎么做房子，怎么建设城市，怎么在“当时的当代”社会各方面背景和需求条件下，去应对施工、设计、造价、做法、空间、结构、形式的问题，当然他们的工作肯定会受制于“当时的当代”。“现在的当代”肯定有很多条件改变了，但是肯定也会有很多东西是不变的。我越来越觉得它既有超越时代的地方，也有超越地域的地方，甚至有超越文化的地方。我们可以把自己现在的所思所想、所作所为，也看成是一种“未来的历史”。

庄：其实，我俩都在面对一个有没有效的问题。

李：这是一种现实，每个人的所思所想、所作所为，只能是一个局部。我可能就是从自己的角度让这个工作变得稍微有效一些，比如说去关注人工和自然的关系，是因为我觉得这里面有当代城市建筑环境失效的某个关键原因。而你就是从另外一种角度了。拿看病救人打个比方，我那个药方可能比较像中药，你这个可能会是一种更切中要害的、要救命的治疗方向。凡是从这样一个基本的角度去思考，它都可能会对建筑学本体的改变有影响。我知道你一直对建筑中的新科技手段很关注，这是不是也是你要应对、改变“失效”的一种方法呢？

庄：技术发展改变空间这件事情，我感觉必然会发生，由上而下，不以我们个人意志为转移。我觉得建筑师有责任去关注，因为他是空间组织者，也是整个建造系统的一分子。建筑师去了解技术，也许能够促成一些更好的使用。

我们对技术的关注，也是从建筑怎么构成开始思考的，跟很多人关注结构构造技术不

tic and “useful”. You retrieve the “useful effect” for those useless things by paying attention to usage, change, and reflecting on temporariness and permanence.

ZS: You especially like traditional Chinese culture. There are many related books in your office and reading room.

LX: I do have a great interest in history, for there are valuables in history and traditions. But now they are hidden, neglected, or misunderstood. If you see it from a bigger scale, they can settle the key issue of human being, the issue of existence. In other words, they solve all that related to architecture. It is a very “useful” system. The problems we are facing now have actually been discovered before and once successfully solved.

Once when we held a meeting at Tianjin University Stadium, I answered Mr. Liu Dongyang that my interest in history was not just about history discipline. I'm also evaluating historical events from a contemporary perspective as a contemporary architect. How did architects back then build and design cities? How did they tackle construction, design, costs, methods, space, structure and form under “their” social contexts and necessary conditions? Of course, their work was also limited by such contexts. Even though many conditions have been changed now, some remain unchanged. It contains elements that transcend time, place and even culture. We can take what we think and do now as a “future history” too.

ZS: Actually, we are both faced with a question of “usefulness”.

LX: This is a reality. People's thoughts and deeds are only partial. For me, I'm just trying to make work more useful from my part. For example, one key reason why the current urban architecture environment is useless might have to do with the relationship between artifacts and nature. For you, it´s from another perspective. For a patient, my prescription is more like traditional Chinese medicine and yours is more like a to-the-point and crucial method. If you view the issue from such a perspective, it would very likely change architectural ontology. I know you have been interested in new science and technologies in architecture; is this another way for you to tackle “uselessness”?

ZS: It is unavoidable that technological development would change how we design space. It is top down and independent of our will. Architects should take the responsibility to pay attention to it since they are spatial organizers and part of the construction system. You can better use it by learning about it.

同，我更关注未来的设备或者相关的科学技术。我们将空间看成是既有的中性空间，它可以一遍遍地改变。在这个时间过程当中结构变得不重要，而空间既有的物质性变得重要，这个物质性就包含了既存现实。由此来看，物质形式的意义相对是短暂的，所以就不会去关注结构、构造这些事情。

当代的建筑在创造一个自身的物理环境时，设备起了关键的作用。比较有代表性的建筑师如福斯特，他的建造体系和设备体系完全是一体化的，只不过中国建筑师讨论得并不是特别多。我们的传统教育中相关涉及是比较基础的，而且设备系统因为本身的改变并不显著，工业化时代之后基本上就定性了，无非是性能的不断提高而已，所以也不太被关注。我们关注它是觉得人造空间里面不仅空间是重要的，物理环境、声光电都是重要的，所以我的直觉是，建筑设备一百多年也没有显著改变过，未来建筑的突破和改变，是不是会由设备的技术突破引发？技术总是组合变化的，并不一定说是整个系统没有了，可能其他的技术会以新的方式介入到我们的系统里来产生一些变化。

我并不是一个技术进化论者。从更深层次的角度来说，未来这些被推动、身不由己的变化，究竟对我们好还是不好？大家都不知道。但是要知道好不好，你就得先去研究它的基本原理，这也起到了一个推动作用。其实我认为我们事务所更偏重于建筑的使用端，我们关心人，因为要研究使用必然要去理解人怎么用，面对一些变化，最核心的当然还是人。

李：是的，而且“使用”这个词是可以有很多扩展含义的，可能那些不同的思考最终本质上都可以归结为“使用”。这其实就涉及我们共同感兴趣的问题，包括鲁安东老师不断提及的“中介”问题。从我的角度来讲，建筑和自然交互实际上是为了人和自然交互，建筑变成了一种人与自然之间的“中介”。传统建筑当然有很多自己独特的中介手段，而到了当代，这种中介手段必然会发生变化，那么很大的变化原因就是当代的技术条件和手段不一样了。

庄：传统的空间观念是很重视中介的，我们在做嘉定博物馆方案［图4］时就用过这种方法。这块基地在城市中一条很热闹的马路边上，背后就是上海名园——秋霞圃。当时一

4　嘉定新博物馆“聆听的视觉”过程方案
"Listening Vision", a process scheme for Jiading New Museum

Our attention to technology also starts from how we construct buildings. Different from the attention many others pay to structural construction technology, we give more focus on future equipment or related science and technologies. We see space as an existing neutral space. It could be changed over and over when putting in a longer time frame. In this process, the structure loses its importance, whereas the existing materiality in space turns to be important. Such materiality includes existing reality. This is a rather relative point of view in which we don't focus on structure or tectonics.

Equipment plays a key role in building a self-contained physical environment for modern architecture. This has always been the focus of architects. Like Norman Forster, he worked with a complete system of construction and equipment. Chinese architects don't talk much about it. What we learn in traditional education in this area is rather basic. Also, since we don't see distinct changes in the equipment system, the changes after industrialization are merely continuous improvements instead of changes in nature. Thus it receives little attention. We focus on it because besides space, physical environment, sound, electricity and light are all important in space. The intuition tells us that construction equipment hasn't significantly changed within a hundred years. In the future, should any breakthroughs occur, would it be in this area? Science and technology are changing together, which doesn't mean that we are to get rid of the whole system. Maybe other technologies would come into the system in a new way and bring about some changes.

I'm not a fan of the technological revolution. From a deeper understanding, would these forced changes be positive or negative in the future? We have no answer. But if we want to find the answer, we have to study its basal principles. I think this is like an engine that pushes us forward. Our studio is directed more towards the user's end of architecture. We care about the usage, targeting at how people use space. In the face of changes, our work has always focused on people.

LX: Yes. Also, the word "use" has extensive meanings, and ultimately all those different thinking would essentially be concluded as "use". Now we touch upon our common interest, the "media" issue raised by Prof. Lu Andong. From my perspective, the interaction between architecture and nature in fact is to facilitate interaction between people and nature. Architecture has become a medium between people and nature. For instance, in the traditions on which I focus, there sure-

个最有意思的方案是考虑在马路上如何能无障碍地看到、感知到园林，成果也是这样做的，因为做到了这个关系，于是建筑也就成为了一个中介。

李：阿那亚金山岭艺术中心[图5]也是这样吗？

庄：是的，对于风景，就是这么操作的，建筑风景被建筑局部"切碎"的同时有串联感，反而激发人的整体性联想。

李：也就是说，使用的手段会不会变，跟"使用"空间所在的环境和人的不同需求密切相关。

庄：其实，这是一个非常独特的中国式的思维方式，众多因素如何能够形成一个良好的关系，建筑在成就这样的关系当中生成。这种方式需要对各种关系足够敏感，抓得住最主要的矛盾，从而统帅解决其他的矛盾，需要发散性的思维，需要有整体的意识[图5]。

5 阿那亚金山岭艺术中心
Art Center of Aranya, Jinshanling

ly are some special media. But now, such media would definitely change mostly due to innovated modern technological conditions and methods.

ZS: Traditionally, medium is emphasized in spatial concepts. When we worked on Jiading New Museum [Fig. 4], we adopted such medium. The site is located near a bustling street and on its back is the famous garden in Shanghai, Qiuxia Garden. The funniest design then was how we could allow the pedestrians to directly see and feel the garden from the road. Then we made it. Since we succeeded in connecting nature and people, the architecture naturally became the medium.

LX: Was it the same case for the Art Center of Aranya, Jinshanling [Fig. 5]?

ZS: Yes. This was how we dealt with the landscape. The architecture landscape is "separated" by the buildings but remains as if they are connected, which actually stimulates the imagination of the whole.

LX: In other words, whether the methods of usage will change or not, highly related to the environment the space of "usage" locates and people's requirements from this space.

ZS: Actually, this is a unique Chinese thinking. How do numerous elements form a good relationship and then lead to a successful architectural relationship? It needs a certain level of sensitivity to all kinds of relationships, an ability to seize the major contradiction in order to solve all other ones, a divergent thinking mode, as well as the ability to sense the whole picture [Fig. 6].

李兴钢

中国建筑设计研究院总建筑师，李兴钢建筑工作室主持人。
主要建筑作品：元上都遗址博物馆、绩溪博物馆、天津大学新校区综合体育馆、唐山第三空间综合体、元上都遗址工作站、北京大院胡同 28 号院改造等。主要著作：《胜景几何论稿》《行者图语》《李兴钢：1991–2020》《静谧与喧嚣》等。曾获得亚洲建筑师协会建筑金奖、ArchDaily 全球年度建筑大奖、WA 中国建筑奖、全国优秀工程设计金 / 银奖、中国青年科技奖、全国工程勘察设计大师荣誉称号等。

6　李兴钢（右）与庄慎（左）对谈中
Dialogue between Li Xinggang (right) & Zhuang Shen (left)

LI Xinggang

Chief architect, China Architecture Design&Research Group. Founder of Atelier Li Xinggang.
His main architectural design works include Xanadu Archeological Site Museum, Jixi Museum, Gymnasium of New Campus of Tianjin University, Third Space Complex in Tangshan, Entrance to the Site of Xanadu, Renovation of No.28 Courtyard Hutong, etc. His main publications include *Essays of Intergrated Geometry and Poetic Scenery* (2020), *Wandering, Walking, Viewing, Living* (2020), *Li Xinggang: 1991–2020* (2020), *Tranquility and Noise* (2015), etc. His practice was honored with design awards, including the Gold Award of ARCASIA Awards for Architecture, the WA Chinese Architecture Awards and ArchDaily Building of Year Award, Gold and Silver awards of the National Excellent Engineering Investigation and Design Awards, etc. He received the China Youth Science and Technology Award and the esteemed National Engineering Survey and Design Master Award.

王方戟 + 张斌 × 庄慎：不统一的设计

WANG Fangji + ZHANG Bin × ZHUANG Shen: Non-Unified Design

时间：2019 年 8 月 31 日
地点：那行文化

张斌（后简称“张”）：你怎么理解你做过的大项目，比如诸暨剧院[图1]？

庄慎（后简称“庄”）：诸暨剧院的设计希望能创造最大的城市开放空间，以避免很多城市都有的类似的情况：剧院往往占有很大的公共用地，但平时又不允许人们轻易进去。因此我们把剧院核心区域——舞台观众厅等独立出来，形成平时不对外开放的可管理区域，把其他的一些功能空间架空，形成环状，这样就使和城市公园融合的日常地面开放空间最大化，被管理、不能进入的空间缩减到最小。

不过之后我觉得，这样策略化地使公共空间尽量多，在一个稠密的既有城市里面会更加有必要，放到刚起步的中国新城里，效果不太明显。需要调整策略的不光是建筑，更要紧的是规划设计。这使得建筑设计策略更像是一个象征。

1　诸暨剧院
Zhuji New Theatre

张：就是说你没法将做小项目的经验投射到这样的项目当中，是吧？

庄：是的，不能直接用。就此可以看到我们要面对的问题，就是方法是否可以普适，以及通过实践验证是否有效的问题。我们在尝试做一些工作，但我觉得现在还比较生硬。

张：这个属于可以共同讨论的话题，大项目还存不存在个人的或者团队的识别性？
你想象一下，不管是在一个新区里还是在一个稠密的旧城区里，有一个大东西要加进来，作为一个有点责任心的建筑师，就会想怎么把这个大东西做得伤害性小一点，那有可能就会把它剥离一下，但实际上还是无法从根本上攻破形成它的背后机制。那怎么理解这个事，这样的项目一旦来了你做不做？或者说这样的项目跟你的个人识别性到底是一个什么关系？

庄：我们会去做的，而且，我们承认不能逃脱大的背景系统。

张：如果用你之前的论述方式，它存在一个如何跨越的问题。所以我觉得你之前讲的“非识别体系”“未来居住研究”，等等，不适合进入这样的语境，因为生产体系不可能被你攻破。这当中存在一个隐含的话题，就是在整个系统存在的前提下，建筑圈在当下应该是什么样子。

Time:　Aug. 31, 2019
Venue:　NEXTMIXING

Zhang Bin (ZB): How do you understand those big projects you have designed, such as Zhuji New Theatre [Fig. 1]?

Zhuang Shen (ZS): For Zhuji New Theatre, we aimed to create the maximum urban open space. Grand theaters in China usually occupy massive public lands, but open a few public access. To avoid this common conflict, we enclosed the core areas, including the stage and audience hall as an independent part of management, combining the rest functions into a grand floating ring. In this way, the ground land was liberated and merging into the surrounding park. Urban open space for everyday use is maximized through minimizing inaccessible space under control.

Later I realized that it may be extremely necessary to strategically maximize the public space in a dense metropolis, while not so effective in a new booming city. Such kind of strategy is not only urgent for architectural design, but also crucial for urban planning and urban design. Therefore, the strategy we applied here is more of symbolic meaning.

ZB: Do you mean you can't apply the working methods you used in small projects to these big ones?

ZS: No, I can't use them directly. We are now facing two problems—Are these methods universal? Is it possible to verify their effectiveness through practice? We are trying something, but they are rather unnatural till now.

ZB: Now we touch a general topic worthy of open discussion—is it possible to keep the identity of an individual designer or a design team in those big projects?
Just imagine, a big project comes, no matter in a new region or a dense old city. As an architect with social responsibility, you may struggle in how to make it less harmful, for example, through separating it from the surroundings, but actually, you can't break the mechanism behind. How will you make sense of this situation? When such a project comes, will you accept it? Or rather, when you accept it, what does it have to do with your own design identity?

ZS: We will do, and we will also admit there is a background macro-system, from which we can't escape.

ZB: Putting this situation in your former narratives, how to transcend remains a problem. That's

庄：好像对我们来说不矛盾，我们的工作建立的前提，包括反思既有建筑学的一部分原因，就是希望我们的工作能建立在理解现有的生产体系之上。按我自己的逻辑，这个情况要求建筑学不要固步自封，要把自己的认知与方式放置在大系统里来评估价值。我们有的问题是问向传统建筑学的，比如非识别体系等，有的做法是从生产角度思考的，比如未来居住等。我们的工作中针对的碎片化的策略等，就是对于这个大系统的一个反应吧，不是试图创造某种全新的范型去重塑世界的革命性的逻辑，而是试图显示出这个系统原始冷静的一面，希望这种认识的角度会给建筑学带来新角度，而这样带来的手段我们发现很多是指向碎片化的。

张：可以理解。好像这个话题更多来自你的调研中对城市和空间的理解，你发现在时间维度上的建成空间和使用是无法匹配的，它展现出一种碎片化或者说是一种消减性，然后你把这个认识平移到你的工作方法当中，形成一个你前面所讲的形式逻辑，是这样吗？

庄：这是城市建筑的改变中可视的一部分，是日常使用中产生的变化，也是我们建筑调研经常能捕捉到的现象。另外一种状态来自生产体系的改变，往往是自上而下的，很多时候是不易察觉的。比如说近现代的建筑，有时候外部还是古典形式的，但它内在的结构与设备已经被新技术替代掉了。我觉得这两种情况可能都是存在的，我们是从这两方面去理解"片段和碎片化的局部"。

张：我觉得这当中有一个悖论。你前面讲过，这十多年来你的工作认知、投射都会有一个途径。这个途径，我感觉它一直有一个潜在的、对抗的东西，可能你要对抗的是当代中国某种僵死的建筑学，而支撑你采取不同方式的背后逻辑，很多情况下来自自下而上的状态。但是你的工作其实是投身于一个自上而下的创造过程，这和你刚刚说的自下而上的形式逻辑，或者说一种主体和系统对抗而产生片段，同时这个片段又是被你制造出来的——这里面是不是有一个悖论？

庄：我的确认为既有的建筑学应该重新审视对于建筑的基础认识，把它放到更大的系统里去，但我不是指把它放到艺术、政治的话语体系里，而是更倾向于放到技术与经济、

why I think the topics such as "unrecognizable system" and "research on future residences" you mentioned before are not suitable for such a context, since you can't break the production system. Here comes an implicit topic: with this existing system, what the world of architecture today should be?

ZS: It doesn't seem contradictory for us. We want to build our work on the comprehension of the existing production system. This is also a driving force for our theoretical reflection on the mainstream architectural discipline. Following my own logic, the architecture discipline should be more open-minded and evaluate their own cognition and method in the big system. Some of our questions are raised for traditional architecture, like the unrecognizable system. Some working methods are created from the perspective of production, like future residences. Our strategies such as on fragmentation are actually the reaction to this large system. It isn't a revolutionary logic that tries to create new paradigms and reestablish the world. On the contrary, it tries to reveal the original face of the system. With this cognition, we hope to bring new perspectives to architecture discipline and found that many methods derived from this lead to fragmentation.

ZB: I can fully understand. It seems to me that this topic is concluded more from your researches and the consequent understanding of city and space. You discovered that those built spaces can't match with everyday usage in the temporal dimension. It presents a sort of fragmentation or reduction. Then you turn this cognition directly into design methodology and create the formal logic you mentioned before. Right?

ZS: Well, this is the visible part of the change of urban buildings, the change resulting from daily use, a phenomenon we can frequently grasp through architectural research. Another aspect is the change generated by the production system, usually from top to bottom, which is barely perceptual. For instance, those contemporary and modern buildings may own a classical appearance, but the internal structure and equipment have been replaced by new technology. I believe both of these two aspects exist, based on this cognition, we tried to understand "fragment and the fragmentized part".

ZB: I think there's a paradox. You've just mentioned that you have had an approach of obtaining working cognition and its projection for over ten years. I feel your approach long has a potential confrontation, maybe you tend to against a stiff Chinese contemporary Architecture. In most situ-

生产的系统里，同时把它放到物质世界的时间性的客观改变里，这会更加中性客观。因此，从这个角度看，自下而上的个体改变并形成整体状态，和体系由上而下的决定作用，可以共同放在一个更基础的话语体系里面思考。

我觉得建筑学目前更像是处于混杂的转换时期，这个时候不管是自上而下还是自下而上，都表现出某种可以类比的局部性。因为不管是技术创新还是设计方法的创新，都不会是凭空突然跳跃到一个新的阶段，它会有一个变化的时期，我感觉我们更像是在这个时期。

张：我理解你讲的，但是我始终会存在一些怀疑。你说现在这个时期属于变化当中不太明确的时期，但从现代化开始，建筑学又有多少时间是稳定的？现代主义被瓦解的前后是不一样的。在现代主义被瓦解之前，建筑学再怎么变，建筑师还是在一个自上而下的环境中工作，虽然建筑师也会来探讨社会变化、生产方式变化、生活方式变化等问题，但他永远是一个造物者的角色，代表着权力，或者让渡权利往下推进。情况在 19 世纪 60 年代末发生了变化，我们看见了大量自下而上的工作方法，当然你可能会问这种工作方法还是不是建筑学？不知道，它可能很偏社会学、人类学，但是它确实产生了另外一种意义上的建筑师。这两个群体在全世界范围内都存在。但是这两个群体其实不共享所谓的共同的建筑学。

庄：但是我觉得这里面的工作方式有一致的地方。其实我也是站在一个自上而下的角度，自下而上不需要建筑师。

张：对，那是社会运动。

庄：但另一方面对于自上而下地决定建筑、预期建筑这件事，我自己的看法有一半是虚无的，我认为预期都是一厢情愿的。但问题是，如果预期都是作废，那工作的价值在什么地方？

张：我想到两个事情：一个是关于要不要变成一个不是自上而下的工作状态，现在很多所谓的自下而上的状态其实会有一个天然设定；另一件事，你刚才讲到建筑师免不了去做一些大尺度的项目，你在决定别人的生活，或者说你要提供一个东西让人家去消化，对那个东西的有效性你是怀疑的，但你又想提高它的有效性。我提的问题是，你怀疑之后，是想提高工作

ations, what supported you to work in a different way is a bottom-up state (the user gives feedback to architects). However, architects are always devoting to a top-down process (they usually serve the client or government), which is totally contradicted with the formal logic you just mentioned. This formal logic needs to be understood in a bottom-up way or through the fragments generated by the confrontation between the subject and the system. If this fragment has been created by yourself, isn't it a paradox?

ZS: I do believe that we should take a fresh look at the basic architectural knowledge when discussing the existing discipline of architecture. We shall review it in a larger system. Instead of reviewing it in the discourse of art or politics, I prefer to put it in the system of technology, economy and production, in the objective temporal change of the material world. This would be more neutral and factual. From this perspective, these two aspects, the bottom-up individual changes and the overall state they achieved, and the decisive effect made by the top-down system, could both be put into a more basic discourse to contemplate.

In my viewpoint, architecture discipline is now entering a hybrid transforming period. In this period, no matter resulting from a bottom-up or top-down process, their outcomes are of partial meaning, that's why they could always be compared. No technological renovation or design method update could jump into a new phase suddenly without experiencing a period of transformation. I feel we are now experiencing such a period.

ZB: I understand what you mean, but I remain deeply suspicious. You have highlighted the present period as a period of uncertainty, but since modernization, how often has architecture discipline been stable? Things are very different before and after the collapse of modernism. Before the collapse of modernism, no matter how architecture discipline changes, architects kept working in a top-down environment. Though they might discuss such issues as social changes, changes in production mode or lifestyle, the architect will always be a creator, who is on behalf of the authority or passing down powers. The situation has largely changed since the late 1960s. Many "bottom-up" working methods appeared. Of course, you may argue whether these methods are still in the category of architecture. There's no answer. Maybe they are quite related to sociology, anthropology, but they do bring in another kind of architects. These two groups of architects exist worldwide. But they never share the so-called common architecture.

ZS: But they work with some common methods. Actually, I am also in a top-down position.

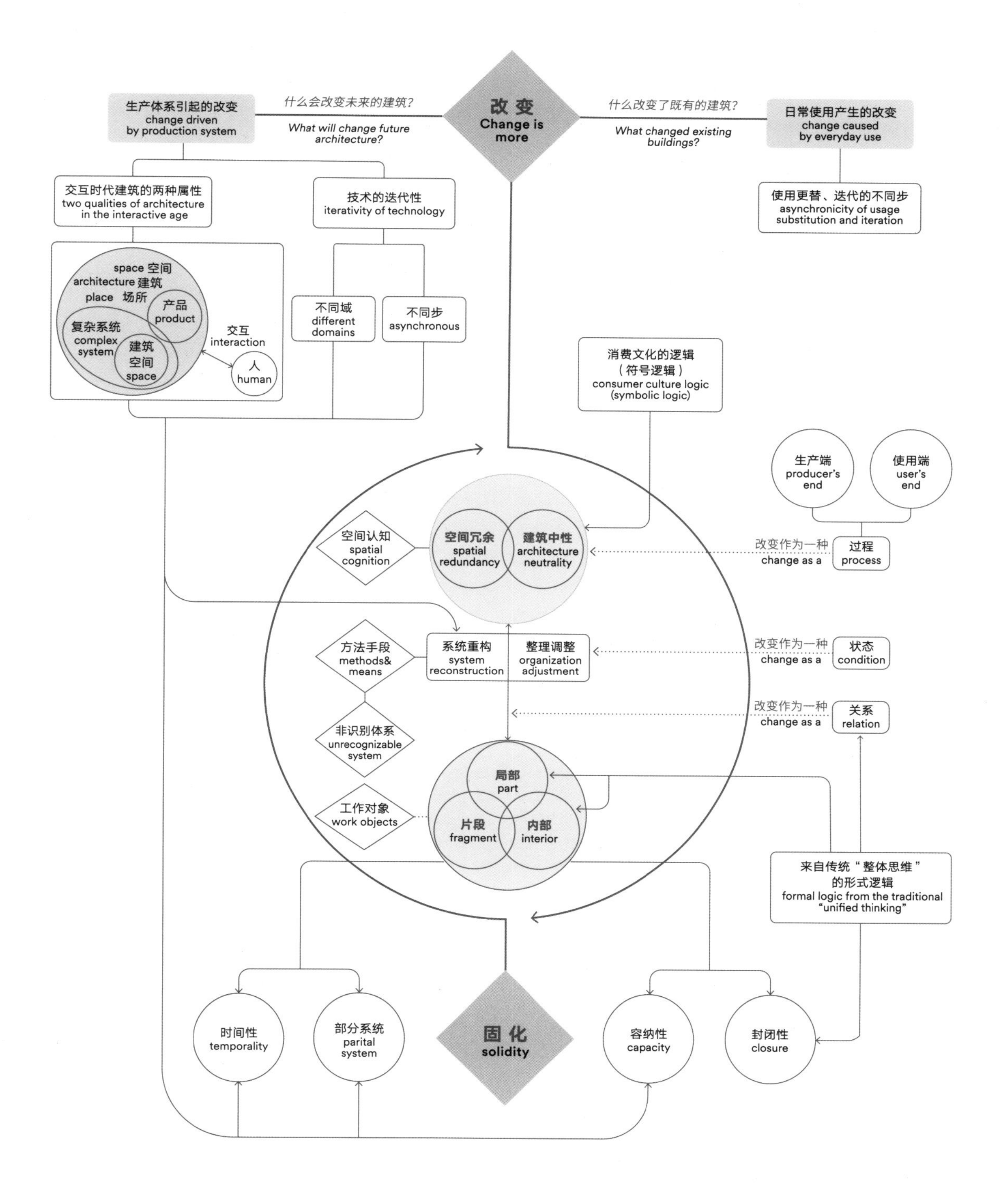

2 设计研究"改变"的理论框架

Theoretical framework for design research on "change"

的有效性，还是说你觉得不可能提高有效性？

庄：我觉得建筑师就是在不断创造冗余。建筑师设计一座城市，就像把一张纸涂满，其中有一部分是正确的，功能是固定的，这部分东西可以被预测，被精心塑造；但是另一部分工作塑造是没有效力的，有可能涂完之后它就会被改变。这里面矛盾的地方在于，你明明知道很多地方是瞎涂的，一方面会想就让它自己去变化好了，心安理得地接受它，变化是一件正常不过的事情，有什么好担心的；另一方面又在纠结我能不能换一种涂法？但是你又找不到一种新的方式，因为你不知道面对的是哪一种可能性。

3 《伊东丰雄：近期作品》
Toyo Ito: Recent Project (A.D.A EDITA Tokyo Co. Lit, Nov. 2008)

张：或者说到底，对于建筑师来说最后都是关心怎么涂。我猜你刚才讲的那种状态，其实是希望把尺度差异过滤掉，去做一种并置。在相对大尺度的项目当中，一定存在着某种系统，但是你又对它不满，然后去并置一个示范性的碎片，是这个意思吗？

庄：跟你说的有点相似。把原来的分离或者并置出一个系统，这两部分我们可能用不同的方式去对待。

张：像商业中心里面公共区域和非公共区域那样？

庄：可能隐含这种想法。

王方戟（后简称“王”）：刚才看了庄慎的这个总结性的图表[图2]，我觉得挺有意思。我也见过其他建筑师在做完这类工作之后，还真的突然之间就有了一种新的认识。2008年秋季，《GA》杂志找伊东丰雄做了一本书[图3]。之前他组织员工搞了三次总结座谈会，对自己以前的思想做了总结。最后总结出三个关键词：边界、洞穴和连续性。我在阅读其他资料时发现，在2001年他跟坂本一成讨论的时候，对自己的三个总结词跟这次的是不一样的。也就是说，他在2008年的时候刻意对过去所有的作品进行了归类。我后来通过这个规律再去研究，觉得他真的在总结之后，对自己的工作按照总结成果有意识地进行了推动。

张：从此他就有个明确的方向。

王：我们目前都把需要设计的建筑看成一个统一的和谐体，这好像是建筑学的基础

Architects are not needed for the bottom-up practice.

ZB: I agree. That's a social movement.

ZS: On the other hand, I am also a half nihilist on designing the building in a top-down way and making expectations. All expectations maybe only the architects' own wish. However, the question lies in: if expectations are invalid, what is the value for design work?

ZB: You remind me of two things. The first is whether it is unnecessary for architects to work with a top-down method. Many so-called bottom-up methods are actually presetted. Second, you just mentioned that every architect will work on some large-scale projects. You decide how other people are going to live, or rather, you have to provide something and let users slowly adapt to it. You suspect whether your work is effective, but you want to make it more effective. So my question is, with this suspicion, whether you prefer to make your work more effective or think it is impossible to improve effectiveness?

ZS: Well, I think architects are constantly creating redundancy. Designing a city is just like painting a whole paper. One part is correctly painted, with fixed functions. It is predictable and can be carefully designed, but the other part of design doesn't work, probably will be changed just after completion. Here comes a contradictory—you clearly know that much space is painted with no logic. On the one hand, you accept the change with peace of mind because change is perfectly normal, and worry is useless. On the other hand, you will think if you can apply a different way of painting. But you fail to find a new way, because you are facing an unknown possibility.

ZB: At the bottom, what architects concern about is how to paint. In the state you just mentioned, I guess you hope to clear away the scale difference. You want to make a juxtaposition. In a rather large scale project, there must be a system reasonable for the current state. But you are not satisfied with that, so you juxtapose a model fragmentation. Is it true?

ZS: Similar to what you said, we will create a new system by separating the original one or juxtaposing. We may deal with these two situations with different methods.

ZB: Like the public area and non-public area in those commercial centers?

ZS: Perhaps it implies such an idea.

Wang Fangji (WF): I have just reviewed Zhuang Shen's summarizing diagram [Fig. 2]. It is quite interesting. I have noticed that other architects achieved a new cognition after this

认识。建筑师总想要做一个有头有尾的东西，从大尺度到小尺度和谐统一。你也不能说它有问题。然而当建筑大到一定程度之后，你还认为它必须是个统一体的话就有问题了。庄慎之前经常跟我们讨论设计问题，通过这些讨论我理解庄慎说的"碎片化"的核心不仅仅是讲一种从下而上的想法，他所讲的"碎片化、片段、局部"等，实际上更是对把建筑当做统一体这类想法的对抗。为什么当它尺度这么大的时候还要维持统一性去做呢？设计完全可以从局部开始，并通过另外一种组织方式去实现各局部之间的关系。

庄：你说的可能是我的一个倾向性的看法：统一的和谐，在城市里根本就做不到，比如说每个建筑师做单体建筑的时候，会把它想象成是一个统一的东西，但是几个建筑师的设计放在一起又不统一和谐了。为什么建筑师每次都会把自己做的东西设计为一个完整城堡？我觉得这是设计的习惯或者说是学科的习惯。

张：王老师讲的和我认为的有点不一样，但我不是完全反对。我觉得非统一性是庄慎需要去为自己提供力量的一个点，他的工作还是在一个蛮传统的路线上，他对所有的事情都会去设定一个系统，只不过是用一种更强调矛盾、碎片、不对等的方式去展开。

王：无论碎片化还是整体化，从建筑学的角度看，它就是一个几何和体验之间的关系。讨论这种关联也就意味着把人的感受问题推到了建筑学问题的前面。庄慎刚才说的，我会认为还是关于系统和体验之间关系的事情。他的整个系统，无论是空间性的系统关系，还是时间性的系统组织关系，其中都涵盖了对人在建筑中的感知方式的讨论。这种感知方式利用了几何跟体验之间的关系，但往往超出了这种关系。在操作的时候庄慎会把它推向一个相对原始的关系。比如说人往往能靠直觉把握那些最原始、简单的形状，并理解自己处于一个什么样的状态。那种纯几何和体验之间的关联，更容易建立起设计与人的关系。庄慎在很多设计中使用的是这种更原始的系统性方式。

张：像我们这样的独立建筑师可能会更倾向于庄慎这样的，用更原始的方式去处理这类事

kind of summarization. In the fall of 2008, the Japanese magazine *GA* has published a book on Toyo Ito, *Toyo Ito: Recent Project* (A.D.A EDITA Tokyo Co. Lit, Nov. 2008) [Fig. 3]. Before this publication, Toyo Ito has held three meetings with his employees to summarize his former design philosophy. Finally, three keywords are concluded: Boundary, Ana (Cave) and Continuity. While I found in other sources, like his conversation with Kazunari Sakamoto in 2001, he had provided three different conclusions. In other words, he intentionally classified all his works in 2008. According to my own research, Toyo Ito does intentionally push his work later according to the summarization in 2008.

ZB: His work was then clearly oriented.

WF: Usually, we tend to see the designed building as a unified and harmonious entity, which seems to be a common sense of architecture discipline. Architects are always expecting a complete work, achieving harmony and unity from large scale to small detail. You can't say this is wrong. But it's problematic if you insist on this unity for a rather big project. Zhuang Shen often discussed with us about design issues before. Through these discussions, I found the core of his "fragmentation" is not restricted to a bottom-up idea. Terms he mentioned such as "fragmentation, segment, part" are sort of confrontation against the idea regarding architecture as a unified body. Why should we maintain unity for such a large scale building? We can absolutely start our design from parts and realize their relation through a different organization.

ZS: What you said perhaps is what I believe: we are actually impossible to achieve unity and harmony in cities. For instance, every architect will treat his single design as a unity, but when putting together, no unity and harmony exist among buildings. Why architects tend to design into a complete castle I believe it has roots in the existed design habit and disciplinary convention.

ZB: Prof. Wang has provided a different viewpoint with which I won't disagree. In my understanding, non-unity could be one point where Zhuang Shen draws his strength, while his work is still based on a traditional foundation in setting a system for everything. Of course, the system he is setting focuses more on a kind of contradiction, fragment or non-equivalent aspect.

WF: Whether fragmentation or integration, from the perspective of architecture, it's all about the relationship between geometry and experience. Discussing such a rela-

情。但是商业公司可能更擅长在一个当代消费层级中处理公共性质的消费空间，剥离掉设计表面的碎片之后，它其实非常基本，不管是专业人士还是非专业人士，一旦进入这个空间就能非常容易地理解它。而且它也能特别清晰地传达自己的意图，所有人都很容易捕捉到这些意图。排除美学的考虑，这些商业设计的方法、操作是很成熟的，甚至不比庄慎所想要达到的状态差。

庄：我一直很佩服商业建筑，因为我觉得它们特别有效、精确，有强大冷静的逻辑，这是由商业本身决定的。我们的建筑学往往缺少这个东西，离人特别远，感觉在自说自话，就是因为不能够非常直接地和整个系统对接。这样的一种内在逻辑性，是我挺想去追求的。

不管商业建筑，还是办公、住宅，有没有可能存在一种更具有建筑学通识的逻辑在里面？我们可以凭着这些逻辑来设计房子。如果是这样，建筑师就更能成为社会的一分子，虽然这点可能要打问号，但是至少有进一步去讨论的可能。如果存在一个这样的建筑学原理，那么它所支撑的技术和理念，应该能够既适用于商业建筑的价值判断，也适用于另一种建筑类型的价值判断。

张：王老师提了一个挺好的话题，关于和谐的问题。和谐其实是一种来自建筑学传统的惯性理解，可能也存在于建筑师想要为自己做的事情自圆其说的时候，想给自己一个理由。像文丘里对于商业、图像、消费、刺激这些事物的理解所产生的某种矛盾、片段、分离，我觉得积累到现在，在特别发达的成熟商业体系里面，它可能还是在制造一种和谐，因为和谐来自于一种对利益的保护。或者像我们刚才提到的特别成熟的商业系统的那种空间设定，它表面不和谐，但其实是和谐的，它有某种容错性，只要某种社会条件不变，它就可以相对和谐地存在下去。

那么根据这个话题，我想问下庄慎，如果我给你一个工作基础，同时你又去叠加一个特殊系统，那个系统可能是你更个人化的一种理解，但当这两个东西叠合在一起之后，可能从认知、美学上它并不和谐，它是一种矛盾。那么你是希望这个房子用这样的系统生产出来进入

tionship means to put people's feelings in a position prior to those architectural problems. I think what Zhuang Shen said is all about the relationship between system and experience. All his systems, either the systematic relationship of space or the systematic organization of time, cover such issues as people's perception inside buildings. This perception takes advantage of the relationship between geometry and experience, even goes beyond. During his operation, Zhuang Shen tends to apply a relatively primitive relationship. For example, just with their instinct, people are able to understand the most primitive and simplest shape and their own situation inside. It is easier to establish a relationship between design and human being through connecting pure geometry and perceptual experience. Zhuang Shen has applied such relatively original systematic method to most of his designs.

ZB: Independent architects like us may be interested in Zhuang Shen's working methods, dealing with these problems in a more primitive way. But commercial companies would prefer to put the public part of shopping space in the hierarchy of consumptions. Leaving alone those surface fragments, this strategy is actually very fundamental. Everyone, no matter professionals or non-professionals, can fully understand the situation when entering such spaces, because the intention could be clearly illustrated and easily conveyed. Put aside aesthetics preference, these commercial strategies and methods are full grown, even could reach the state Zhuang Shen has expected.

ZS: I really admire commercial buildings, because they are extremely effective, accurate and with strong calm logic. This is the result of commercialization. There is no such thing in our architectural discipline. Architecture stays far away from people, satisfying self-talking, due to its failure to connect directly with the whole system. Such kind of internal logic is what I desire.

Is it possible to find a general architectural logic, no matter in commercial, office or residential buildings, based on which we can complete our design? In that case, architects are easier to be a part of the society, though not absolutely sure, at least we can further discuss it. If such an architectural principle does exist, the technology and philosophy it contains should be adaptable to the value judgment of both commercial buildings and other types.

ZB: Prof. Wang has presented a good topic about harmony. In the category of architecture discipline, we have used to implant the idea of harmony into our design. Perhaps architects need

到社会环境后，它去促进一种永久性的和谐？还是希望制造问题，让它不和谐？

庄：这个话题很有意思。我知道在目前这样一个系统里面，我们是很难挣脱的，一切看似不同的意义都可以被消解。但正因为如此，从另外一面来说，建筑学、建筑师就可以完全不必拘泥，愿意尝试哪方面就去做好了。不必要的拘泥会成为一种自我制造出来的价值清规戒律，既妨害了不同类型的建筑实践方式的统一，也妨害了学科研究时的探索范围与价值目标的设定。

王：还是刚才那个话题，比如说一个大房子，它其实有一个基本成型的系统。你跟它的关系是什么？

张：我可以尝试帮他回答一下。庄慎在做这种改造的时候，他是把已经存在的东西作为一个条件，是一个即存物，然后往上叠加一个他自己的东西，一般来说会有个并置。

如果你做一个新的东西，你的逻辑会是怎么样的呢？千岛湖码头小镇那个项目很典型，如果换一个建筑师，那新系统最好是由空间逻辑、建造逻辑、结构材料产生。但是庄慎是把绝大部分花钱最多的部分做成一个通用系统，这个通用系统可能在不同项目里会不一样，通用系统弄好之后他再去加一个自己的东西，平台或者楼梯，然后将它们并在一起。

诸暨剧院也是一样，一千多座位的剧院、舞台这类内核不能变，然后他把公共空间和一些可以灵活组合的东西放在另一个系统里。但这个项目里另外的系统和主体在建造意义上没有太大的差异，可能也是因为设计得比较早，他并没有去做建造意义上的多样性。如果让他现在再做一遍，他可能会把外面的那个环形在建造意义上做得也不一样。我觉得在某种意义上，当你进入一个大尺度项目的时候，总归有一些东西没法被攻破，或者说你没法指望通过一个你所描述的新图景来替代掉。

庄：对。我一直没有试图去做一个全新的范式，我个人觉得那种典型性是做不到的，也没有意义。用一个新的方式做出一种新的系统性的范式来，本身我就觉得这种命题不存在。我做的那些东西，可能就是对于旧玩意或者被定义为“旧有”的一些改变、穿透、叠加、破坏，好像我更倾向于用这种方法。

to offer themselves a reason, which is actually a satisfactory explanation for their design. Taking Robert Venturi as an example, in a full and very developed business system, maybe the contradictory, fragment and separation formed on the basis of his understanding of commerce, image, consumption and stimulation are still creating harmony. Because harmony might root in protecting certain interests. Or rather, as we just mentioned, the spatial design of a fully developed commercial system is seemingly disharmonious, but actually in harmony. Since such design is open for faults, it is able to survive as long as the related social condition remains unchanged.

Under this topic, I'd like to ask Zhuang Shen a question. "If I provide you a work foundation, on which you can add a special system based on your own understanding. When these two put together, they become contradictory, look disharmonious from the perspective of cognition and aesthetics. What would you expect? After putting into the social context, will the building created on these systems promote an eternal harmony, or cause troubles and lead to disharmony?

ZS: Well, this topic is really interesting. I admit that we are all in this system, which to some extent, is difficult to escape. All seemingly different significances can be cleared up. But just because of this, on the other hand, architects really don't have to restrain themselves. They shall do what they really want to do. Instead, those unnecessary restrictions will turn into self-produced conventions, which not only prevents the unification of various practice methodologies, but also limits setting exploration range and value goal in disciplinary research.

WF: Let's continue the former topic. For instance, there is a big building, and it is actually an almost completed system. What exactly is your relationship to it?

ZB: Let me try to answer for him. I strongly believe that when Zhuang Shen starts a renovation, he will take the existed situation as a precondition, an existing object, then add his own design on it. The strongest possibility is to apply a juxtaposition; for me, this is the most impressive part.

If you need to design a new building, what is your working logic? East Dock Town in Thousand Islands Lake (Hangzhou) is a typical project. Other architects may put the new system on spatial logic, tectonic logic or material issue. However, Zhuang Shen has turned the part which may cost the most budget into a general system. This general system may vary in different projects. When getting this done, he will add his new system, like a platform or a stair, and then combine these two systems together.

Zhuji New Theatre is a similar case. The core that the theatre has, an auditorium of more than one

张：我觉得这种方式在你身上应用得还蛮有效的，好像一个特别复杂的问题会被你用这样的方式整理出条理，哪些是不用去触碰的，哪些是在现有资源下面有信心可以搞定的。

庄：我基本上会快速地分析一下各种因素和矛盾，然后把它自动地分解为一些类型，或者分解为一些体系、一些矛盾，然后再把某一组拿出来，有点像拆解组合的方式。

张：这东西我觉得跟类型学有关。

庄：一个任务，一旦放到具体的情况里面，因为涉及很多因素，就会和这个任务产生多线的关系或矛盾，我觉得我们更多是在理清这些东西，循着一个个实际的矛盾因素开展设计，这好像不太像类型学的方法。

王：从一个第三者的角度来看这些关系，其实都有一个非常强的简单体验在里面。我觉得这也是一个线索，并不仅仅是把这些经验叠加在一起，而是把人的体验考虑进去，这样加完之后会带来一个特定的新的体验方式。

庄：对，当你分析任务的时候，它会具体化。比如说你得到一个任务书，其实那个任务书写得再详细也还是抽象的，因为它没有时间、地点、周围的状态。

王：关于诸暨剧院，你刚才好像是说因为人要从较远的地方过来，所以对它的有效性保持怀疑。但实际上即使他要走那么远过来，因为有其他活动以及公园的背景，还是产生了相应的场域的，所以我觉得这个策略还是非常有效的。

对于大型建筑，我感觉最大的特点是可以动用更多的资源。动用的资源可以让你的这些片段、局部以某种方式或者以更大尺度的方式实现。

张：我觉得庄慎的工作方法和某些观念在早期实践中就存在了，感觉得到他不是很喜欢做特别统一的东西。阿科米星初期探索的那条线，可能只是强化了这样一个事实。后来通过对城市的认识，把操作上的不统一性、矛盾性放大了，是不是可以这样理解？

庄：可能是这样的，因为我确实会喜欢不一样的东西，习惯于把建筑放到各种关系中去看，我觉得最早有可能是这样产生的。

张：其实说到底还是组织。

thousand seats and a stage, is impossible to change. So he combines those less relevant public spaces and flexible programs into another system, though in this project, there is no major difference between this new system and the main building in tectonic sense. Maybe as it was his early work, he hasn't pursued construction diversity then yet. If he was asked to design this theatre again, he would build the ring differently. Nevertheless, when you are dealing with a large project, there's always something you can't break through, or replace with your own vision.

ZS: You are right. I haven't been trying to make a brand new model. In my viewpoint, it's impossible and meaningless to pursue such typicality. It is even a false proposition that you can create a new systematic paradigm with a new method. I prefer to apply old strategies such as change, penetrate, juxtapose and break, which may be defined as "old".

ZB: I found you have successfully applied this method. An extremely complicated problem will be classified into those untouchable and those you own the confidence to handle with available resources.

ZS: Basically, I will make a quick analysis on all kinds of factors and contradictions, classify them into several types, systems or contradictions, and then highlight one aspect, just like disassembling and combining.

ZB: I think it relates to typology.

ZS: Put a project into a specific situation, since many factors are involved, and multiple relationships or contradictions will occur with the program. We are more likely to clear these issues and kick-off the design following real conflicts. I don't think it has something to do with typology.

WF: From the perspective of a third party, all these relationships contain strong but simple experiences. For me, this is a clue. It is not only a simple combination but also taking human experience into consideration. After adding a new system, a specific fresh experience is achieved.

ZS: Yes! The project becomes more concrete when you analyze the program. For example, you are given a commission. No matter how detailed the program is written, it is still abstract, because no specific time, place or surrounding is included.

WF: Respect to Zhuji New Theatre, you have doubted its effect because people need to travel a long journey to reach this place. In my understanding, even when they come

庄：是，因为以前讲“组织”时，觉得它是一种消费逻辑，继而是一种形式操作方法。后来就不太去说了，因为我觉得自己在形式方面可能比较灵活，认为形式不是重要的，所以就有点偏执，反过来回避这件事情。我现在回过头来想一想，觉得这个事情不俗，只不过你不要把它看成是纯形式的一种方式。

张：某种自己的自反。

庄：有可能吧。我现在有点觉得我们的员工不太适应我的方式，我说这个事情不是很好做吗？后来发现其实是因为我自己做起来比较灵活。

张：为什么你会觉得做一些分离，或是赋予一种新的子系统，或去构建各种系统的关系是很容易的？这里有几个前提：一是你形式操作能力很强；还有你的场景体验，或者是某种叙述性、戏剧性这种直观层面上的感觉也比较敏锐。但是我觉得刚才那个话题当中其实存在类型学。我们把一个念头推演到最终的结果总是形式，这个形式是靠一种内在结构方式构建的。比如说它到底是什么，你把它讲清楚，这不就是一个类型吗？

庄：如果像你这么说的话可能就是这样。因为形式确实可以有模式和类别。我自己考虑形式的话，肯定会关联地去想某种形式可能是某种模式，或者它像什么。

张：早期你们讲“组织”时，都是寻常的系统，拿来的素材只是空间不同，但整个系统是比较均质化的。到现在，我觉得你其实还是靠组织在工作，我的理解是它还是组织，只不过现在你的素材发生了变化，你会有更不一样的东西。我觉得可能你在弱化“组织”这个词，但是其实好像还是一脉相承的。

庄：对，我觉得这个词很好，事实上最近我对它有些新的认识，只不过可能它更容易被简单理解为一种操作方法，所以现在总结系统关键词时不大提及了。像我很早以前做过的路桥改造项目［图4］就是组织的最早、最原始的方式，它一直影响到后来的很多项目，比如常州棉仓的设计［图5］。

张：类似这种工作方法，或最早出现的这种状态是来自于哪些先例？

庄：早期可能就是当年硕士论文［图6］时候留下来的形式操作方法，对关系十分重视与

4　台州路桥商业建筑
Taizhou Luqiao Commercial Building

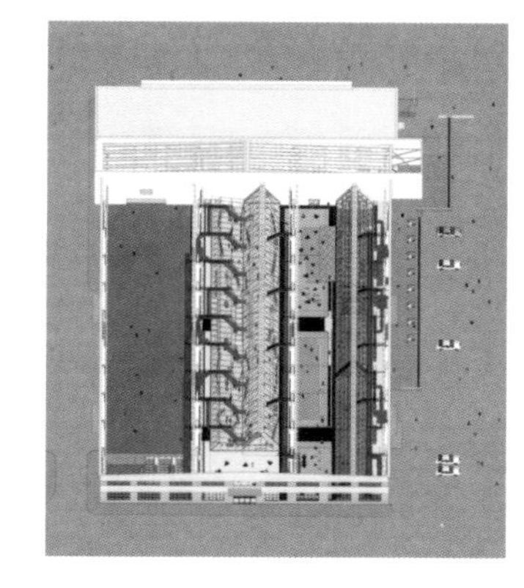

5　棉仓城市客厅
Cotton Lab Urban Lounge

from far away, this place is still special because of other activities and the adjacent park. So this strategy does work well.

For a big project, you can take advantage of more resources to implement the ideas of fragments and parts in some way or on a larger scale.

ZB: According to my understanding, the core of Zhuang Shen's design method or concept was actually born in his early practice. He was never satisfied with producing unified schemes. It has been strengthened by the establishment of Atelier Archmixing. Later, through urban researches, this approach aiming at non-unification and contradiction is highlighted. Am I right?

ZS: Possibly like what you are saying. I do like different things, and I am used to thinking about architecture in various relationships. That's how my working style was developed.

ZB: In fact, it's organization.

ZS: Yes. I used to see "organization" first as a consumption logic, then a formal and operational approach. Later, I didn't talk much about "organization" any more, because I might be flexible in form creation, so I don't think it is important. I became even a bit stubborn in trying to avoid "form". Looking back, this is actually not so conventional, if you cease to take the organization as a way only for pure form.

ZB: Kind of self-confrontation.

ZS: Maybe. I have a feeling that our employees have difficulties in fitting into my working style. I always insist that design is an easy job, but later I find that it's only because I am flexible in practice.

ZB: Why do you take it easy to make some separations, create a new subsystem or establish relationships among various systems? There are several preconditions. The first is that you are pretty strong in the formal operation, and the second is that you are also strong in scene experience or intuitive sense, something narrative and dramatic. But I insist that the topic mentioned contains typology. A concept will materialize eventually in some form, which is established on an internal structure. What on earth is it? If you can explain it clearly, isn't it a type?

ZS: Maybe you are right because the form can be classified as models or types. When I am thinking of form, I will definitely relate it to a certain model or think about what it looks like.

ZB: In the early stage, the organization you mentioned was a general system. Although those borrowed elements bring in different spaces, the whole system is rather homogeneous. Now I still

敏感，还有对事情习惯性的正反双向思维，所以在设计中总是用更多方式去组织关系。然后这个方式在研究消费逻辑时得以加强、补充，因为开始把形式和意义剥离开来看待，于是形式组织方式就变得很丰富。我觉得这是我们工作的形式操作基因。

现在我看到城市里面的一些建筑、改造项目，会觉得特别亲切，因为它很多地方是符合那个基因的。原来的房子改了一点点，不就是所谓的依附吗？它们是各自成就的，我一点都不会觉得违和，特别能够理解这件事情，然后我就会愿意去做局部，让我改个立面也不会觉得不爽。所以最早的那些意识影响了形式操作的方式。

张：我觉得可能不只是形式操作，你刚才讲的东西我也挺认同，因为我能理解你为什么是这个样子，主要还是从对空间的想象出发的。

王：按照我的理解，首先是要确立新和旧，哪怕是你自己创造出来的新和旧，那么这两者之间的关系就成了一种体验或组织的系统，然后很多可能性就可以从这个关系开始。

庄：每个设计都是对既有环境的改变，都会产生新问题，一旦有了新问题就会有新的方式再去解决它，这是一个类似递归的子系统。只要你抓住一个问题，它就会产生一个关系。

王：一个项目有它具体的条件，你用什么物质方式去呈现，是不是可以超越具体性，会有一个抽象的想法去建立建筑的基本结构吗？或者说你对于一个以单一的、超越具体性的抽象存在开始的设计方法，会怎么去评判？

庄：我应该没有超越这个具体性。比如说在一堆关系里面拿出一个关系来，其实先就设定了一个具体的目的。为了达到这个目的，肯定会涉及具体的材料、空间景象或者具体的功能。这中间可能存在完全不一样的变化，但是其实目标是一样的，所以这就使它变得更具体。

王：对，现在没有超越具体性。因为无论是关系还是最后的落实，都还是比较具体的。你是否有以一个抽象的系统，将想法落实到场地中的工作方法或者说工作意识？

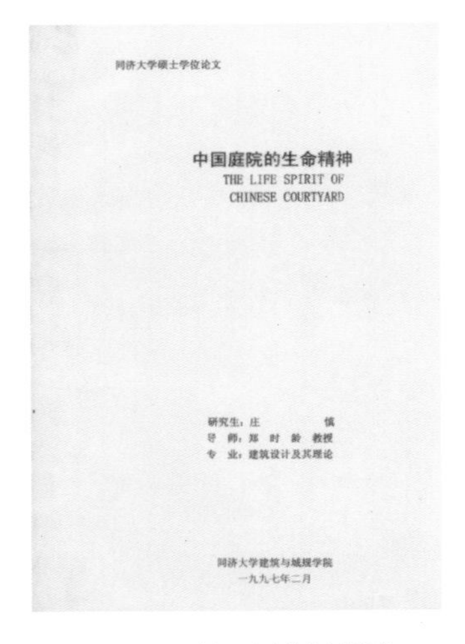

同济大学硕士学位论文

中国庭院的生命精神
THE LIFE SPIRIT OF
CHINESE COURTYARD

研究生：庄　　慎
导　师：郑 时 龄 教授
专　业：建筑设计及其理论

同济大学建筑与城规学院
一九九七年二月

6　庄慎硕士论文《中国庭院的生命精神》
Zhuang Shen's masters' thesis "The Life Spirit of Chinese Courtyard"

believe you are working by organizing. At least, I still take it as an organization. But you've got different elements to achieve different effects. Perhaps you no longer want to highlight the word "organization", but actually, it has been inherited.

ZS: Yes, "organization" is a good term. In fact, recently I had a new understanding. The reason why I seldom mention it as a keyword in our recent summarization is that it's easily taken as an operational method. In the very early renovation project in Luqiao, Zhejiang Province [Fig. 4], I have applied this "organization", in a very primitive way. Many recent projects can still trace back to this idea, including Cotton Lab Urban Lounge in Chang Zhou [Fig. 5].

ZB: Do you have any precedents for your working method? Or what kind of previous references have led to your early application?

ZS: The earliest case is the form operation method resulting from my master thesis [Fig. 6]. Through that research, I started to focus on the relationship and became very sensitive to it. I developed a habit of two-way thinking, which means I was able to apply more methods in organizing relationships in design. These methods were further strengthened and complemented when we examined the consumption logic. Through liberating the form from its significance, we developed various formal organization methods. This is the gene of our formal operations.

Now I feel blessedly familiar with urban buildings and renovation projects, because they are close to our working gene in many aspects. Isn't the little change of the existed building close to the so-called "attachment"? They are mutually supporting. I never feel strange and can truly understand this situation. Then I am willing to design a part, happy to make a façade renovation. In short, those earliest cognitions have led to our formal operation method.

ZB: I don't think it's only about the formal operation. I quite agree with what you said. You are what you are, mainly because of your strong capacity in spatial imagination.

WF: According to my understanding, first we have to establish what is "new" and what is "old", even if they were created by yourself. Then the relationship between them will become a system of experience or organization. Many works can start from such a relationship.

ZS: Every design intends to change the existing environment and brings new problems. New methods will be created to solve these problems, like a recursive subsystem. Once a problem

庄：我没有想以一个抽象的方式或者某种显著的标示来展开设计，但是我觉得会有一些固定的东西和自己的原则，我会更倾向于找到一些关联，把建筑变成关系中的一分子。

张：这可能跟我们读书时候的状态有关系。我从读书开始就从来不会设想一个均质、统一的内在结构，我好像总是抗拒做那种纯粹的设计，我发觉自己做的设计也是一种矛盾的叠加并置，在这一点上我跟你比较像，但是我们具体的操作方式又不太一样。
庄慎会把它剥离得很清楚，他会把一个复杂的问题简化掉，简化为清晰的可以被体验、阅读的两种东西，然后把它们轻松地结合，实现出一种状态。他这种做法相对来说更灵活一点，但是我们在思维习惯上是有某种共性的，我在抽象层面上也是天然抗拒纯粹性。这种状态是和个人相关，还是和某种教育系统有关？

王：对于这一点我觉得很理解。做设计的时候有些人一上来就是左右手互博，也就是抓一系列最核心的问题，在希望同时解决这些问题的努力之中推进。不同建筑师的设计方法是不同的，这也是对于建筑是什么的观念的差别。有的建筑师也许一上来就把建筑设定为一个独立而稳定的系统，然后将这样一个系统维持到最后。每个项目在进展的过程中都有一些具体的问题不断出现，很难在一开始就找到平衡点。也许是不得不在一种互搏的过程中逐渐将系统稳定下来吧。如果一上来就是互搏的话，就是将建筑中各种条件之间逐渐达到的关系默认为一个前提。
刚才张斌也问到一个类型学的问题。我发现有一些建筑师不管场地什么样，项目具体如何，都能扔个类型上差不多合适的东西过来，然后再去调整关系。有意思的是，通过这种方法也能达到一种最后稳定的状态。

庄：以前在同济上学的时候，老师不会先教你一个强势逻辑，然后叫你用这种强势逻辑把功能搏进去。我不知道是不是因为上海这种城市的氛围。

张：也有可能，我觉得可能跟我小时候对于这座城市的体验也有关。上海给我的感觉就是一个无法统一的城市，脑子里都是片段，各种片段。

庄：而且它又很务实[图7]。

is held, a relationship will occur.

WF: Each program has its specific conditions, how will you present it materially? Will you apply abstract concepts instead of concrete considerations to create a basic structure for the building? Or what do you think of a design method starting from a single and abstract idea beyond concrete issues?

ZS: I haven't transcended this concreteness. To pick a relationship from a dozen of relationships is actually to set a specific goal. Specific materials, space sceneries or specific functions are indispensable in the process of achieving this goal. There might be totally different changes, but the goal remains the same, thus it is made more specific.

WF: Yes, now you haven't transcended the concreteness. No matter in terms of the relationship or the final execution, it is relatively specific. Can you apply your idea to the site with an abstract system? Did you ever find such a working method, or are you conscious of working like this?

ZS: I have never thought of designing with an abstract method or some prominent identity, but I confess that I have some fixed philosophy and my own principals. I'd be far more inclined to find some relationships and turn architecture into one of them.

ZB: Perhaps it has something to do with our college life. In my case, there is something in common, although now our concrete operations are rather different. I have never imagined a homogeneous and unified internal structure since I entered the university. It seems that I have always been resisting those pure designs. My designs always consist of a contradictory juxtaposition and combination. On this point, we are alike, but we work with different specific methods.
Zhuang Shen will clearly separate it. He will simplify a complex problem, turning it into the combination of two clear, perceptible and legible situations that are accomplishable. Relatively, his method is more flexible, but we share some common thinking habits. I can't think of those pure things, at an abstract level, pureness doesn't exist in my dictionary. Is this state related to an individual or some education system?

WF: I completely understand. You start the design like fighting your left hand with your right hand, which means catching a series of key problems at the very beginning, and developing the design with the hope of solving these problems simultaneously. Different architects use different methodologies, which roots in their varied comprehension of what architecture is. Some architects may see architecture as an independent and stable system and keep it unchanged to completion. With the development of each project, specific problems constantly appear, so it's very difficult to find a balance at the start. Perhaps the system has to be gradually stabilized with this self-fighting. Starting from a self-fighting means taking the relationship evolved gradually between architectural conditions as a prerequisite.

Zhang Bin just asked about typology. I found that some architects are able to apply a roughly suitable type of design no matter what the site is or how specific the project is. Then they start to adjust these relationships. Interestingly, they can also achieve a stable state through this method finally.

ZS: When we studied architecture at Tongji University, professors wouldn't teach you a strong logic at first and then apply it and implant functions. I don't know whether it is due to the urban atmosphere in Shanghai.

ZB: Possibly, I think maybe it has something to do with my experience in this city in childhood. For me, Shanghai is a city that can't be unified, full of fragments, a variety of fragments.

ZS: And the city is also very practical. [Fig. 7]

王方戟

同济大学建筑与城市规划学院教授，上海博风建筑设计咨询有限公司主持建筑师。
《时代建筑》杂志兼职编辑，《世界建筑》杂志编委，《建筑师》杂志特邀学术主持，《西部人居环境学刊》通讯编委，南京大学建筑学院研究生建筑设计课课程教授。其建筑设计作品大顺屋、带带屋、桂香小筑、七园居等在国内杂志上多次发表，并多次参加国际建筑展。

张斌

致正建筑工作室创始合伙人，同济大学建筑与城市规划学院客座教授。
他的研究与实践涵盖城市、建筑、室内和景观多个领域。至今已完成多项重要作品，包括同济大学建筑学院 C 楼、同济中法中心、青浦练塘镇政府办公楼、东岸望江驿、崇明东滩湿地科研宣教中心等。参与展览：2017 上海城市空间艺术季、2017 深港城市 / 建筑双城双年展、2019 年未知城市：中国当代建筑装置影像展等。

7　王方戟（右）、张斌（中）与庄慎（左）对谈中
Dialogue among Wang Fangji (right), Zhang Bin (middle) and Zhuang Shen (left)

WANG Fangji

Professor of College of Architecture and Urban Planning, Tongji University. Principal Architect, TEMP Architects. Part-time Editor, *Time+Architecture* Magazine, Shanghai. Member of Editorial Committee, *World Architecture* Magazine, Beijing. Guest Moderator, *The Architect* Magazine, Beijing. Corresponding Editor, *Journal of Human Settlement in West China*, Chongqing. Studio Advisor, School of Architecture, Nanjing University. His works Dashun Pavilion, Daidai Pavilion, Laurel Fragrance Follé and Hotel Septuor are published in Chinese journals and exhibited in international architectural exhibitions.

ZHANG Bin

Co-founder of Atelier Z+, Visiting Professor of Collage of Architecture and Urban Planning, Tongji University.
He established Atelier Z+ with Zhou Wei in 2002. His field of research and practice covers the city, architecture, interior and landscape. Atelier Z+ has so far accomplished several significant projects such as Building C of College of Architecture and Urban Planning of Tongji University, Sino-French Centre of Tongji University, Liantang Town Hall in Qingpu, Shanghai, River View Service Stations, East-bund, Pudong, Shanghai, Wetland Research and Education Center, Dongtan, Chongming, Shanghai, and has participated in several exhibitions: 2017 SUSAS, 2017 Bi-city Biennale of Urbanism/ Architecture, 2019 Unknown City, etc.

作品
Selected Projects

决定我们工作的几个因素
Several Factors that Define Our Work

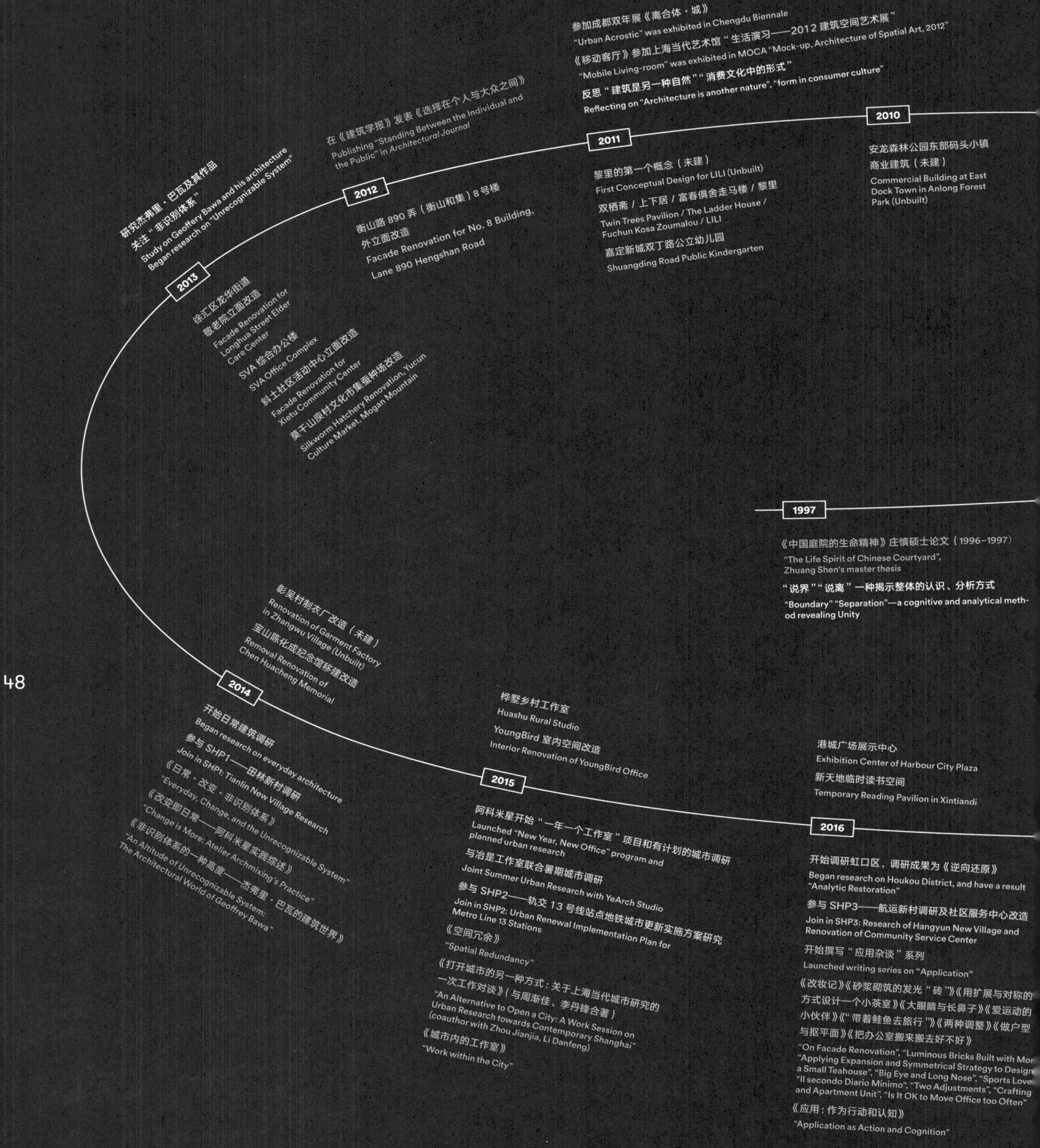
参加成都双年展《离合体·城》
"Urban Acrostic" was exhibited in Chengdu Biennale
《移动客厅》参加上海当代艺术馆"生活演习——2012 建筑空间艺术展"
"Mobile Living-room" was exhibited in MOCA "Mock-up, Architecture of Spatial Art, 2012"
反思"建筑是另一种自然""消费文化中的形式"
Reflecting on "Architecture is another nature", "form in consumer culture"
2010
安龙森林公园东部码头小镇商业建筑（未建）
Commercial Building at East Dock Town in Anlong Forest Park (Unbuilt)
2011
黎里的第一个概念（未建）
First Conceptual Design for LILI (Unbuilt)
双栖斋 / 上下居 / 富春俱舍走马楼 / 黎里
Twin Trees Pavilion / The Ladder House / Fuchun Kosa Zoumalou / LILI
嘉定新城双丁路公立幼儿园
Shuangding Road Public Kindergarten
在《建筑学报》发表《选择在个人与大众之间》
Publishing "Standing Between the Individual and the Public" in Architectural Journal
2012
衡山路 890 弄（衡山和集）8 号楼外立面改造
Facade Renovation for No. 8 Building, Lane 890 Hengshan Road
研究杰弗里·巴瓦及其作品
关注"非识别体系"
Study on Geoffery Bawa and his architecture
Began research on "Unrecognizable System"
2013
徐汇区龙华街道敬老院立面改造
Facade Renovation for Longhua Street Elder Care Center
SVA 综合办公楼
SVA Office Complex
斜土社区活动中心立面改造
Facade Renovation for Xietu Community Center
莫干山庾村文化市集蚕种场改造
Silkworm Hatchery Renovation, Yucun Culture Market, Mogan Mountain
1997
《中国庭院的生命精神》庄慎硕士论文（1996-1997）
"The Life Spirit of Chinese Courtyard", Zhuang Shen's master thesis
"说界""说离"一种揭示整体的认识、分析方式
"Boundary" "Separation"—a cognitive and analytical method revealing Unity
彰吴村制衣厂改造（未建）
Renovation of Garment Factory in Zhangwu Village (Unbuilt)
宝山陈化成纪念馆移建改造
Removal Renovation of Chen Huacheng Memorial
2014
开始日常建筑调研
Began research on everyday architecture
参与 SHP1——田林新村调研
Join in SHP1: Tianlin New Village Research
《日常·改变·非识别体系》
"Everyday, Change, and the Unrecognizable System"
《改变即日常——阿科米星实践综述》
"Change is More: Atelier Archmixing's Practice"
《非识别体系的一种高度——杰弗里·巴瓦的建筑世界》
"An Altitude of Unrecognizable System: The Architectural World of Geoffrey Bawa"
桦墅乡村工作室
Huashu Rural Studio
YoungBird 室内空间改造
Interior Renovation of YoungBird Office
2015
阿科米星开始"一年一个工作室"项目和有计划的城市调研
Launched "New Year, New Office" program and planned urban research
与冶是工作室联合暑期城市调研
Joint Summer Urban Research with YeArch Studio
参与 SHP2——轨交 13 号线站点地铁城市更新实施方案研究
Join in SHP2: Urban Renewal Implementation Plan for Metro Line 13 Stations
《空间冗余》
"Spatial Redundancy"
《打开城市的另一种方式：关于上海当代城市研究的一次工作对谈》（与周渐佳、李丹锋合著）
"An Alternative to Open a City: A Work Session on Urban Research towards Contemporary Shanghai" (coauthor with Zhou Jianjia, Li Danfeng)
《城市内的工作室》
"Work within the City"
港城广场展示中心
Exhibition Center of Harbour City Plaza
新天地临时读书空间
Temporary Reading Pavilion in Xintiandi
2016
开始调研虹口区，调研成果为《逆向还原》
Began research on Houkou District, and have a result "Analytic Restoration"
参与 SHP3——航运新村调研及社区服务中心改造
Join in SHP3: Research of Hangyun New Village and Renovation of Community Service Center
开始撰写"应用杂谈"系列
Launched writing series on "Application"
《改妆记》《砂浆砌筑的发光"砖"》《用扩展与对称的方式设计一个小茶室》《大眼睛与长鼻子》《爱运动的小伙伴》《"带着鲑鱼去旅行"》《两种调整》《做户型与抠平面》《把办公室搬来搬去好不好》
"On Facade Renovation", "Luminous Bricks Built with Mor "Applying Expansion and Symmetrical Strategy to Design a Small Teahouse", "Big Eye and Long Nose", "Sports Love "Il secondo Diario Minimo", "Two Adjustments", "Crafting and Apartment Unit", "Is It OK to Move Office too Often"
《应用：作为行动和认知》
"Application as Action and Cognition"

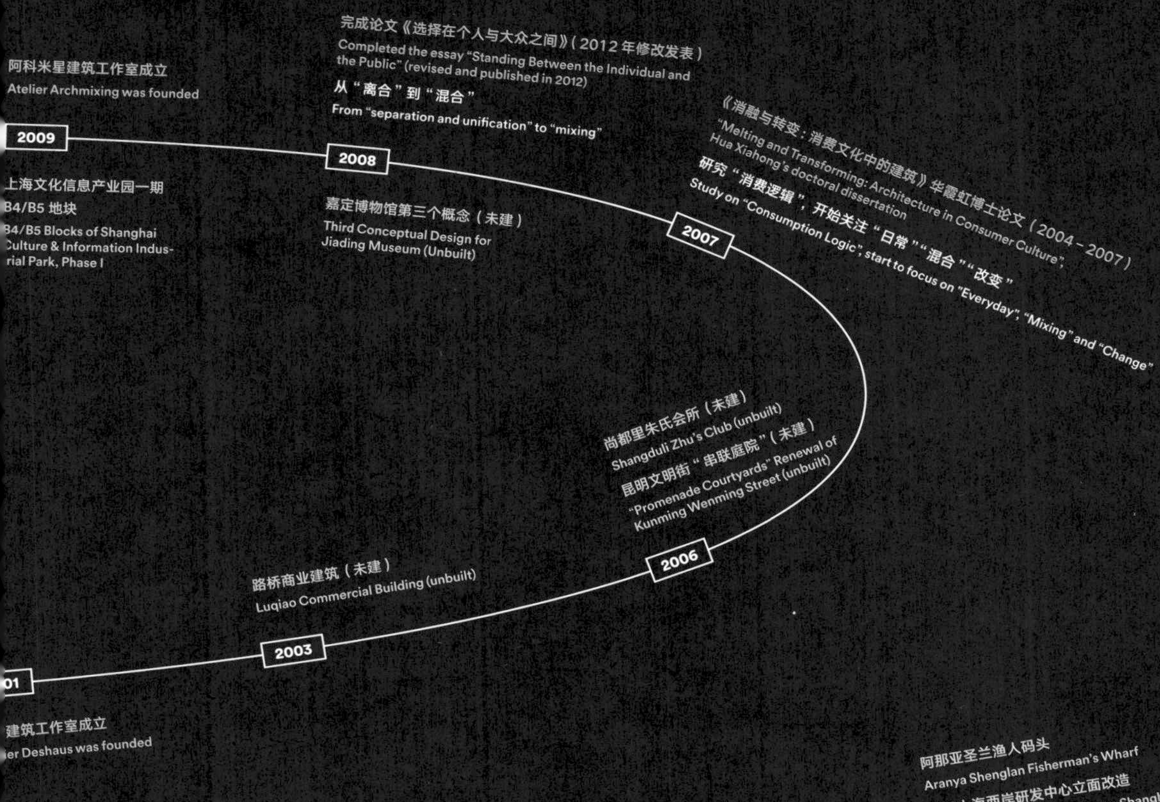

阿那亚圣兰渔人码头
Aranya Shenglan Fisherman's Wharf
网易上海西岸研发中心立面改造
Facade Renovation for Netease Shanghai West Shore Development Center
申威达厂房改造
Renovation of Shenweida Workshop Space
永嘉路口袋广场
Pocket Plaza, Yongjia Road
宝山绿心幼儿园
Greenheart Kindergarten in Baoshan District
昆山实验小学西侧地下车库改扩建
Expansion and Renovation of Kunshan Experimental Primary School

上海南汇新城中法学院
Sino-French Institute, Nanhui City
智慧欧莱托育园
Wisdom Olion Nursery
舟山青龙山公园入口改造
Qinglong Mountain Park Entrance Renovation
阿那亚金山岭接待中心
Reception Center at Aranya, Jinshanling
徐汇养老基地项目
Xuhui Elder Care Center
棉仓城市客厅
Cotton Lab Urban Lounge

宝山贝贝佳欧莱幼儿园
Baoshan Beibeijia Olion Kindergarten
成都麓湖生态城 D6D7 地块小学幼儿园(未建)
Primary School & Kindergarten on Plots D6D7 in Chengdu Luhu Ecological City (Unbuilt)
悦阅书店
Yueyue Bookstore

2017

2018

2019

开始调研徐家汇——城市的内部空间
Began research on Xujiahui—the Inner Space of the City

开始"空间冗余"杂谈系列
Launched writing series on "Spatial Redundancy"

《局部》《内部》《有关冗余的笔记 1:poché》《有关冗余的笔记 2:老白的实验室》《有关冗余的笔记 3:<火星救援>与二手宇宙》
"Parts", "Interior", "Note on Redundancy 1: poché", "Note on Redundancy 2: Water White's Laboratory", "Note on Redundancy 3: *The Martian* and Used Universe"

整理观念、策略、设计方法的线索
Summarizing concepts, strategies and design methods

改变 Change
建造 | 使用 Construction | Use
冗余 Redundancy
中性的建成环境 Neutrality of Built Environment
非识别 Unrecognizable
创造 | 整理 Creation | Organization
创新 | 调整 Innovation | Adjustment
系统 | 片段 System | Fragment
性能 | 效能 Performance | Effect
整合 | 局部 Whole | Part
外部 | 内部 Exterior | Interior

《向日常学习》(《建筑研究季刊》英文论文)
"Learning from the Everyday World" (English Paper featured in *Architectural Research Quarterly*)

《〈火星救援〉与二手宇宙》
"*The Martian* and Used Universe"

《逆向还原:城市空间演变的日常逻辑 / 对话庄慎与华霞虹》
"Reverse Restoration: Daily Logic of Urban Space Evolution/ Interview with ZHUANG Shen and HUA Xiahong"

《走向城市建筑学的可能:"虹口 1617 展览暨城市研究"研讨会评述》
"Possibilities towards an Urban Architecture: Notes on the Seminar of 'Hongkou 1617 Exhibition and Urban Study'"

《整体的回响:来自内部的反向思维》
"The Echo of the Unity: Reverse Thinking from the Interior"

《日常,改变,非识别体系》(《建筑设计》英文论文)
"Everyday, Change, Unrecognisable System" (English Paper featured in *Architectural Design*)

《看不见的改变:论使用端空间技术化的可能》
"Invisible Change: Technologizing Space at the User's End"

《柔性倍增——基于人工智能、移动机器人、智能立库的未来居住空间概念研究》
"Flexible Multiplication: Spatial Concept on Future Living Integrating Technology of AI, VGA and Intellectual Elevator-Shaft Garage"

《城市游牧工作室》参加深圳坪山"未知城市:中国当代建筑装置影像展"
"Urban Nomadic Office" was exhibited in "UNKNOWN CITY: China Contemporary Architecture and Image Exhibition", Shenzhen

《内的宣言》联合展览
"Manifestos on Interiority" joint exhibition (Archmixing/ Temp Architects/LanD Studio)

开始研究"设计面对未来复杂人工世界的方法"
In search of "the Methodology of Design Towards Future Complicated Artificial World"

开始"移动办公模式"
Launched "Mobile Office Mode"

《阿科米星工作室》
"Atelier Archmixing"

在我们自己看来，形成阿科米星工作特征的因素主要包括以下五个方面：

1. 隐性的形式逻辑
2. 消费逻辑带来的冷静
3. 非识别的日常城市与建筑
4. 变化中的不变积累
5. 被忽视的使用端与改变

这些先后发现并逐渐融合形成整体的因素，不约而同都带有某种隐性的特征。

1 隐性的形式逻辑：影响所有设计思维方式的产生

阿科米星不是一个风格式的事务所，我们的设计没有形式上的固定趣味，也不追求形式上的固定趣味。

建筑空间与形式的设计被我们视为用来解决实际问题的手段。因为实际问题的多样性与不确定性，我们倾向不受形式、空间的既有使用方式与含义的约束，自由灵活地组织应用设计手段。我们对建筑的看法倾向于某种实用主义、相对主义，不希望把建筑师的工作强调为一种“创作”。

使用需求、环境气候、物质技术、社会文化等条件，在城市、建筑以及建筑师建立新空间的过程中，通常相互形成种种矛盾。我们的设计方法有赖于发现和分析这些矛盾的敏锐度与经验值，找出对设计最重要的矛盾作为工作的契入点，围绕解决主要矛盾这一核心目标来建立所有要素之间的关系，使所有要素最终形成整体性。这种整体性建立在矛盾的平衡关系中，一旦出现新的变化力量或者因素，所有的关系就会被拆散，再重新整合。这也是我们的设计没有固定形式趣味，变化多端的方法和原因。自然我们也不会把建筑视为最佳固化状态的“作品”。

然而，在这样的设计方式中，的确存在十分固定的东西。那是一种形式生成的逻辑——形式在共同形成整体性、揭示整体性，以及成就其他形式成立的过程中，自身得以存在。这

The factors that define Atelier Archmixing's practice mainly include:

1 Implicit formal logic
2 Calmness brought by consumption logic
3 Unrecognizable everyday city and architecture
4 Continuous accumulation in change
5 Neglected user's end and change

These factors are discovered consequently, then gradually integrated. They spontaneously carry some implicit features.

1 Implicit Formal Logic: The Initial Thinking Mode Affecting All Designs

Atelier Archmixing is not a design studio with a notable aesthetic style. Our design doesn't carry fixed formal feature, nor pursue a fixed formal feature.

We always see the design of architectural space and form as a tool to solve practical problems. Thanks to the diversity and uncertainty of real world problems, we tend to organize and apply design methods flexibly, free from the limitation of conventional methodology and significance about form and space. We would rather take the position of architectural pragmatism or relativism, instead of highlighting architectural design as a "creation".

In cities and buildings, in the process of constructing new spaces, numerous conflicts would arise among usage requirements, environmental climate, material technology and social culture, etc. Our design method relies on how observant and experienced we are to discover and analyze these conflicts. More specifically, it starts from exploring the essential conflicts, and the design process focuses on problem solving through establishing relations among all essential factors and integrating them as a unity. Such integration is implemented by balancing all conflicts. Once new factors arise, the old structure should be deconstructed and reorganized. This is the reason as well as the methodology for our flexible and diverse formal design. Consequently, we refuse to see architecture as the solidified "work" of best performance.

However, our design method does contain a fixed issue, which is a logic of formal emergence. Form exists only in the process of being integrated, revealing the whole and shaping other forms. This is a formal logic of a sense of wholeness, of relativity, of reverse thinking and of concepts. In our architectural design, it has turned into an invisible formal gene, a formal feature. More

是一种整体性意识下的形式生成逻辑，一种相对性的形式生成逻辑，一种反向思维的形式生成逻辑，一种观念性的形式生成逻辑。更为重要的是，在我们的实际设计中，这不仅是一种隐性的形式基因，一种形式感，而且在常年的工作里形成了一种超越形式操作的思维方式。在《改变：阿科米星的建筑思考》一书中选入的《选择在个人与大众之间》《整体的回响：来自内部的反向思维》两篇文章都阐述分析了这一主题。

在我们的工作中，包括嘉定新城文化信息产业园 B4/B5 地块、诸暨剧院、双栖斋、莫干山庾村文化市集蚕种场改造、宝山陈化成纪念馆移建改造、桦墅乡村工作室、YoungBird 室内空间改造、悦阅书店、舟山青龙山公园入口建筑、阿那亚金山岭艺术中心、昆山实验小学西侧地下车库改扩建[图1]、黎里、台州路桥商业建筑、昆明文明街串联庭院、安龙森林公园东部码头小镇商业建筑、前滩休闲公园 4、5 号建筑、彰吴村制衣厂改造[图2]、昆山森林公园会所改造等众多实施项目与设计都体现了这一设计思维方式。

1　昆山实验小学西侧地下车库改扩建
Expansion and Renovation of Kunshan Experimental Primary School

2　消费逻辑带来的冷静：对既有建筑学学科反思的开始

从研究消费逻辑开始，对于建筑设计在社会生产—消费系统中价值的理解，以及建筑师在该体系中的位置的重新审视，推动我们解决了两个认识上的问题。

一方面，形式与符号之间的关系不再固定，我们也不再将两者的固定关系视为严肃问题去对待。这帮助我们在设计应用手段上打破了传统价值认同上的界限，开始自由地组织形式与符号。同时，也第一次把另一个领域的原则——最初是消费逻辑——平移转用到既有的建筑学方法里。这种跨界组合的习惯后来也被我们应用在把生产端的技术思路转用到建筑设计的方法里。

另一方面，我们第一次从生产体系的角度去审视过去学到的建筑学的每一项价值判断与方法原则，并对于什么是更有价值的实践形成了自己的判断。显然，很多被称为有价值的实践与做法，经过这样的审查后，在我们看来不再具有真正的价值，而是一种包装出文化光环的符号现象，各种看似丰富的个性也只是一种“边缘性差异”。同时，我们也认识理解到消费

importantly, with years of practice, it has converted into a mode of thinking exceeding formal operation. The theme of “implicit formal logic” has been elaborated and analyzed in two articles collected in the book *Change is More: Architectural Thinking by Atelier Archmixing*: “Standing between the Individual and the Public” and “The Echo of the Unity: Reverse Thinking of the Interior”.

Many of our designs have illustrated this mode of thinking, including B4/B5 Blocks of Shanghai Culture & Information Industrial Park, phase I; Zhuji New Theatre; Twin Trees Pavilion; Silkworm Hatchery Renovation, Yucun Culture Market, Mogan Mountain; Removal Renovation of Chen Huacheng Memorial; Huashu Rural Studio; Interior Renovation of YoungBird Office; Yueyue Bookstore; Qinglong Mountain Park Entrance Renovation in Zhoushan; Art Center of Aranya, Jinshanling; Expansion and Renovation of Kunshan Experimental Primary School [Fig. 1]; LILI; Taizhou Luqiao Commercial Building; Promenade Courtyards, Renewal of Kunming Wenming Street; Commercial Building at East Dock Town in Anlong Forest Park; No. 4 & No. 5 Buildings in Qiantan Park; Renovation of Garment Factory in Zhangwu Village [Fig. 2]; Renovation of Kunshan Forest Park Club, etc.

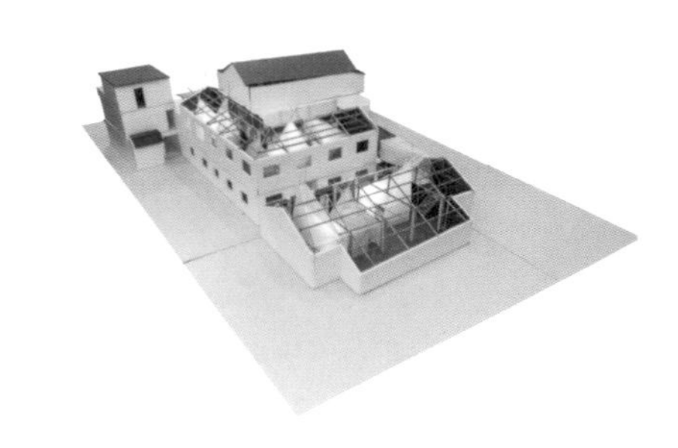

2　彰吴村制衣厂改造
Renovation of Garment Factory in Zhangwu Village

2　Calmness Brought by Consumption Logic: Start to Rethink the Current Architectural Discipline

Starting from examining the consumption logic, we have obtained a better understanding of architectural design’s value and architects’ position in the social production-consumption system. This new judgement has led to two cognitive breakthroughs.

On the one hand, since we know that the relation between form and symbol is no longer fixed, we stop to take the solid connection between form and sign seriously. That’s why we start to break the traditional values, break the boundaries between different fields and organize forms and symbols more freely. Meanwhile, we apply for the first time a principle from another field, initially the consumption logic, to existing architectural methodology. With these experiences of trans-boundary organizations, later we are able to apply technical methods used at production end to architectural design.

On the other hand, for the first time we review every value judgment, methodology and principle in the architectural field we have learned before from the perspective of the production system, so that we can judge what a more valuable practice is. Obviously, after this inspection, many

3　衡山路890弄（衡山和集）8号楼外立面改造
Facade Renovation for No.8 Building, Lane 890, Hengshan Road

4　成都麓湖水镇民宿岛——酒店B
Homestay Island Hotel B, Luhu, Chengdu

形成的舒适性和美学等吸引建筑学眼光的东西，并由此开展了反思。我们希望自己的工作在学科里能实现真正的差异。因此我们尝试尽量避开流行话题，按自己的方式慢慢工作。

研究消费文化带来的反思让我们看到了设计背后生产体系冷峻的力量，这反而使我们觉得建筑设计这个工作更有真实感了，也使我们对设计与实践价值的判断态度变得客观冷静。《选择在个人与大众之间》一文阐述分析了这一认识与思想的转变。

在我们的工作中，衡山路 890 弄（衡山和集）8 号楼外立面改造[图3]、阿那亚金山岭艺术中心、成都麓湖水镇民宿岛——酒店 B[图4]、昆山实验小学西侧地下车库改扩建、台州路桥商业建筑、昆明文明街串联庭院、安龙森林公园东部码头小镇商业建筑等实施项目与设计中都体现了这样的思想方法。

3　非识别的日常城市与建筑：一个近在咫尺的最佳范本

我们为何会关注日常城市与建筑？

相对于富有文本与符号意义的建筑创意世界，我们很早就发现日常城市的物质世界里有一种更真实的力量。

这个看法首先源于我们所受的专业训练。日常城市的物质生成方式中包含大量具有应用潜力的、现实的智慧，专业训练使我们具有找出它们的分辨力。

其次，关注日常的城市建筑应该也是价值观所驱使的。研究消费逻辑使我们“厌倦”投身于被符号包裹或者充满深刻道理的空间形式的创作中，不管这些形式是时尚的还是质朴的，商业的还是学术的。我们更愿意接近剥离意义的使用状态，而日常城市里面的很多东西正是如此，是一种只为使用而来的真实状态。关注日常城市、感兴趣于各种丰富的城市大众文化、亚文化，与反思权威、学院式、正统精英式的建筑方式有关，与反思识别性及非识别性有关。

最重要的，关注日常的城市建筑与我们想要实现探索新的理念的愿望有关。专业的直觉与生活的经验都让我们相信，中国的日常城市有其与众不同之处。很多例子告诉我们，特别

of the so-called valuable practices and methods have lost their value. They are merely symbols polished by culture. Their seemly rich individual characters are just a sort of “marginal difference”. Besides, we begin to realize and understand those aspects catching architects’ eyes, the comfort and aesthetics brought by consumption and start to rethink. In order to achieve real disciplinary differences through practice, we try to work slowly at our own pace without being interfered with by trending topics.

After the research on consumer culture, we have discovered the stern power of production system behind the design, which makes the architectural design more real for us. We have become more objective and coolheaded when judging the values of design and practice. The article *Standing between the Individual and the Public* has elaborated and analyzed this transformation in recognition and thinking.

Many of our practices and designs have given expression to this mode of thinking, including Facade Renovation for No.8 Building, Lane 890, Hengshan Road [Fig. 3]; Art Center of Aranya, Jinshanling; Homestay Island Hotel B, Luhu, Chengdu [Fig. 4]; Expansion and Renovation of Kunshan Experimental Primary School; Taizhou Luqiao Commercial Building; Promenade Courtyards, Renewal of Kunming Wenming Street; Commercial Building at East Dock Town in Anlong Forest Park, etc.

3　Unrecognizable Everyday City and Architecture: A Perfect Sample in Close Proximity

Why did we pay close attention to everyday city and architecture?

We have noticed very early that compared with architectural creation rich in texts and symbols, the everyday urban world embodies more authentic power.

Firstly, this idea roots in professional training of architecture, which has gifted us the insight to discover applicable and pragmatic wisdom from the materialization of everyday urbanism.

Secondly, the attention we pay to everyday urban architecture is driven by values. Having studied the consumption logic, we get bored with devoting ourselves to creating spaces of symbols or profound truth, no matter fashionable or simple, commercial or academic. We would rather stay close to the user’s condition free from meanings. In fact, many things in the everyday city were born for this real user mode. Paying attention to the everyday city, following rich urban mass cultures and subcultures with interest, they are related to our reflections on definitive, ac-

的典型的现象可能带来新的认知，比如拉斯维加斯、纽约、东京，让文丘里、库哈斯、犬吠工作室得到了关于城市建筑的新的认知。与此相似，中国的典型大城市也是这样的案例。我们选择观察中国快速城市化形成的物质空间本身的变化，因为这更接近于我们想要获得的操作物质空间的方法。

观察中国当下城乡建成环境和开展日常城市研究所得是阿科米星工作室这十年来最具现实性、丰富性与启发性的收获。我们创造了“空间冗余”“中性的建筑”“调整”“有效建造”“片段”“局部”“内部”等直接可应用于实践的概念。我们也将城市建筑的典型特征看作是一种“非识别体系”的样板。在《改变：阿科米星的建筑思考》一书中选入的《改变即日常——阿科米星实践综述》《日常、改变、非识别体系》《城市中的工作室》《空间冗余》等文章都阐述分析了该主题。

在我们的工作中，嘉定新城文化信息产业园 B4/B5 地块、衡山路 890 弄（衡山和集）8 号楼外立面改造、莫干山庾村文化市集蚕种场改造、宝山陈化成纪念馆移建改造、桦墅乡村工作室、YoungBird 室内空间改造、悦阅书店、棉仓城市客厅、宝山贝贝佳欧莱幼儿园、永嘉路口袋广场［图5］、智慧欧莱托育园、徐汇区龙华街道敬老院立面改造、申威达改造、昆山实验小学西侧地下车库改扩建、富春俱舍走马楼、城市内的工作室、昆明文明街串联庭院、安龙森林公园东部码头小镇商业建筑、彰吴村制衣厂改造、昆山森林公园会所改造等实施项目与设计中都应用了从日常城市研究中发现的经验和策略。

5　永嘉路口袋广场
Pocket Plaza, Yongjia Road

4　变化中的不变积累：用时间来研究空间的工作方法

我们的工作与研究对象是快速与高度压缩的城市变化过程，采用的研究方式却是一个长期慢速积累的过程。

众所周知，中国几十年的快速城市化在时空压缩后定有其特征，那是什么？如何去发现？我们找到了与之相适应的方式：一方面依靠时间的积累。经过团队，尤其是主持建筑师们 20 多年的实践积累，与快速城市化过程相伴随形成的经验，持续形成了对城市空间变化

ademic, orthodox elitist architectural methods, as well as the reflections on recognizability and unrecognizability.

Most importantly, paying attention to everyday urban architecture is closely related to our desire to explore new working philosophy. Based on professional intuition and life experience, we believe that everyday cities in China are different from the rest. Many examples have taught us that we may acquire new cognitions from particularly typical phenomena. Like Robert Venturi, Rem Koolhaas, Atelier Bow-Wow, they have acquired such new cognitions respectively from Las Vegas, New York and Tokyo. Typical large cities in China must embody the same possibilities. We have decided to observe the change of physical space caused by China's rapid urbanization to discover design methodology.

For Atelier Archmixing, the most practical, abundant and enlightening work in the last ten years results from our observation of Chinese urban and rural built environments and research on everyday cities. We have invented concepts such as “Spatial Redundancy”, “Neutral Architecture”, “Adjustment”, “Efficiency”, “Fragment”, “Part”, “Interior”, which can be applied to practice. Those typical characters of urban architecture are seen as a sample of “unrecognizable system”. This topic is elaborated, analyzed in four articles collected in the book *Change is More: Architectural Thinking by Atelier Archmixing*, including “Change is More: Atelier Archmixing's Practice”, “Everyday, Change and the Unrecognizable System”, “Work within the City” and “Spatial Redundancy”.

The experience and strategies we found in everyday city research have been applied to many of our practices and designs, including B4/B5 Block in Shanghai Culture & Information Industrial Park, Phase I; Facade Renovation for No.8 Building, Lane 890 Hengshan Road; Silkworm Hatchery Renovation, Yucun Culture Market, Mogan Mountain; Removal Renovation of Chen Huacheng Memorial; Huashu Rural Studio; Interior Renovation of YoungBird Office; Yueyue Bookstore; Cotton Lab Urban Lounge; Renovation of Baoshan Beibeijia Olion Kindergarten; Pocket Plaza, Yongjia Road [Fig. 5]; Wisdom Olion Nursery; Facade Renovation of the Longhua Street Elder Care Center; Renovation of Shenweida Workshop Space; Expansion and Renovation of Kunshan Experimental Primary School; Fuchun Kosa Zoumalou; Work within the City; Promenade Courtyards, Renewal of Kunming Wenming Street; Commercial Building at East Dock Town in Anlong Forest Park; Renovation of Garment Factory in Zhangwu Village; Renovation of Kunshan

6 诸暨规划展示馆和科技馆
Zhuji Urban Planning Exhibition Center & Science and Technology Museum

7 新天地临时读书空间
Temporary Reading Pavilion, Xintiandi, Shanghai

的认知，我们的设计实践与研究一直与之相伴并持续开展。我们坚持的工作原则是知行合一，可以很慢，可以间断，但不要停下来。第二方面，则是我们发现了形成新的典型性的方式。城市建筑的典型性并非源于一两个经典案例，而是由数量与局部共同积累构成的，这几乎是一种涌现，相应地，阿科米星的实践也更重视由数量与时间积累构成的成果与特征。

在《改变：阿科米星的建筑思考》一书中选入的《空间冗余》《走向城市建筑学的可能——“虹口 1617 展览暨城市研究”研讨会评述》等文章记录了上述的研究过程和发现。在我们的实践中，嘉定新城双丁路公立幼儿园、诸暨剧院、YoungBird 室内空间改造、宝山贝贝佳欧莱幼儿园、永嘉路口袋广场、申威达改造、昆山实验小学西侧地下车库改扩建、诸暨规划展示馆和科技馆[图6]、新天地临时读书空间[图7]、昆明文明街串联庭院设计等实施项目与设计都融入了对变化中不变积累的思考。

5 被忽视的使用端与改变：未来的工作界面

多年的国内工作经验与对于城市建筑的研究让我们关注“使用”引起的改变，与传统建筑学关注的设计建造产生端不同，使用端一直未被认为是建筑学的研究领域，而我们认为这是可能获得建筑学新见解的研究与实践领域。

在日常的城市里，大量自下而上的“使用”改变了建成环境，提示了使用端的一种状况。但是仅仅局限于现实领域，会使使用端的研究陷入到“都市考现学”式的具体而琐碎的汪洋中，由数量构成的典型性有某种危险，难以萃取，甚至会近乎抽象，无法固化。这样的使用端缺少另外一块拼图，那就是生产体系由上而下的推动与决定，由生产与技术变化带来新的使用端的引导与迭代。这就是改变的两个问题：什么改变了既有的建筑？什么会改变未来的建筑？我们最终确定的改变的领域是一个总领，使用端的聚焦点是一种集成，均旨在指导未来的实践。

在《改变：阿科米星的建筑思考》一书中选入的《火星救援与二手宇宙》《看不见的改变：论使用端空间技术化的可能》两篇文章记录了上述的研究过程和发现。我们的实践工作中，

Forest Park Club, etc.

4 Continuous Accumulation in Changing: Space Research Along with Time

We have been working and researching on rapid and highly compacted urban change with a method of long and slow accumulation.

With decades of rapid urbanization, we all know that there is a time-space compression in China's built environment. But what exactly is that and how can we discover it? We have found two corresponding methods. Firstly, it takes time. With more than twenty years of practice and experience along with rapid urbanization, our team, especially the principal architects have achieved independent cognition on the change of urban space, based on which our design practice and research keep developing. We insist on integrating cognition with practice, which could be slow and interrupted, but never stopped. Secondly, we have discovered the methods to make our work typical. Individual classic cases can't represent the typicality of urban architecture. There is an emergence of such typicality piled up by quantity and parts. Correspondingly, Atelier Archmixing has attached more importance to the results and features piled up by quantity and time.

The research process and discovery mentioned above have been recorded in the articles such as "Spatial Redundancy" and "Possibilities towards an Urban Architecture: Notes on the Symposium of 'Hongkou 1617 Exhibition and Urban Studies'" collected in the book *Change is More: Architectural Thinking by Atelier Archmixing*. Many of our practices and designs have discussed continuous accumulation in changing time, including Shuangding Road Public Kindergarten; Zhuji New Theatre; Interior Renovation of YoungBird Office; Renovation of Baoshan Beibeijia Olion Kindergarten; Pocket Plaza, Yongjia Road; Renovation of Shenwei da Workshop Space; Expansion and Renovation of Kunshan Experimental Primary School; Zhuji Urban Planning Exhibition Center & Science and Technology Museum [Fig. 6]; Temporary Reading Pavilion, Xintiandi, Shanghai [Fig. 7]; Promenade Courtyards, Renewal of Kunming Wenming Street, etc.

5 Neglected User's End and Change: the Future Battlefield

With years of domestic work and research on urban architecture, we have noticed the change caused by usage. Different from the production end always highlighted by the mainstream academic field, the user's end has always been neglected in architectural research. However, we

悦阅书店、棉仓城市客厅、智慧欧莱托育园、徐汇区龙华街道敬老院立面改造、阿那亚金山岭艺术中心[图8]、城市内的工作室、昆明文明街串联庭院、安龙森林公园东部码头小镇商业建筑、彰吴村制衣厂改造等实施项目与设计都体现了对使用端与改变的思考。

8 阿那亚金山岭艺术中心
Art Center of Aranya, Jinshanling

hold the opinion that new architecture philosophy could be discovered in this research and practice area.

In everyday city, lots of bottom-up use changed the built environment and revealed the existing state of the user's end. However, the research on the user's end could easily be overwhelmed if only limit in the real world. Quantitative typicality is too abstract to be extracted and fixed. The last puzzle for the user's end is the promotion and decision made by the top-down production system. Production and technological breakthroughs will lead to a new direction and upgrade for usage. Then we raised two questions about change: What has changed the existing building? What will change the future building? The changing area and the focus of the user's end we finally defined are guidance and integration, which will direct future practice.

The research process and discovery mentioned above have been recorded in articles such as "*The Martian* and Used Universe" and "Invisible Change: Technologizing Space at the User's End" collected in the book *Change is More: Architectural Thinking by Atelier Archmixing*. Many of our practices and designs have recorded our thinking on user's end and change, including Yueyue Bookstore; Cotton Lab Urban Lounge; Wisdom Olion Nursery; Facade Renovation for Longhua Street Elder Care Center; Art Center of Aranya, Jinshanling [Fig. 8]; Work within the City; Promenade Courtyards in the Renewal of Kunming Wenming Street; Commercial Building at East Dock Town in Anlong Forest Park; Renovation of Garment Factory in Zhangwu Village, etc.

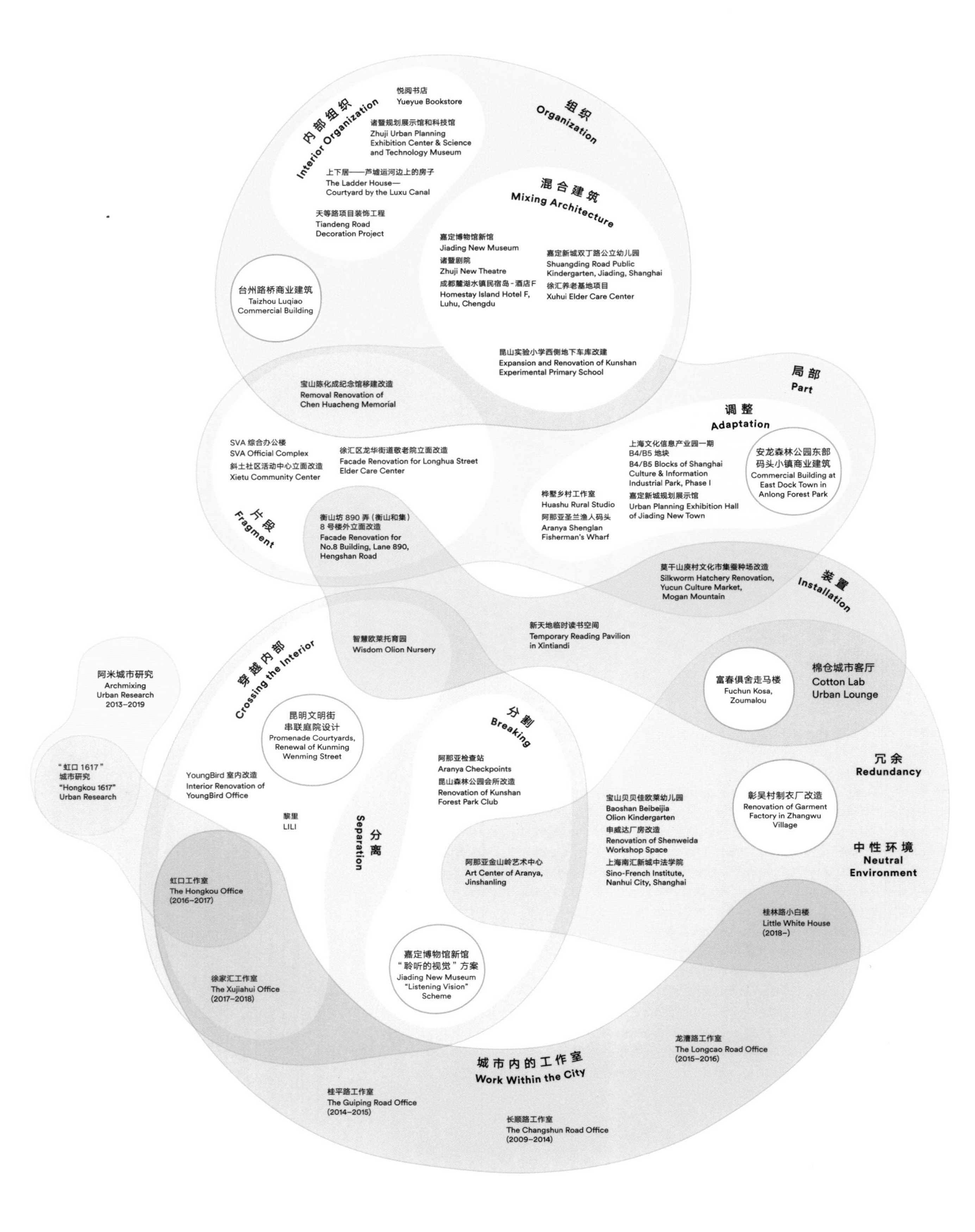
组织
Organization
内部组织
Interior Organization
悦阅书店
Yueyue Bookstore
诸暨规划展示馆和科技馆
Zhuji Urban Planning Exhibition Center & Science and Technology Museum
上下居——芦墟运河边上的房子
The Ladder House—Courtyard by the Luxu Canal
天等路项目装饰工程
Tiandeng Road Decoration Project
混合建筑
Mixing Architecture
嘉定博物馆新馆
Jiading New Museum
诸暨剧院
Zhuji New Theatre
成都麓湖水镇民宿岛 - 酒店F
Homestay Island Hotel F, Luhu, Chengdu
嘉定新城双丁路公立幼儿园
Shuangding Road Public Kindergarten, Jiading, Shanghai
徐汇养老基地项目
Xuhui Elder Care Center
台州路桥商业建筑
Taizhou Luqiao Commercial Building
昆山实验小学西侧地下车库改建
Expansion and Renovation of Kunshan Experimental Primary School
宝山陈化成纪念馆移建改造
Removal Renovation of Chen Huacheng Memorial
局部
Part
调整
Adaptation
SVA 综合办公楼
SVA Official Complex
斜土社区活动中心立面改造
Xietu Community Center
徐汇区龙华街道敬老院立面改造
Facade Renovation for Longhua Street Elder Care Center
上海文化信息产业园一期 B4/B5 地块
B4/B5 Blocks of Shanghai Culture & Information Industrial Park, Phase I
嘉定新城规划展示馆
Urban Planning Exhibition Hall of Jiading New Town
安龙森林公园东部码头小镇商业建筑
Commercial Building at East Dock Town in Anlong Forest Park
桦墅乡村工作室
Huashu Rural Studio
阿那亚圣兰渔人码头
Aranya Shenglan Fisherman's Wharf
片段
Fragment
衡山坊 890 弄(衡山和集)8 号楼外立面改造
Facade Renovation for No.8 Building, Lane 890, Hengshan Road
莫干山庾村文化市集蚕种场改造
Silkworm Hatchery Renovation, Yucun Culture Market, Mogan Mountain
装置
Installation
新天地临时读书空间
Temporary Reading Pavilion in Xintiandi
智慧欧莱托育园
Wisdom Olion Nursery
阿米城市研究
Archmixing Urban Research
2013–2019
穿越内部
Crossing the Interior
昆明文明街串联庭院设计
Promenade Courtyards, Renewal of Kunming Wenming Street
分割
Breaking
富春俱舍走马楼
Fuchun Kosa, Zoumalou
棉仓城市客厅
Cotton Lab Urban Lounge
"虹口 1617" 城市研究
"Hongkou 1617" Urban Research
YoungBird 室内改造
Interior Renovation of YoungBird Office
阿那亚检查站
Aranya Checkpoints
昆山森林公园会所改造
Renovation of Kunshan Forest Park Club
冗余
Redundancy
彰吴村制衣厂改造
Renovation of Garment Factory in Zhangwu Village
黎里
LILI
分离
Separation
宝山贝贝佳欧莱幼儿园
Baoshan Beibeijia Olion Kindergarten
申威达厂房改造
Renovation of Shenweida Workshop Space
上海南汇新城中法学院
Sino-French Institute, Nanhui City, Shanghai
中性环境
Neutral Environment
阿那亚金山岭艺术中心
Art Center of Aranya, Jinshanling
虹口工作室
The Hongkou Office
(2016–2017)
桂林路小白楼
Little White House
(2018–)
嘉定博物馆新馆"聆听的视觉"方案
Jiading New Museum "Listening Vision" Scheme
徐家汇工作室
The Xujiahui Office
(2017–2018)
龙漕路工作室
The Longcao Road Office
(2015–2016)
城市内的工作室
Work Within the City
桂平路工作室
The Guiping Road Office
(2014–2015)
长顺路工作室
The Changshun Road Office
(2009–2014)

六个超级模型
Six Super Models

阿科米星的工作对象主要是城市建筑。

在以往的工作中，有一些方案对我们持续的实践与研究具有重要意义。这些设计，首先都是在现实具体的需求、复杂的城市环境下开展的。它们都发现了隐蔽在现实条件之后的关键矛盾，并提出了解决问题的有效方法。它们也都是准确、灵活地应用基本设计原理与技巧的结果。另一方面，这些案例的设计方式也反映了阿科米星及其主持建筑师彼时的建筑设计研究内容，它们既受到当时理论思考的影响，也会在其后被不断地反思与应用，这是一个实践与认知相互交织的过程。

因此，这些解决方案就成为某种典型的设计试验模型，也成为阿科米星工作中具有指导性、参考性的原型，超越了通常的设计项目个案的级别。这些方案大部分未曾实施，有的设计可能并非该类模式中最早的尝试，却是最典型的早期范本。这些有意思的模型包括 2005 年的台州路桥商业建筑未实施方案、2014 年的樟吴制衣厂改造未实施方案、2006 年的昆明文明街串联庭院未实施方案、2008 年的嘉定博物馆新馆未实施方案、2012 年的安龙森林公园东部码头小镇商业建筑未实施方案和 2011 年的富春俱舍走马楼。没有这些超级模型所代表的设计方法的思考，阿科米星将不可能实现棉仓城市客厅、昆山实验小学西侧地下车库改扩建等综合性项目。

1 台州路桥商业建筑

2005 年，主持建筑师庄慎参与浙江台州路桥的旧城更新设计[图1]。由于地处传统建筑风貌协调区，小商业用房项目之前形式较为当代的一版设计被当地管理部门否决了。新设计被要求尽量做成传统式样，以便与街坊的整体仿清式风格相配合。此外，确定施工的是一支专做商业化仿清式古建筑的队伍，设计需照顾其习惯的形式施工工艺。由于这些因素，这个小房子第一次采用了一个不常见的设计方法——“规划”一个建筑。这个不足 200 平方米的房子被“拆”成两部分，一部分是由建筑师决定空间位置并详细设计具体形式的墙体，以及古

1 台州路桥商业建筑
Taizhou Luqiao Commercial Building

Atelier Archmixing targets chiefly urban architecture.

In our past working experience, some schemes are of great significance to our continuous practice and research. These designs all start with realistic and concrete needs in the complicated urban environment. They all have revealed the key conflicts hidden in real-life conditions and raised effective solutions to those problems. They also result from the accurate and flexible application of basic design principles and techniques. In another aspect, the design approach of these projects also reflects the architectural design research of Archmixing or its principal architects back then. These designs have been influenced by concurrent theoretical thinkings and will be continuously reflected upon and put into application. Such is an interwoven process of practice and cognition.

As a result, these solutions have become some experimental design prototypes, or instructive and referential work models for Archmixing, which have transcended those general individual design cases. Most of these projects have not been implemented. Some might not be the earliest attempts in models of the same kind, but the most typical early examples. These intriguing models include Taizhou Luqiao Commercial Buildings (2005, unbuilt); Renovation of Garment Factory in Zhangwu Village (2014, unbuilt); Promenade Courtyards, Renewal of Kunming Wenming Street (2006, unbuilt); Jiading New Museum (2008, unbuilt); Commercial Building at East Dock Town in Anlong Forest Park (2012, unbuilt) and Fuchun Kosa Zoumalou (2011, completed). Without these symbolic design methods and thinkings from these supermodels, Archmixing couldn't have realized comprehensive projects such as the Cotton Lab Urban Lounge in Changzhou and the Expansion and Renovation of Kunshan Experimental Primary School.

1 Taizhou Luqiao Commercial Buildings

In 2005, the principal architect Zhuang Shen was involved in a renovation program in Luqiao, a historical town in Taizhou city, Zhejiang Province [Fig. 1]. Since the site located in a coordinating area of traditional buildings, those modern schemes of the earlier stage for this commercial case were vetoed by the local administration. The new design was expected to be in a traditional style to merge with the whole neighborhood that mimics the architectural style of the Qing Dynasty. In addition, the designated construction team is specialized in duplicating ancient Qing-style commercial buildings, and the new design should take this into consideration. Due to these factors,

建建设的平面轮廓；另一部分是建筑师规划后交由专业古建修缮公司负责深化、施工的仿古建筑。现代的前一部分也作为仿古的后一部分的空间限定基础。这样的设计方式产生了满足各方需要的新旧融合的新建筑。

对这个小商业房来说，设计的关键首先在于设计方法本身的设计，其次具体在于不同建造方式的混合组织。虽然最终没有落地实现，但这个设计是最早的一个应用“组织”方式的混合设计模型。这个模型显示了一种跳出完整建筑式的思维习惯，尝试局部自由组合的思维。这个最早的组织建筑的模型中有着之后在阿科米星的实践里逐步清晰的“局部”“调整”思想的影子，它影响了后来阿科米星的很多设计，包括新建项目中的嘉定博物馆新馆实施方案、诸暨剧院、嘉定新城双丁路公立幼儿园、成都麓湖水镇民宿岛——酒店F、徐汇养老基地等；也包括改造与建筑内部设计中的悦阅书店、上下居——芦墟运河边上的院子、天等路项目装饰工程、宝山陈化成纪念馆移建改造等。

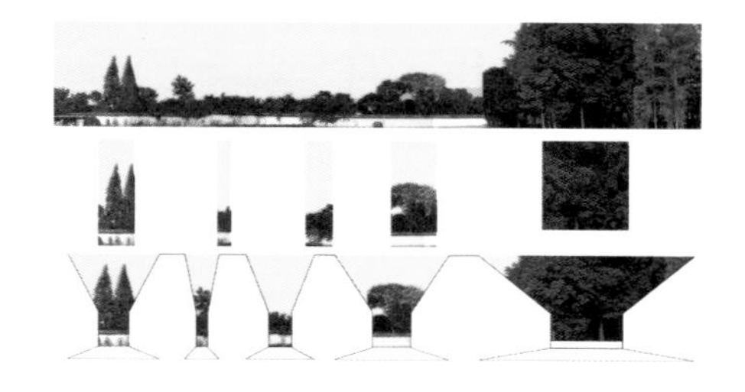

2–1 嘉定博物馆新馆过程方案“聆听的视觉”，叙事概念
"Listening Vision" narrative concept in the process plan for Jiading New Museum

2 嘉定博物馆新馆“聆听的视觉”方案

2008年，主持建筑师庄慎设计了嘉定博物馆新馆，在数轮方案中，有一稿未建成方案，是一个重要的设计原型。拟建博物馆的秋霞圃西侧地块隶属嘉定老城州桥历史文化风貌区的协调区域，一墙之隔即有着500多年历史的秋霞圃——上海五大古典名园之一。

这稿方案的最大特征是自身没有特别的形象，而是将新建筑作为“中间者”，希望通过它重新建立秋霞圃和当代城市之间的关系。通过新建筑的间隙，原本被博物馆遮挡、在街道上看不到秋霞圃的城市中的行人，他们的视线可以不断地触及被切割成片段的秋霞圃美丽的西墙与林木，并会将这一次次的片段连缀成想象中的完整画面。为了保证城市与秋霞圃之间的视线畅通无阻，连接展厅的通廊经过了巧妙的设计：最西段二层的连廊向外悬到人行道上空，即行人头顶之上；靠近秋霞圃的一层的连廊被藏到了传统形式的院墙之后，于是，行人与园林之间再无他物。

2–2 嘉定博物馆新馆过程方案“聆听的视觉”，沿街邂逅园林
"Listening Vision" encounter garden along the street for Jiading New Museum

an unusual design method was adopted in this small project, "to plan" a building. The building with less than 200m^2 plot area was divided into two parts. One part is the concrete walls defining the courtyards and the outline for the commercial buildings, for which, the architect carefully arranges the layout, the form and tectonic details. The other part is a mock ancient style building, planned by architects and then refined and implemented by the construction company. The modern part also defines the spatial boundary for the traditional part. Such a design method results in a new building uniting ancient and modern elements to satisfy the needs of all parties.

For this small commercial building, the key lies in firstly working out the design method itself, and then mixing different construction styles. Although unfortunately not realized, this scheme is our first mixing design model applying the "organization" approach. This model has shown fresh thinking, which has jumped out of the conventional scope of complete building and tried to combine parts flexibly. This earliest model of organizing architecture has embodied such concepts as "part" and "adaptation", which will turn clearer in later practices of Archmixing. Its trace is notable on many new projects including Jiading New Museum; Zhuji New Theatre; Shuangding Road Public Kindergarten; Homestay Island Hotel F, Luhu, Chengdu; Xuhui Elder Care Center, as well as some renovations and architectural interior designs such as Yueyue Bookstore, The Ladder House—Courtyard by the Luxu Canal; Tiandeng Road Decoration Project; Removal Renovation of Chen Huacheng Memorial in Baoshan District.

2 Jiading New Museum "Listening Vision" Scheme

In 2008, the principle architect Zhuang shen designed Jiading New Museum. Among all five proposals, an unimplemented one is also an important design prototype. The museum was planned west to Qiuxia Garden of more than 500 years history, one of the five classic gardens in Shanghai. This is also part of Zhouqiao historical and cultural preservation area in this ancient town of Jiading. Qiuxia Garden is separated from the new site only with a wall.

This scheme aimed to use featureless buildings as "a mediator" to reconnect Qiuxia Garden with the current urban life. Through the repeated gaps in the new museum, pedestrians strolling along the street could constantly encounter with beautiful pieces of the west wall and tall trees of Qiuxia Garden and weave them into a united picture in their mind. To guarantee the free connection between the street and the garden, we have deliberately designed those galleries in the new

嘉定博物馆的这个模型，重要性在于示范了一种反向建立自身价值的设计思路与手法，它采用“打断”“分离”整体的手法，反而加强了对于整体的感知，在这一过程中，促成这种关系的新建筑也得以成就自身的价值。这个超级模型为我们的设计增加了一种建立关系的新原型。后来的阿那亚金山岭艺术中心、阿那亚检查站、昆山森林公园会所改造等方案都采用了同样的模式。

3 樟吴制衣厂改造

2014 年，阿科米星在参与浙江樟吴村的乡建集群设计时，被分配到设计改造樟吴制衣厂。该项目拟将村里一组空置的制衣厂房改为一处大学建筑教学的实践基地，主要接待暑期来此开展设计工作营的师生。由于存留的厂房构造简易、空间大小错综复杂，要做整体完善性的修复与改造代价巨大，条件不允许。同时，民间自建的既有房屋也没有建筑施工图纸等作为改造的依据。因此针对怎么设计、如何施工，阿科米星的设计方法绕过了正式的设计文件编制，只用概念设计即可施工；设计思路也跳过了整体设计的概念。

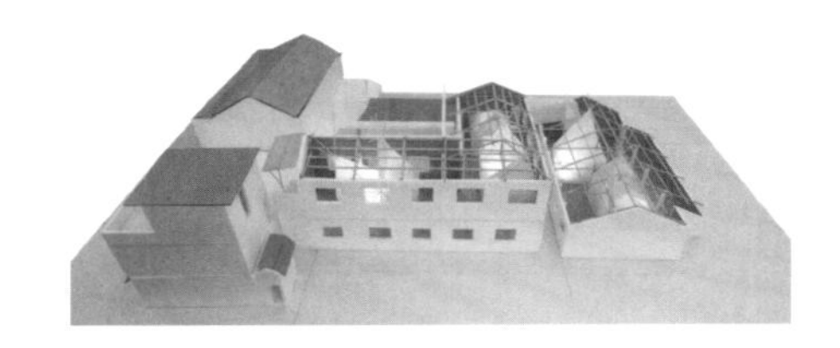

3 彰吴村制衣厂改造
Renovation of Garment Factory in Zhangwu Village

我们首先采用整理的方式，将原来建筑群里破败、临时的东西清除干净，将原有的房屋门窗拆除、留下洞口，这样就将原来的房屋整理为一处仿佛天然存在的空间骨架和覆盖物，就像“自然环境”一样。在这个空间里，采用现场定位、灵活搭建的木构与织物围护的临时“小建筑”形成一个局部的环境，“小建筑物”内部接入分体空调，开敞的原有房屋的其他剩余空间保留自然通风。

这个设计不同于以往完整的图纸型设计，更像是一份指导在现场直接工作的指导书，其重要性在于这是阿科米星第一次在实际设计中将现有建筑视为建成的“既有中性环境”的设计，是第一次利用“冗余空间”的设计。同时在这个设计里，“内部”、关注“使用端”性能、“屋中屋”模式等设计研究概念也已明确地体现了出来。在其后的工作里，棉仓城市客厅、宝山贝贝佳欧莱幼儿园、申威达厂房改造、上海南汇新城中法学院、阿科米星桂林路小白楼

museum: on the second floor, the gallery was hanging above the sidewalk and the passersby, and on the first floor, the gallery clung to the garden wall in the same traditional style. Nothing stands between the pedestrians and the garden.

The model of Jiading Museum is important because it has shown design logic and strategy of building value reversely. The sense of unity was strengthened through breaking and separating the unity. During this process, the in-between new building has achieved its own value. This super model has equipped us with a new prototype to build connections. The same model was applied in later projects such as Art Center of Aranya, Jinshanling; Anaya Checkpoints, and Renovation of Kunshan Forest Park Club.

3 Renovation of Garment Factory in Zhangwu Village

In 2014, when participating in the group design of rural renovation in Zhangwu Village, Zhejiang Province, Archmixing was designated to renovate Zhangwu Garment Factory. The project aimed to turn a group of vacant factory buildings into an architectural education base, mainly for summer design studios. Since the existed buildings were of simple structure, varied-size and complicated spaces under poor conditions, it would be uneconomical and unrealistic to carry out an entire renovation. Meanwhile, the existed buildings were mainly informal constructions without any drawings.

As a result, Archmixing decided to guide the construction on conceptual design instead of formal design documentation. An overall design logic was also abandoned. We first applied a method of organization, cleaning up those shabby and temporary parts, removing the doors and windows to leave the openings. The original buildings were then sorted into a spatial skeleton and coverings like a “natural environment”. Inside this space, temporary “small buildings” with timber structure and fabric covering were positioned on-site and constructed flexibly, constituting a partial environment. The “small buildings” were equipped with split air conditioner units. The rest spaces were remained unchanged to allow natural ventilation.

Different from a complete conventional design based on drawings, this design is more like an instruction manual to guide work on site. Its importance lies in the point that this was the first time for Archmixing to see a current building as an “existing neutral environment” in real practice. It was also its first time to use “spatial redundancy” in design. Meanwhile, design research concepts

办公室等设计都可以视为该原型的变体。

4　昆明文明街“串联庭院”

2006年，主持建筑师庄慎在昆明文明街旧城更新项目中，曾产生一个“串联庭院”的设计构想。文明街是昆明的传统老街坊，包含很多云南传统民居——“一颗印”合院。文明街的空间结构是由街道—巷道—院落组成的，新的商业开发改造将导致的巨大突变，与原本的建筑格局与风貌、原有的市井生活必然产生冲突。因为现有街坊结构中的巷道是一个由传统合院的实体院墙组成的封闭结构，而城市商业开发需要更多打开的面，因此会破坏实墙面。怎样才能实现既不破坏原有传统建筑的风貌，又满足大规模商业功能的诉求？串联庭院的方案就是一个解决方案，采用了“穿越”建筑群内部的方式。我们分析了原有建筑群的结构，利用民居“柔软”开放的内在庭院，打通本来隔绝的独立院落，把它们连接成一体，将商业的能量引入建筑群落的内部，形成了街巷与串联庭院两套平行的、具有不同形态与功能的步行组织体系，其中串联庭院形成商业动线，原有街巷则保持旧有风貌。

方案采用的“切割”“破坏”方式，改变了内部原有的结构，创造了内部的另外一个维度。这个超级模型的重要性在于揭示了“内部”这一概念需要在改变中呈现出来，使“内部”的概念具体化为一种与改变手段关联在一起的方法，也成为可以应用的原型。阿科米星此后完成的宝山贝贝佳欧莱幼儿园、黎里、YoungBird室内改造、阿科米星虹口办公室等都是这样应用的。

5　安龙森林公园东部码头小镇商业建筑

2012年，阿科米星参与了安龙森林公园东部码头小镇设计项目，着手设计的一组建筑包含商业、餐饮与部分小客房等混合业态。组团坐落在山坡地上，面向千岛湖的水面。这个设计要处理的主要是未来商业功能的不确定性与希望获得一个具有明确特征的建筑形象之间的

4–1　昆明文明街串联庭院设计
Promenade Courtyards, Renewal of Kunming Wenming Street

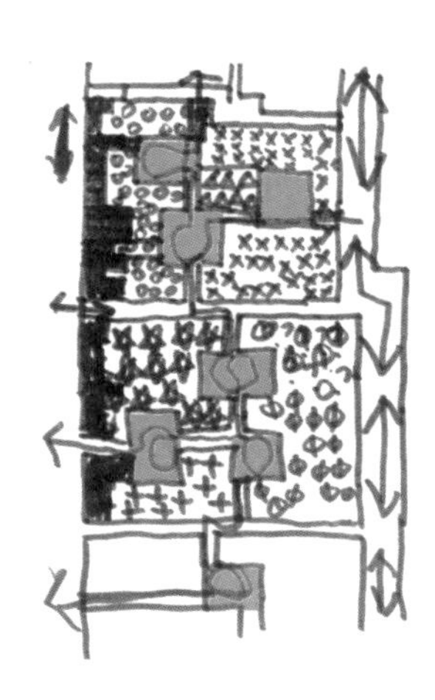

4–2　昆明文明街串联庭院设计，草图
Drawing of Promenade Courtyards

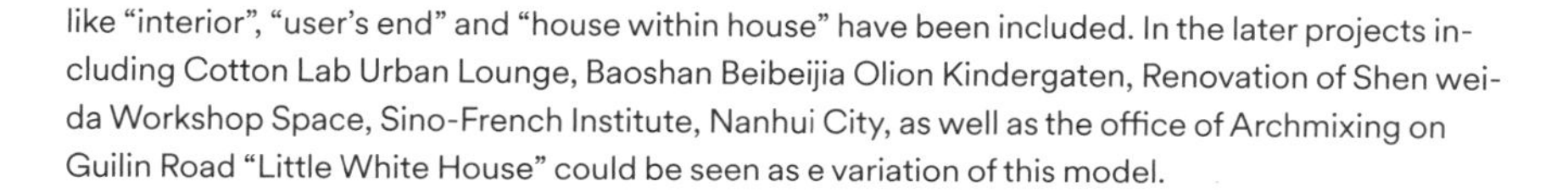

like “interior”, “user’s end” and “house within house” have been included. In the later projects including Cotton Lab Urban Lounge, Baoshan Beibeijia Olion Kindergaten, Renovation of Shen weida Workshop Space, Sino-French Institute, Nanhui City, as well as the office of Archmixing on Guilin Road “Little White House” could be seen as e variation of this model.

4　“Promenade Courtyards”, Renewal of Kunming Wenming Street

In 2006, the principle architect Zhuang Shen conceived an idea to build “promenade courtyards” in the renovation project of Wenming Street in the old city of Kunming. Wenming Street belongs to a historical neighborhood constituted with vernacular courtyard houses called “Yikeyin”. The original urban fabric was a network of street-alley-courtyard. The mutation resulting from commercial development and renovation will largely challenge the original architectural structure and form and the local lifestyle. Since the vernacular urban structure is closed, alleys are constituted with those solid walls of the courtyards. While commercial space needs more openings, which may destroy the original solidity. How can we avoid destroying the appearance of original buildings while satisfying the commercial needs on a large scale? One of the solutions is a system of promenade courtyards, connecting through the interior of the buildings. After having analyzed the original structure, we decided to make the best use of the open “soft” courtyards, integrating the previously independent courtyards, and introducing commercial vitality into these traditional dwelling complex. Thus two parallel pedestrian walks in distinct form and function were formed in the alleys and connected courtyards. When the promenade courtyards support commercial circulations, the alleys could remain the original state.

Through “separating” and “destroying”, the original interior structure was changed and another dimension was created inside. The importance of this supermodel lies in its revelation of how we can only present “interior” through changing. The abstract concept of “interior” turns into a concrete methodology of change, as well as an applicable prototype. This model was adopted in later cases like Renovation of Baoshan Beibeijia Olion Kindergarten , LILI, Interior Renovation of YoungBird Office, Archmixing office on Hongkou district.

5　Commercial Building at East Dock Town in Anlong Forest Park

In 2012, Archmixing participated in the project of East Dock Town in Anlong Forest Park,

5　安龙森林公园东部码头小镇商业建筑
Commercial Building at East Dock Town in Anlong Forest Park

矛盾，必须思考不确定与确定、临时与永久、改变与不变之间的关系并提出解决方案。我们的方案将建筑的构成分为两部分：一部分是处于下方的实体空间，稳定坚固，主要用于商业、餐饮和客房功能；另一部分是上方的木构平台，灵活轻巧，提供受欢迎的观湖商业平台。设计特别之处在于，下部看似稳定的空间形式，实际使用时由于商业功能的缘故，它的功能与外观在未来将是不确定与变化的；而看似脆弱、临时的木构平台，反而因为功能的明确而不会再改变。这个平台就是一个相对的局部，结构上依附在稳固的建筑上，但又与它相对立，具有自己独立的形式，强势的形式统领控制了整个建筑。

这个方案的重要性在于将"局部"的概念与意义显示出来，对局部的自治性，调整对于整体改变的作用，调整在获得有效性上的作用等问题作出思考。此前我们设计的嘉定城市规划展示馆改造，上海文化信息产业园一期 B4/B5 地块，此后的桦墅乡村工作室、陈化成纪念馆移建改造等都是这个模型思路的不同实践。其中比较典型的是徐汇龙华街道老人院立面改造，局部调整的做法成为了整体系统更新的关键措施。更综合的发展出现在 2018 年设计的昆山实验小学西侧地下车库改扩建里，在一些看似难以协调的市政、公共空间、学校扩建的复杂条件下，一个跨越通常的部门边界、城市规划管理边界的综合性的局部调整方案解决了难题。

6　富春俱舍走马楼

2012 年，阿科米星在改造富春江边的富春俱舍走马楼时，需要将原本因为天井而开敞底层的四合院房屋封闭起来，形成可用空调的室内环境。新加的封闭体的边界、位置和方式需要与精致的传统四合院构造相协调。我们的解决方案是做一个活动装置，这个装置由钢结构、塑料膜、竹子装饰构成。为了与室内原有结构、细节、氛围协调，装置被设计为独立柱伞状结构，顶部的塑料膜布可根据需要打开，雨水汇集则采用了漏斗式的中心汇排水。

这个超级模型的典型之处是采用"装置"的方法，生成了一个具有临时性与可变性的局

in charge of designing a mix-used program of retails, restaurants and hotels. The complex lies on a hill facing the Thousand-Island-Lake. The main challenge is how to settle the conflict between the uncertain commercial functions and a fixed distinctive appearance. We had to consider the relationship between certainty and uncertainty, temporariness and permanence, changing and non-changing and then propose a solution. We divided the building into two parts: the lower part is a solid volume, which is stable and strong to accommodate commercial, dining and guest-rooms; the upper part is a wooden deck, which is flexible and light to serve as a sightseeing platform. The uniqueness of the design is that the seemingly stable lower part will be functionally and formally uncertain and changeable to fit for commercial use, while the seemingly fragile and temporary platform will keep stable with a clear and lasting purpose of sightseeing. The platform is relatively attached to the building, with its structure standing on the stable volume, but on the other hand, it also constitutes a contrast with its independent shape which strongly dominates the whole building.

The importance of this design lies in exhibiting the concept and meaning of "parts". We have considered over such issues as how we can maintain the autonomy of parts, how adjustment can change the whole and achieve efficiency. Varied practices of this model can be seen in the renovation of Urban Planning Exhibition Hall of Jiading New City; B4/B5 Blocks of Shanghai Culture & Information Industrial Park, Phase I; Huashu Rural Studio and the Removal Renovation of Chen Huacheng Memorial. A typical case could be the Facade Renovation for Longhua Street Elder Care Center. Partial adaptation turns into a key strategy in updating the whole system. More comprehensive development of this model could be seen in the project of Expansion and Renovation of Kunshan Experimental Primary School in 2018. It seems hard to coordinate the complicated conditions of municipal, public space and school expansion, but finally a comprehensive partial adjustment that crosses over the conventional boundaries between different departments and urban planning administrative scopes has solved the problem.

6　Fuchun Kosa Zoumalou

In 2012, when renovating the Zoumalou for Fuchun Kosa by Fuchun River, we needed to enclose the courtyard to build an interior space with air conditioners. The edge, position and style of the new addition should coordinate with the delicate structure of this traditional courtyard. We

部，这个原型体现了对时间的考虑。阿科米星在此前设计的莫干山庾村文化市集蚕种场改造，此后的新天地临时读书空间、宝山贝贝佳欧莱幼儿园、柔性倍增——城市未来居住空间等实践中，都运用了这种解决问题的方式。

6 富春俱舍走马楼
Fuchun Kosa Zoumalou

built here a changeable installation composing of steel structure, plastic film and bamboo decoration. To coordinate with the original structure, details and spatial atmosphere, this installation was designed as an independent column in an umbrella shape. The plastic sheet could be open on demand. A funnel-like centralized drainage was installed to collect rainfalls.

What features this supermodel is the application of "installation", and a temporary and changeable part is thus formed. This prototype shows our consideration on time. Similar strategies were applied in the Silkworm Hatchery Renovation, Yucun Culture Market, Mogan Mountain before and will later be used in the Temporary Reading Pavilion in Xintiandi, Shanghai, the Renovation of Baoshan Beibeijia Olion Kindergarten and the "Flexible Multiplication: Future Urban Living Space".

上海文化信息产业园一期 B4/B5 地块
B4/B5 Blocks of Shanghai Culture & Information Industrial Park, Phase I

01

悬挂的庭院

“悬挂的庭院”是B4/B5地块的核心设计概念。建筑师对使用者的定位是从事信息科技和文化创意行业的年轻白领，需要户外活动和公共交流。设计将原应落地的庭院升至二、三、四层，一方面扩大了基地范围内的开放空间；另一方面，仅靠调整“挑院”位置，就将标准化的矩形办公单元转变为一户一型的内部空间，通过单元组合又形成丰富的外部空间。

大部分规格统一为宽8.8米，出挑5米，高3.6米的空中庭院采用钢结构悬挂方式固定在外墙上，一是避免粗笨的混凝土梁板造成空间压抑；二是保证足够大的出挑距离，加上足够高的围护界面，内部使用时具有较为私密的庭院感，外部则形成足够的体量感和空间分隔效果。

明确的阴影

挑院的地面采用防腐木地板开缝锚固在钢梁上，阳光可以穿透缝隙。包覆材料经过长时间的协商比较才选定。最初方案采用双层金属扩张网板，后来曾考虑过木条、藤编等各种可能，最终决定使用较耐久、性价比较高的方孔钢板，包覆挑院的三个外立面和底面。为了强调金属板的质感，与光滑的实墙面形成对比，板面开孔参考了江南园林建筑中木格窗的比例，5厘米见方的孔，间隔宽3厘米，以便在50米开外也能感知到穿孔效果，同时在挑院内及外墙上投出明确的阴影。

在这里，庭院脱离了地面漂浮到空中，镂空花墙失去了厚度贴附着钢结构，浓重的砖木材料被替换成轻薄且带光泽的钢板网，所有的空间要素、界面形态和材料质感似曾相识却又面貌新颖，整体是一派简约的现代办公场所意象。然而，在首层形态不一、忽明忽暗的庭院间漫步穿行时，开孔率约60%的方孔钢板好似一层薄纱，若有若无地限定并分割着天空和建筑，外墙上洒落串串光格和斑驳树影；坐在办公室或空中庭院时，四面围合产生的内省感，透过方孔隐约可见的树姿，都饱含江南园林特有的恬静气息。

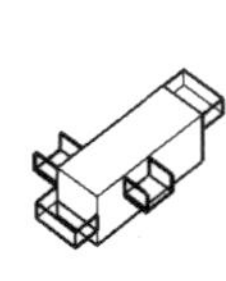

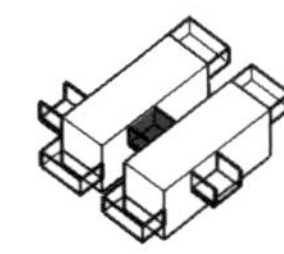

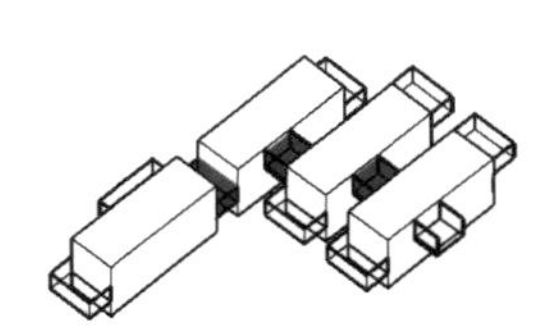

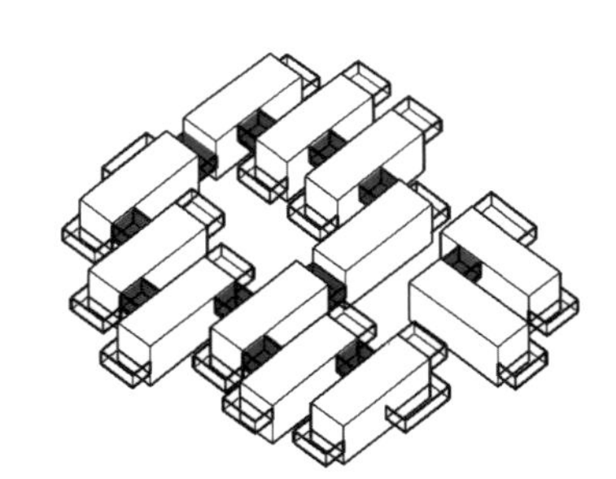

Hanging Courtyard

"Hanging courtyard" is the key design concept of Block B4/B5. This project is designed for young white-collars who are dedicated to the IT industry and cultural creation. These groups of young people need public activities and communication. The courtyard usually located on the ground floor was now elevated to the second, third and fourth floor. A huge ground space was thus released. By hanging the courtyards, the standardized rectangular office units were transformed into units with different flat surfaces. A rich outer space was formed with the composition of simple units.

Most of the courtyards are unified, 8.8m wide, 5m outreach and 3.6m high. They are fixed on the exterior wall by a suspended steel structure. On the one hand, the structure can avoid space depression caused by awkward concrete beams and slabs. On the other hand, it ensures a large cantilever. With high enough enclosure interface, the inner space possesses a private courtyard and the outer space achieves notable volumes and space segregation.

Clear Shadow

The hanging courtyard adopts anticorrosive wood flooring. The floors are slatted and anchored on the steel beam, between which sunlight goes through. The coating material is selected after a long discussion and comparison. The initial proposal adopts double-layer expanded metal mesh. Later, other possibilities like the wooden strip and rattan plaited materials are discussed. Finally, we choose the durable and economical square-hole steel plate to cover three facades of and underneath the courtyard. To emphasize the texture of the metal sheet and contrast with the smooth solid wall, we take reference from those wooden lattice windows in traditional Chinese gardens in the south Yangtze River Delta area. Square holes of 5cm side length are drilled 3cm apart from each other, so that the holes can be sensed 50m away, and clear shadows can be observed inside the hanging courtyard as well as on its exterior wall.

Here, the designer elevates the courtyard, makes the hollow lattice wall stick to the steel structure, and selects light shiny expanded metal mesh instead of thick brick and wood. All the space elements, interface shape and material texture seem familiar and renovated. A simple image is created to describe modern office space. However, when walking in courtyards with shadows and sunshine on the first floor, the square-hole steel plate with an aperture ratio of about 60% turns to be a gauze, by which sky and buildings are uncertainly defined and separated. The exterior walls are scattered with light grids and mottled shadow of the trees. When sitting enclosed in the office or the hanging courtyard, people can feel a sense of introspection. Besides, Trees might be seen indistinctly through the square hole. All these demonstrate a sense of tranquility once featured in traditional garden in this region.

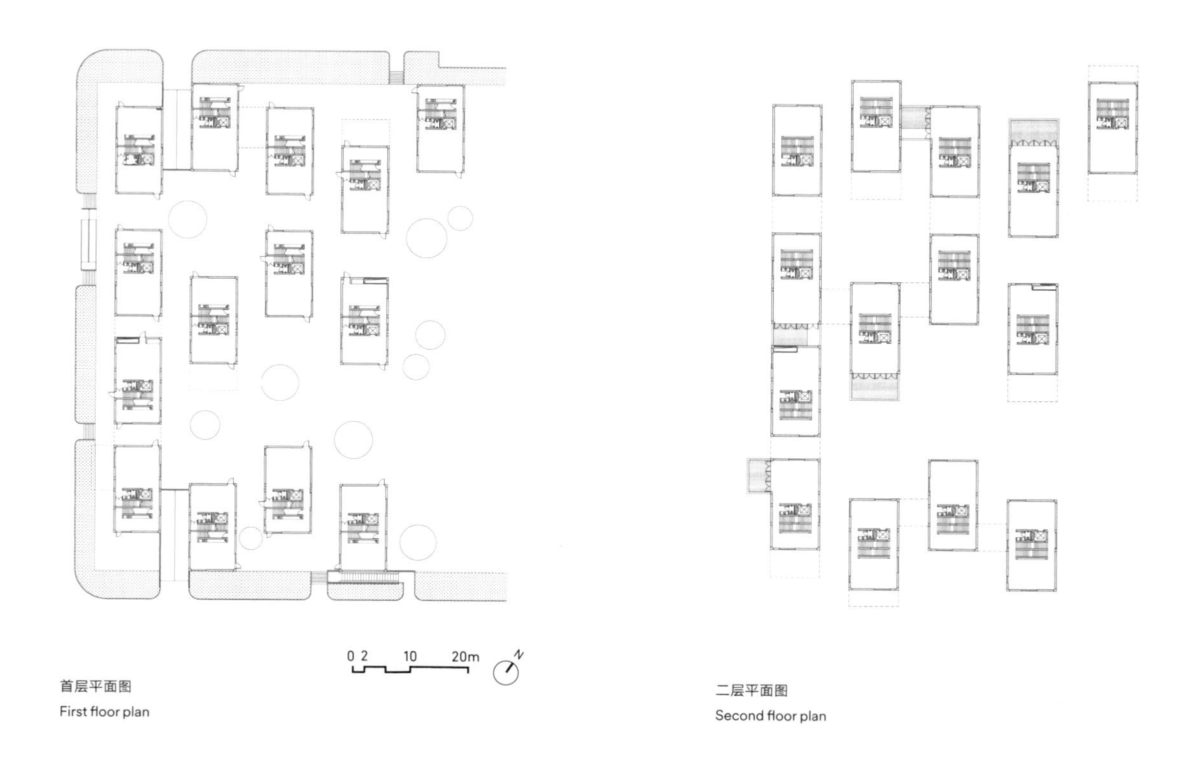

首层平面图
First floor plan

二层平面图
Second floor plan

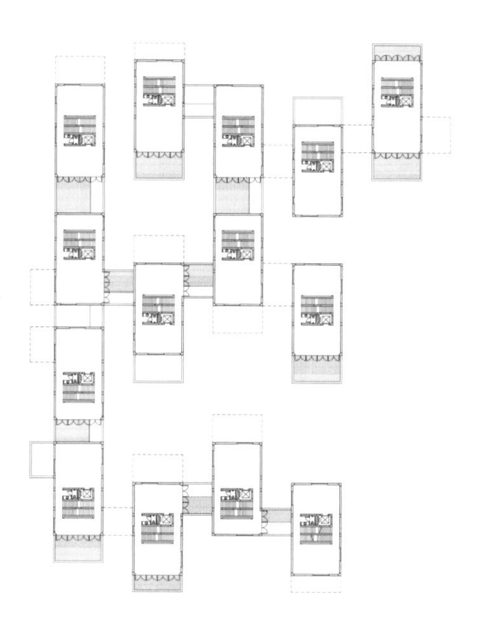

三层平面图
Third floor plan

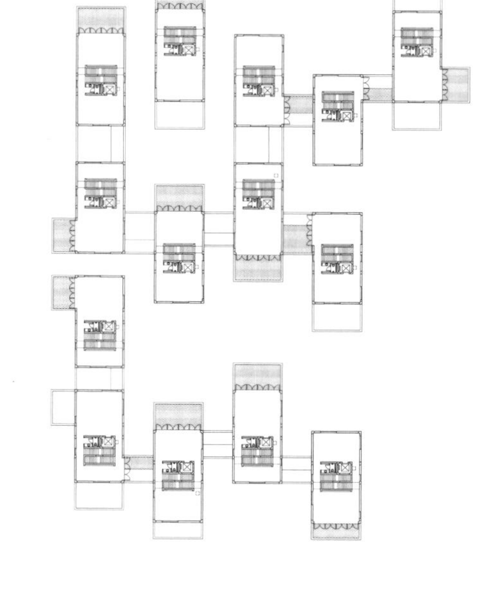

四层平面图
Fourth floor plan

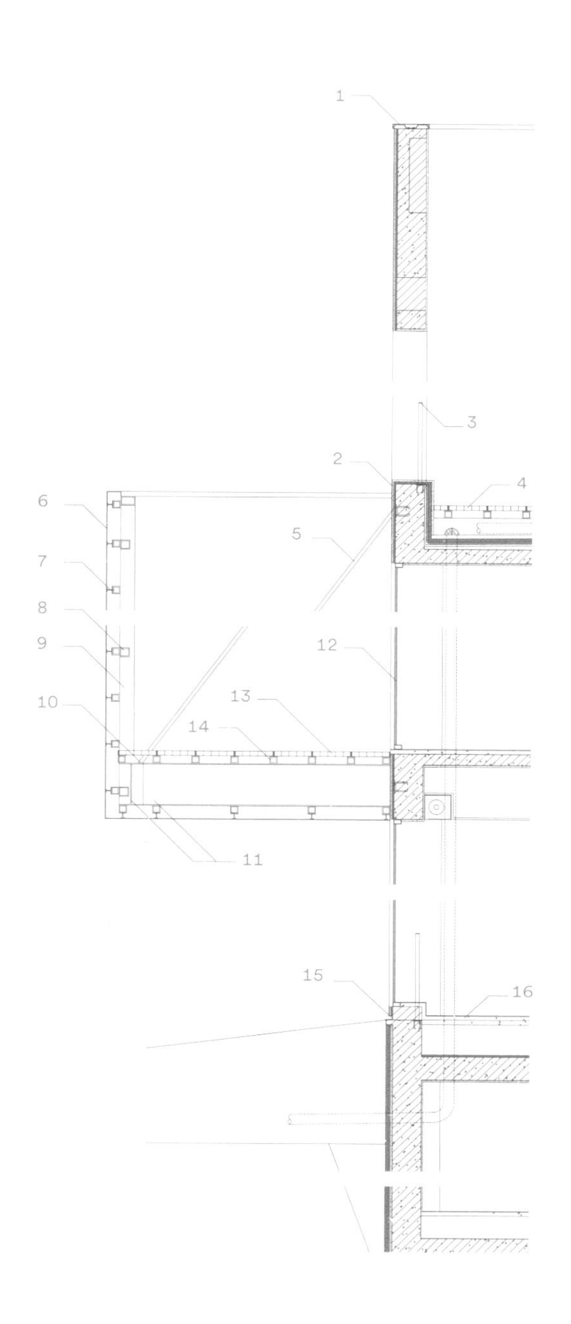

1	钛锌板压顶	Titanium-zinc coping
2	丙烯酸弹性高级涂料墙面	Acrylic elastic advanced coatings wall
3	5/40 扁钢氟碳喷涂	5/40 flat steel fluorocarbon paint
4	成品不锈钢篦子	Stainless steel drainage grate
5	圆钢拉索	Round steel cable
6	3 厚金属穿孔板	3mm perforated metal plate
7	幕墙挂件	Curtain wall connection
8	幕墙横梁	Curtain wall beams
9	幕墙立柱	Curtain wall column
10	圆钢拉索节点	Round steel cable joint
11	H 型钢梁	H-section steel beam
12	断热低辐射铝合金门	Thermal insulation Low-E aluminum alloy doors
13	400/400 不锈钢篦子	400/400 stainless steel drainage grate
14	方管	Square tube
15	钢凹槽	Steel grooves
16	环氧树脂自流平楼面	Epoxy self-leveling floor

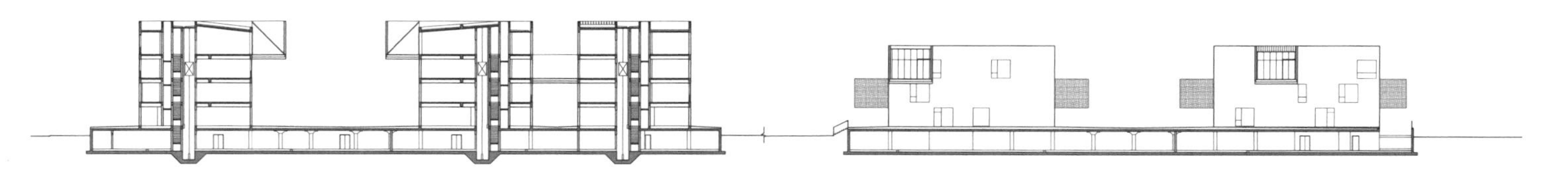

40

嘉定博物馆新馆
Jiading New Museum

02

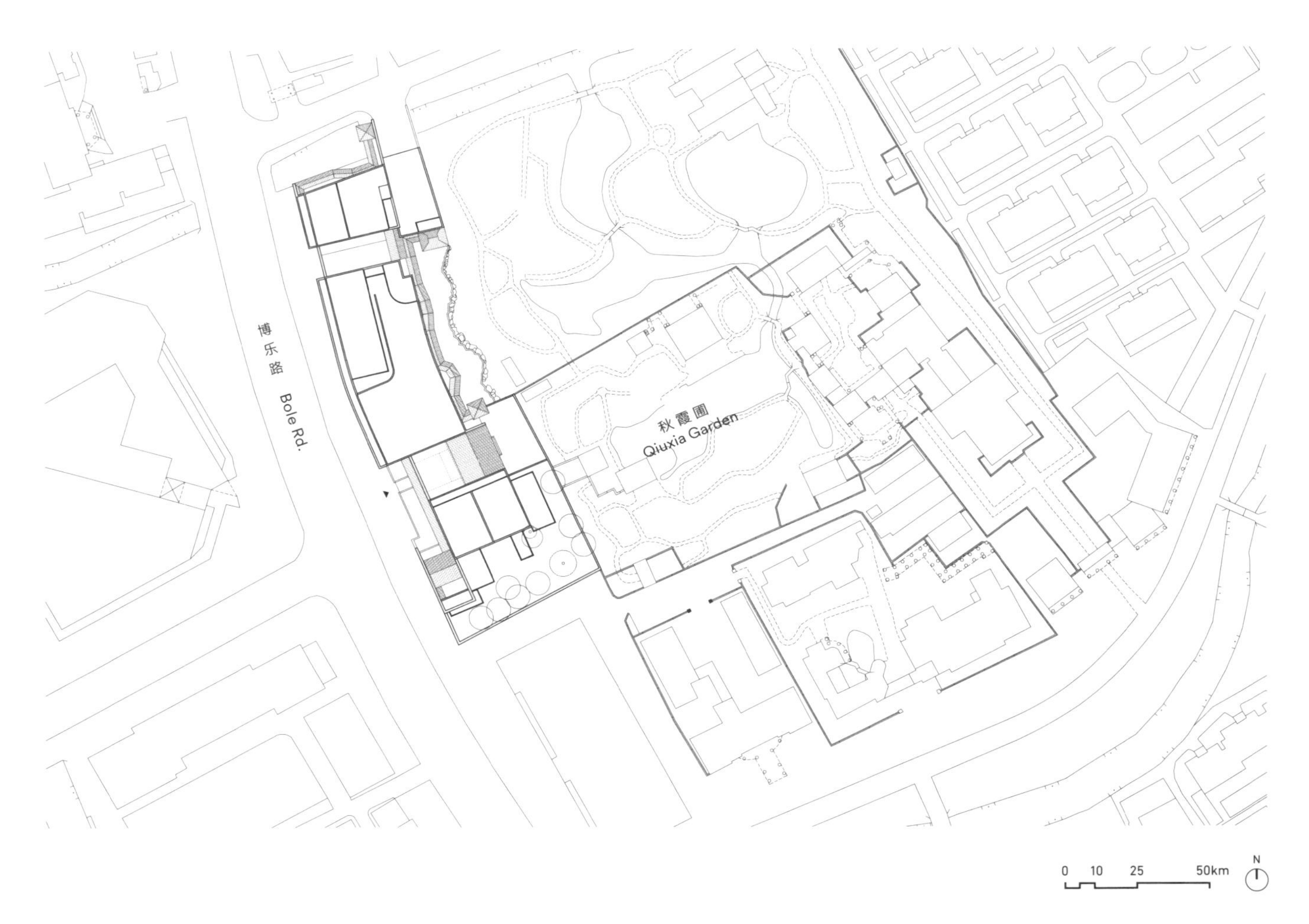
博乐路
Bole Rd.
秋霞圃
Qiuxia Garden
0 10 25 50km
N

新旧并置

秋霞圃西侧地块隶属嘉定古城州桥历史文化风貌区的协调区域，西临入城主干道博乐路，西南方向有老城主要的历史建筑和街区，包括孔庙、汇龙潭、州桥老街和法华塔，当然最主要的是一墙之隔，有着 500 多年历史的秋霞圃——上海五大古典名园之一。这种特殊的地理位置使其成为一个敏感的地块。从城市设计到最终的实施方案，我们经过了五轮的设计和调整。

结合在西侧地块为秋霞圃增加新入口的要求，我们采用了混合策略，即创造新旧截然不同的两种形式和空间，通过两者的并置和交替，形成对比强烈的体验。新建的传统园林作为秋霞圃的西入口与古典园林相接，博物馆屋顶部分被设计成现代地景园，既作为室外展场，又作为多功能空间，从屋顶还可以用新的视角俯瞰传统园林。新旧两部分空间保留了可分可合的可能，新旧形式和空间形成一定对比。

透漏的立面

在形式上，中国传统美学喜欢文质彬彬的和谐；在建筑上，则偏爱蕴含文质和虚实意象的透漏形式。基于对传统美学的研究，博乐路沿街立面采用透漏材质，幕墙采用从传统建筑纹饰发展而来的多种开孔方式，形成细腻典雅的文质效果。

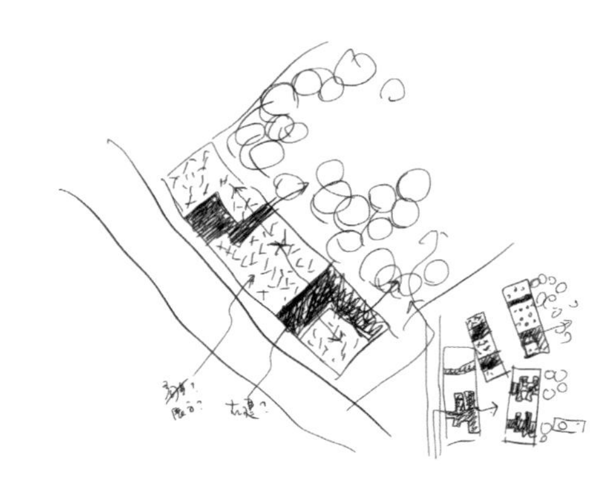

五轮比较设计方案模型

Five-round comparative design scheme models

Juxtaposition of the New and the Old

The west area to Qiuxia Garden is part of the historical and cultural district of Jiading Zhouqiao Ancient Town, facing the main road (Bole Road) on the west, connecting the town with major historical buildings and blocks to the southwest, such as Confucius Temple, Huilong Pond, Zhouqiao ancient alley and Fahua Pagoda, most importantly Qiuxia Garden just beyond the wall, one of the five reputed classic gardens in Shanghai with more than 500 years of history. Such location has rendered the place an extremely sensitive site. We have gone through five rounds of urban design and adjustment before finally settling down the plan.

In order to open a new entrance to the west of Qiuxia Garden, we adopt new strategies to create two distinct forms and areas. A sharp contrast is presented through aligning and intertwining these two. The newly built traditional style garden connects the classic Qiuxia Garden and its west entrance. The rooftop of the museum has been designed into a modern landscape garden, used both as an outdoor display area and for multi-purposes. From the rooftop, visitors can take in a fresh bird's eye view of the classical garden. Although the new and the old areas keep flexible in separation and combination, their form and space constitute notable contrast in character.

Translucent Facade

Traditional Chinese aesthetics pursue harmony in exterior etiquette and interior cultivation. In architecture, this philosophy and the image of solid and void combination are frequently conveyed through penetrating and perforating forms. Based on our research on traditional aesthetics, we have applied penetrating and perforating materials to achieve a transparent/opaque effect on the facades along Bole Road. The curtain wall has adopted various opening patterns developed from traditional buildings, resulting in the effect of exquisite and elegance.

嘉定博物馆
JIADING MUSEUM

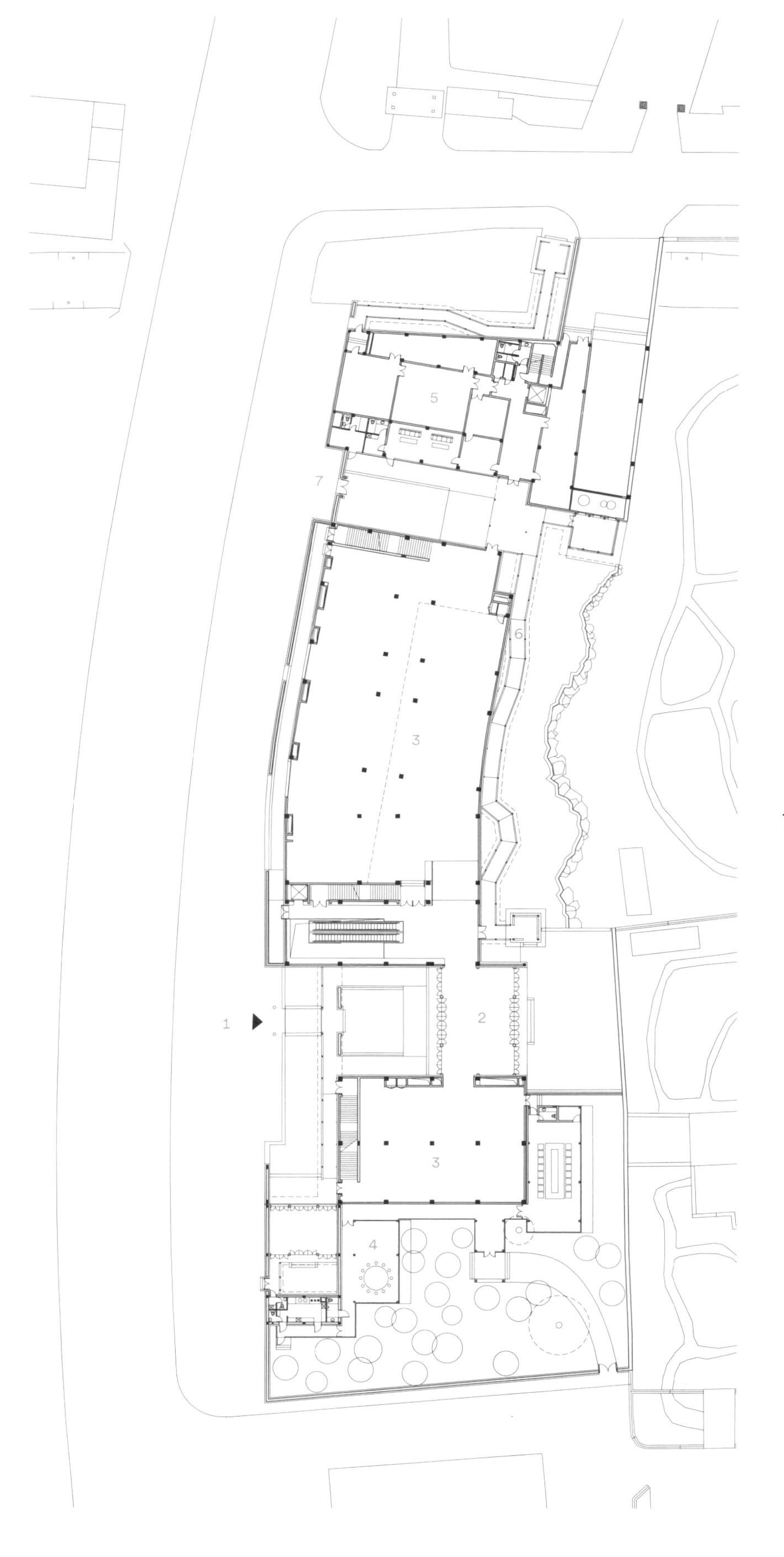

1	主入口	Main entrance
2	序厅	Vestibule
3	展厅	Exhibition hall
4	接待院落	Reception yard
5	库房	Storage
6	游廊	Veranda
7	次入口	Logistics entrance

首层平面图

First floor plan

嘉定新城双丁路公立幼儿园
Shuangding Road Public Kindergarten, Jiading New City, Shanghai

03

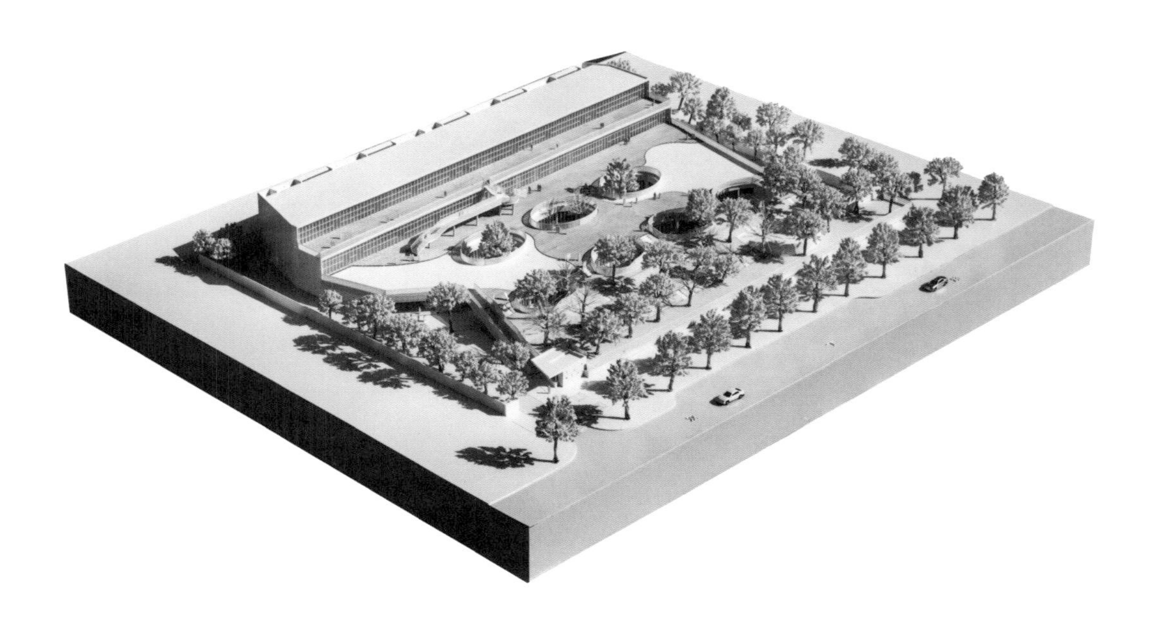

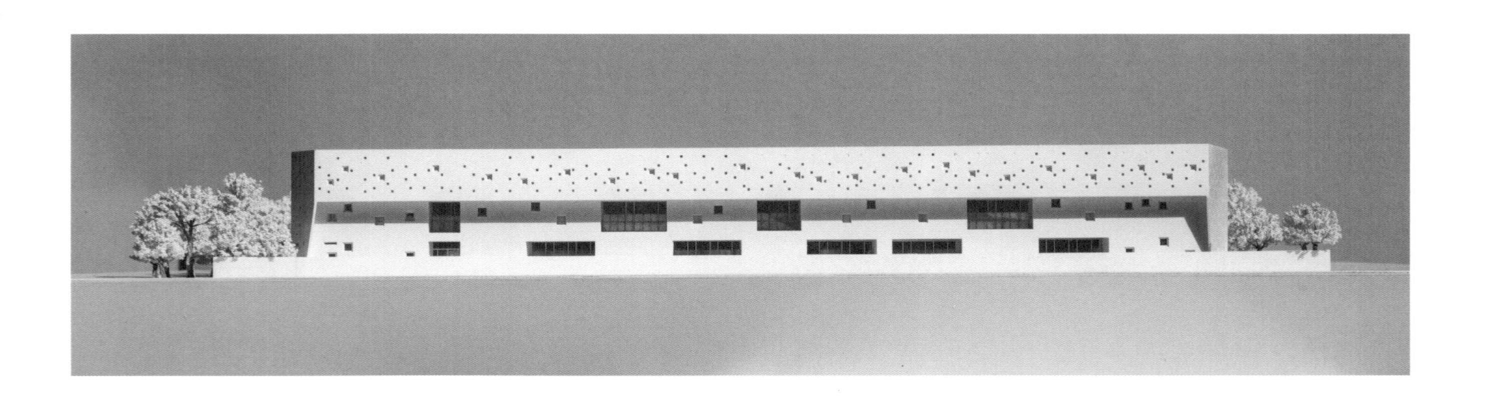

户外活动场地最大化

上海嘉定新城双丁路幼儿园是一座 15 班的标准公立幼儿园，用地按市政规划最小指标配置。为了让户外活动场地最大化，采用了建筑、场地与景观整体融合式设计策略。三层高的主体建筑贴紧基地北边布置，首层设置各类专用教室、教师和服务用房，二、三层是 9 个幼儿班全部朝南的活动单元，每层南面通过退台形成分班活动场地。主体建筑南面贴着宽扁的一层体量，用以布置 5 个托儿班、1 个早教班和 1 个多功能活动室，低幼儿可方便进出教室及其直接毗邻的椭圆形庭院，首层教室屋顶成为全园集中活动场地。这片场地优点至少有三：面积是采用建筑场地分离的做法能留出的最大场地的 2.43 倍；因为处于中间高度，跟上下两层仅需一层楼梯就可顺畅连接；与所有室外活动场地直接毗邻的建筑界面均高一层，尺度亲切。

微地形

设计者还创造了一片连续而有趣的微地形：建筑体量南面的场地在竖向上被平分成三组，每组有一斜一平两片东西向延展的草坡，用于种植四季树木，布置沙坑、水池、旗杆、种植园和其他趣味活动设施。草坡顶部设计与二层硬质场地无缝连接，底部与双丁路北面的绿化隔离带连为一体。整座幼儿园体量和空间从北面最高处层层跌落，直到城市道路，在不同楼层的孩子视线可以穿过层层叠叠的彩色圆钢栏杆和疏密交织的树枝，一直抵达城市的车水马龙和远处的风景。通常相当封闭的幼儿园获得了罕见的开放性和公共性。场地全部抬高则保证了安全。

场地如同一块完整的画布

在施工过程中，一方面由于高质量的彩色外墙涂料价格较高，另一方面想强调活动场地完整性这一核心概念，最后色彩方案也采用了融合式的设计。主体建筑外墙面和门窗框料分别保持浅灰和中灰，庭院内壁、场地面层和栏杆则采用彩色，并且不同颜色的分区是超越材质和界面的。同一庭院内壁墙面和上面的金属栏杆采用同一颜色，以形成统一的空间和体量感。为了与椭圆形庭院边界协调，二层集中场地的塑胶场地采用弧线划分出两个区块，颜色根据被圈入的庭院颜色确定。塑胶地面跟木地板标高一致，不同材质的场地在空间上是连贯的。从三层平台通到二层的钢楼梯也被刷成统一色。从各层室外平台望出去，那些彩色的庭院和透空的栏杆、楼梯就像虚虚实实的大型彩色玩具一样，空间隔而不断，场地宛如一块完整的画布。颜色的融合式设计不仅实现了空间整体统一的效果，一些施工粗糙和误差问题也不那么突兀了，甚至反而在统一颜色的主宰下，形成一些微妙的光影，颇为有趣。

Maximize the Playground

Shanghai Shuangding Road Public Kindergarten in Jiading New City is a standard public kindergarten with fifteen classes. Its building area is allocated according to the minimum standard of municipal planning. To maximize the open playground, the kindergarten was designed to integrate architecture, site and landscape. The three-story main building is attached closely to the north. Various functional classrooms, teacher's rooms and service rooms occupy the first floor. On the second and third floor, there are south-facing activity units for nine nursery classes. On each floor, south terraces are stepping backwards to form activity space for each class. At the south side of the major building, there is a wide and flat one-story building where five infant classes, one early education class and one multi-functional activity room are accommodated. Young children can get easy access to their classrooms and affiliated oval gardens. The roof of this building is designed as a central activity area. There are at least three advantages of this activity space on the first floor roof. Firstly, 2.43 times of the maximum open space is achieved compared with a conventional separating plan. Secondly, by putting the main platform in the middle (on the second floor), people can get easy access to the roof garden by climbing only one floor high. Thirdly, the space is user-friendly since all the buildings around the outdoor activity space are one story high.

Micro-Topography

The designer has also created a continuous and interesting micro-topography. The southern area is vertically and averagely divided into three parts. Each part includes a piece of slope grassland and a piece of flat grassland, extending from east to west. This site is designed for planting trees, setting sand pit pools, flagpoles, plantations and for other interesting activity facilities. The top of the grassland is seamlessly connected with the second floor, while its bottom is connected with the green belts at the north of Shuangding Road. The building volume goes down from the highest point in the north to the urban road in the south. Children's sight can rest upon the busy urban life and distant landscape by looking through colored round steel railings and dense branches. Kindergartens are usually established in a highly closed place, while this one is more open and public. Since its site is elevated, it remains safe.

Playground like a Unified Color Canvas

During the construction, since high-quality coating for exterior walls is very expensive and the designers intend to emphasize the integrity of the activity site, a fusion design is adopted. The exterior walls of the main building, the door and window frames are respectively kept light-grey and medium-grey, while the interior walls of the courtyard, the ground floor and the railings are kept colored. Different materials or areas might be colored in the same. The walls and the metal railings in the same courtyard are painted with the same color so as to unify the sense of space and volume. Since the plastic floor and the wooden floor are of the same height, two sites with different flooring constitute a continuous space. Steel staircases between the third floor and the second floor are all painted in the same color. Consequently, looking out from the outdoor platforms on each floor, those colorful courtyards, railings and staircases seem like huge colored toys. The spaces are distinguished but not separated, which makes the whole site a complete canvas. The color fusion design not only achieves perfect spatial integrity, but also helps to alleviate shoddy construction and construction errors. It has even achieved a subtle light and shadow effect.

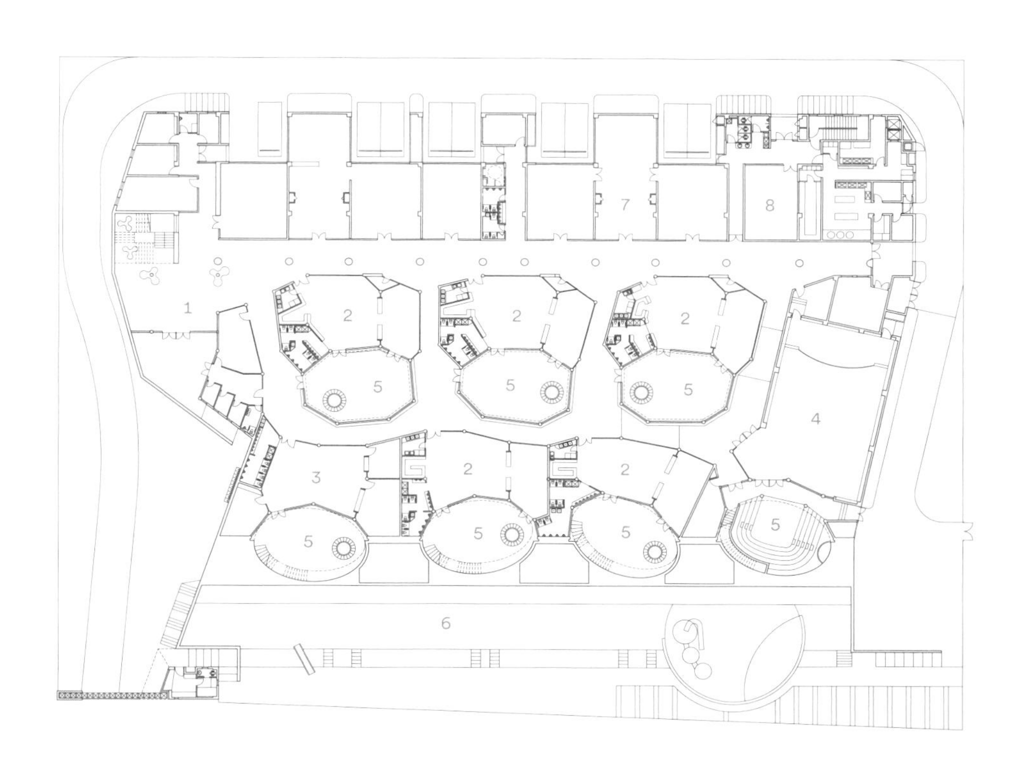

首层平面图

First floor plan

1	门厅	Lobby
2	托儿班活动室	Nursery activity room
3	早教指导中心	Guidance center
4	多功能活动室	Multi-functional room
5	庭院兼分班活动场地	Courtyard activity space
6	草地活动区	Grassland activity area
7	专用活动室	Special activity room
8	教师餐厅	Teachers' restaurant

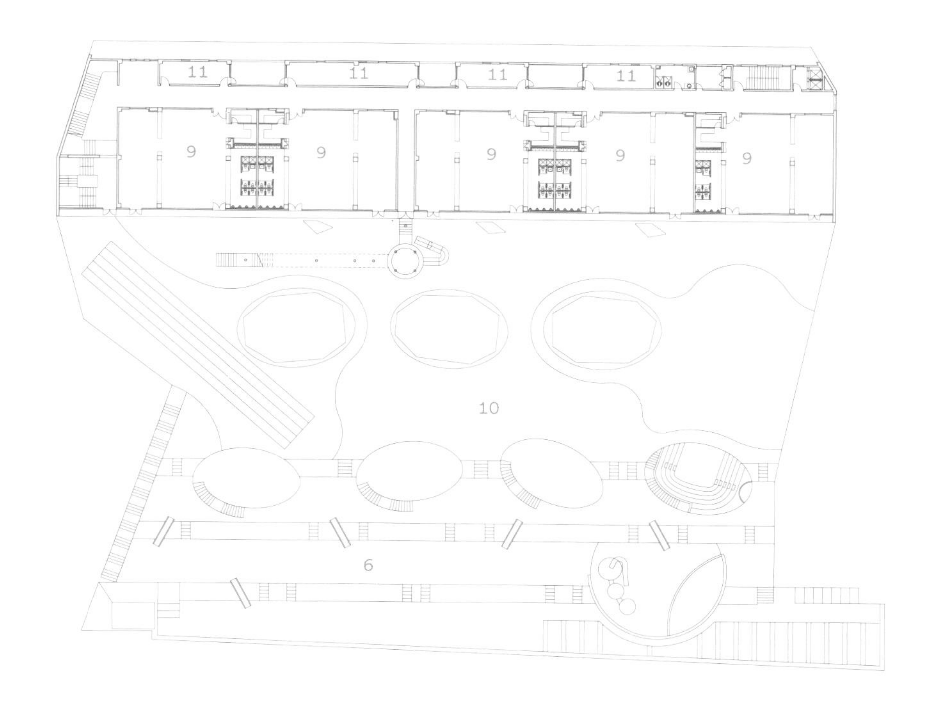

二层平面图
Second floor plan

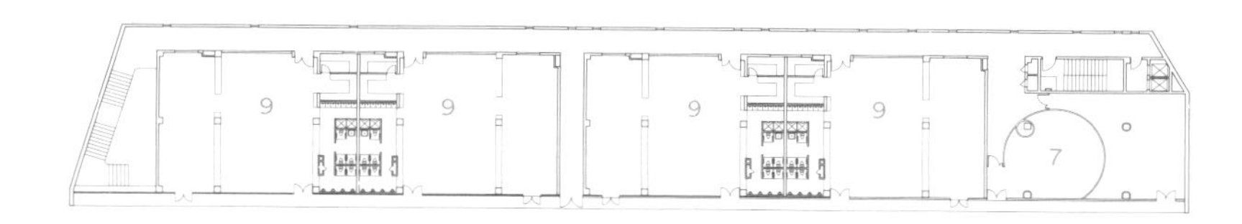

三层平面图
Third floor plan

6	草地活动区	Grassland activity area
7	专用活动室	Special activity room
9	幼儿班	Classroom
10	活动场地	Playground
11	办公区域	Office

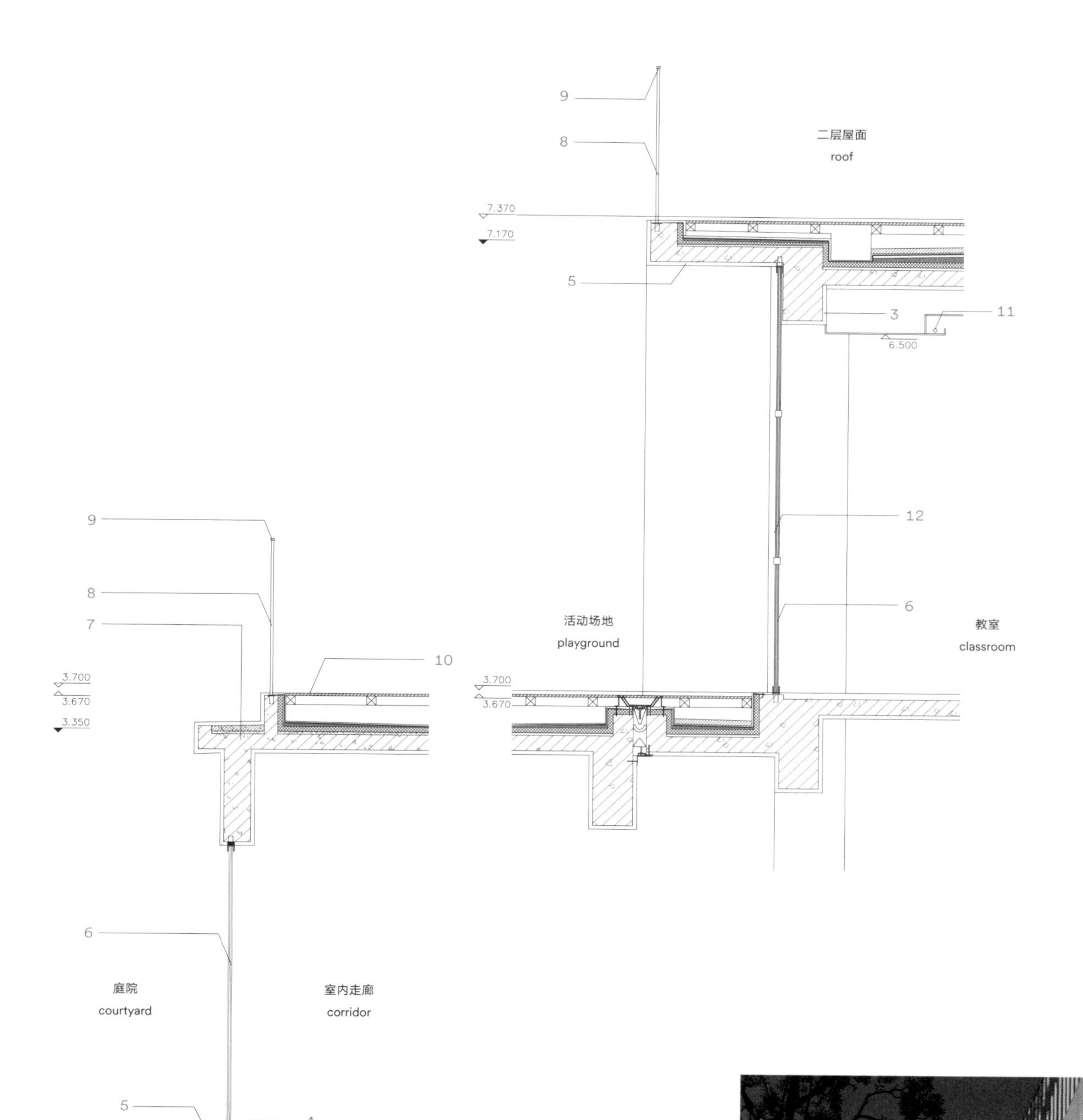

内庭院的垂直剖面
Vertical section of the inner courtyard

1	2 厚 PVC 卷材地面	2mm PVC floor
2	2 厚 PVC 卷材踢脚	2mm PVC baseboard
3	20 厚无机保温砂浆	20mm inorganic thermal insulation mortar
4	软包墙面	Artificial leather wall skirt
5	30 厚无机保温砂浆	30mm inorganic thermal insulation mortar
6	断热铝合金遮阳型玻璃窗	Thermal insulation aluminium window with sunshading
7	刚性防水屋面（防水涂料面层）	Rigid waterproof roof (waterproof coating)
8	3 厚 Φ20 钢管	3mm Φ20mm round bar
9	40/15 扁钢	40/15 flat steel
10	950/4000/19 防腐木	950/4000/19 anticorrosive wood boarding
11	日光灯槽	Lamp chamfer
12	铝扣板	Aluminum gusset sheet

恒森大厦
MOON HOTEL
1963

诸暨剧院
Zhuji New Theatre

04

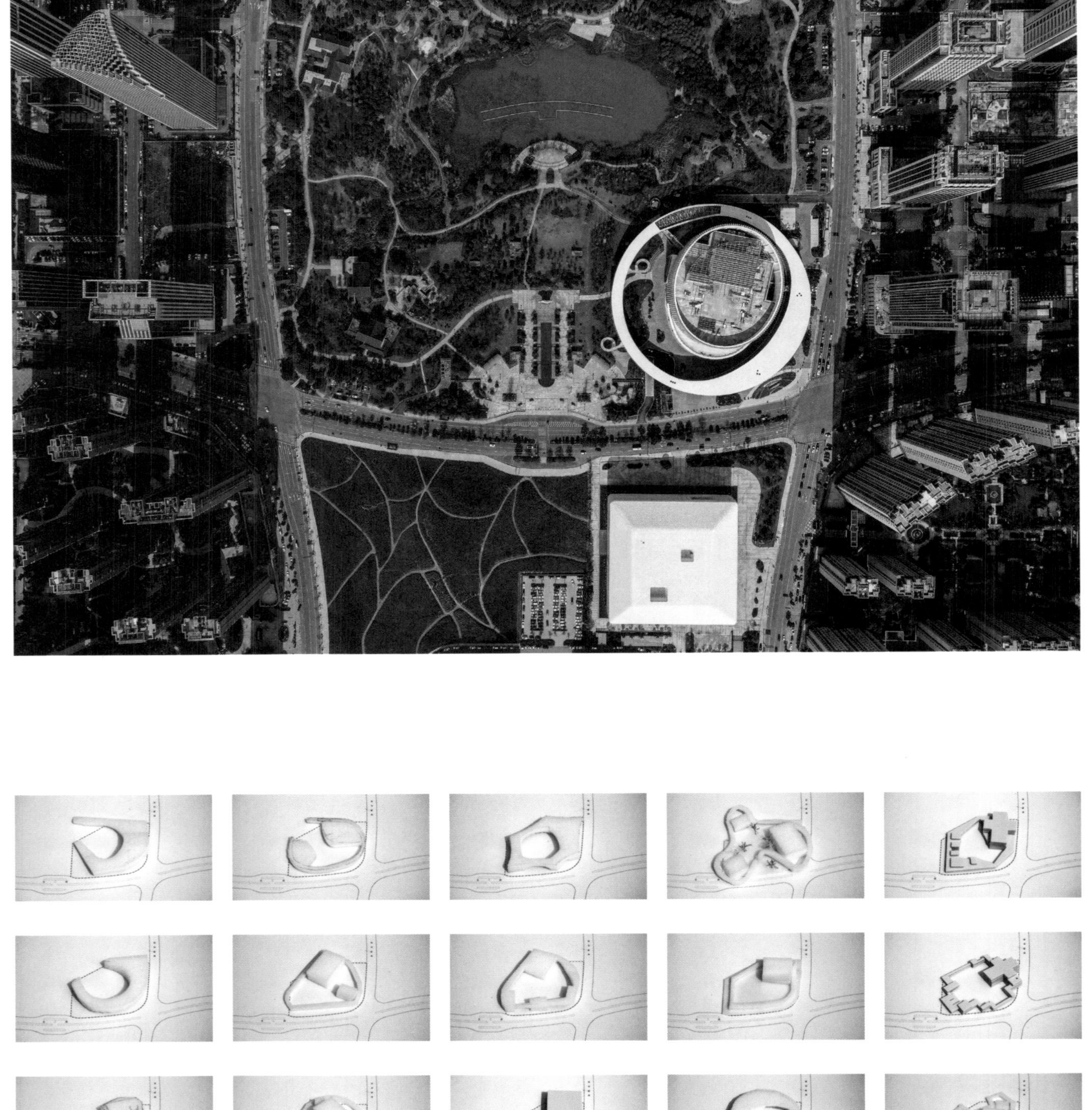

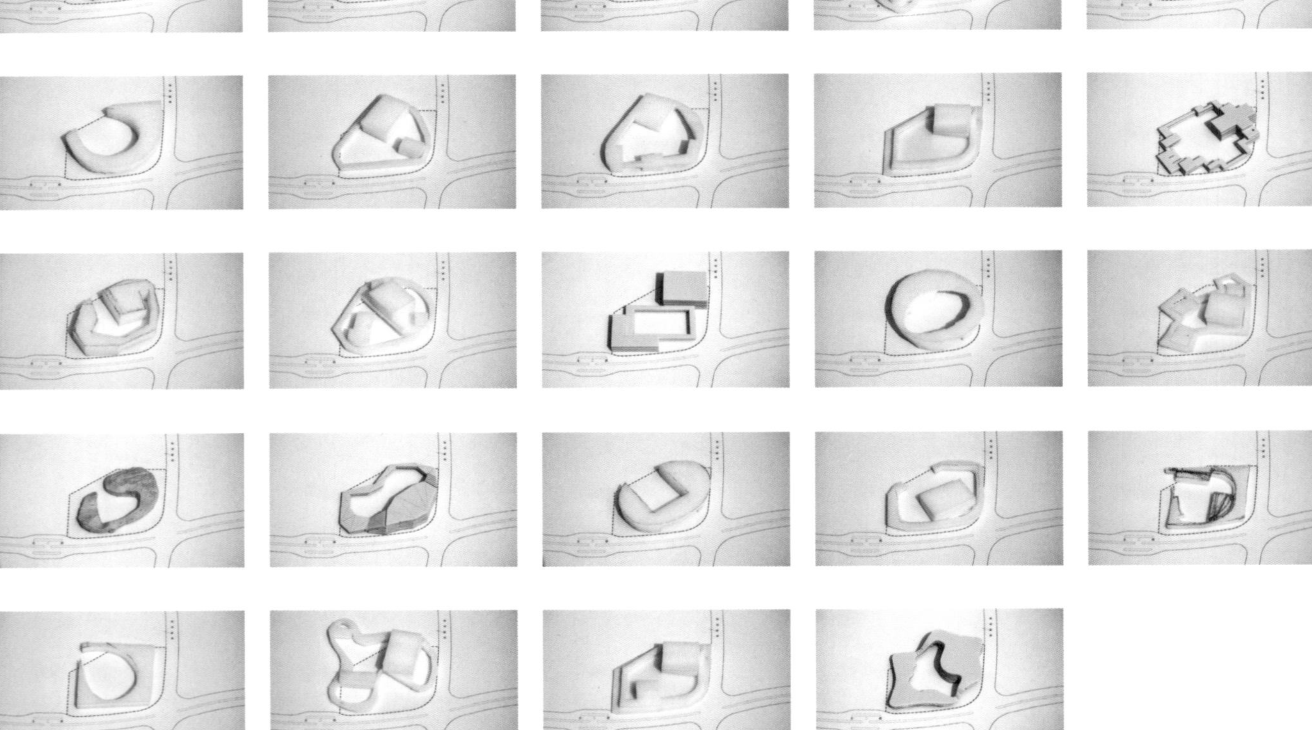

开放的剧院

在 2000 年前后，为了满足传播城市形象和提升生活品质的需求，中国不少直辖市、省级城市开始兴建大剧院、规划展示馆、体育场等文化设施。这些剧院通常占有大面积的新城中心用地，但由于管理限制，其公共空间未能得到充分利用。诸暨剧院位于浙江省诸暨市城东区市民公园的东南角，业主希望将其与南侧同时建造的规划展示馆一起打造成为城市文化新地标，并能与富有地方文化特色的人文戏曲因子相关联。由于选址位于大型城市中心公园的一角，设计师考虑将建筑封闭管理的边界尽量缩小并后退到极限，其他空间尽量开放，与公园融合，可供市民自由享用。

城市光环

建筑布局根据功能分为两部分：中心偏北是椭圆造型的主体建筑，内含需封闭式管理的 1380 座观众厅，配备了标准的升降乐池、左右侧台、后转台的舞台，其台口宽 17.6 米、高 10 米；其他所有功能被放入外围圆环形建筑中。外环建筑像一道光环，架空漂浮起来，使地面空间充分对外打开，两个建筑之间形成开放式内外广场与绿化。这样的布局一方面减轻了庞大的建筑体量对周边城市空间的压迫感，另一方面也使公园绿地、城市空间自然融合。外环建筑在立面上呈现出自然起伏的造型，其南向有 32 米大跨度的挑空，既突出了建筑主入口的鲜明意象，也为观赏建筑提供了更好的角度和视野。材料选择上，主体建筑以 S 形白灰色穿孔铝板覆面，外环建筑表面采用弧形灰色半透明玻璃幕墙，两者相互映衬烘托。在这道城市光环下，市民们可以自由地徜徉在公园的自然美景和剧院的人文盛宴之间。

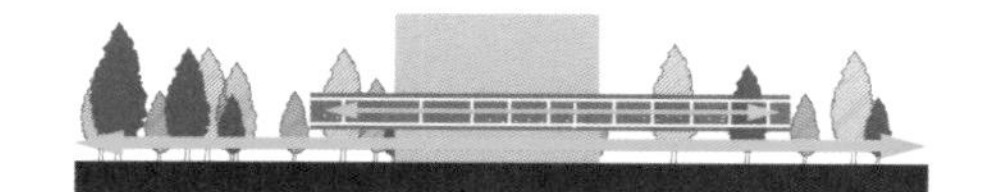

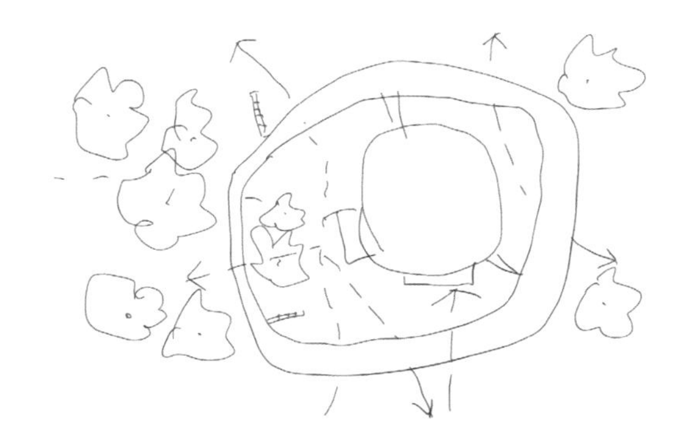

An Open Theatre

Around 2000, to promote the image of urban culture and improve the urban living quality, there is a trend for Chinese provincial cities to erect public facilities like grand theatres, urban exhibition halls and grand stadiums. Although most of these theatres occupy large land in the new city center, they are poorly accessible as an urban public space under strict management. Zhuji New Theater is located in the southeast corner of Citizen's Park in Chengdong Area of Zhuji City, Zhejiang Province. The client hopes that together with Zhuji Planning Exhibition Hall on the south, it can become a new cultural landmark also showcasing the local heritage in the traditional drama, and creating an urban space that can be actively used. Since it is located at a corner of a large central park, the designer has decided to minimize and bring back the boundaries of the close structures under the full management to the uttermost, while making other spaces more integrated with the park and open for free use.

A Bright Ring in the City

According to the functional requirements, the building is divided into two parts. The main structure with oval shape sits in the north center, which contains the minimum controlled facilities, a 1,380-seat auditorium and a stage with a standard lift orchestra pit, left and right side stages and a rear rotary stage, whose proscenium is 17.6m wide and 10m high. An outer circular ring-like structure accommodates all the other functions, just like a floating light right in the city. An open interior and exterior plaza with greenery are formed between these two structures. On the one hand, it successfully reduced the pressure on surrounding urban space caused by the tremendous volume of the theatre. On the other hand, the natural undulating form of the outer ring structure and the 32m large span of cantilevered space not only highlight a vivid image of the main entrance, but also bring in excellent urban scenery. The main structure is cladded with S-shaped white and grey perforated aluminum plates, and the outer ring structure covered with curved grey translucent glass curtain wall to form a harmonious unity. Under this grand city light ring, people can wander between the natural scenes of the central park and the cultural performance in the theatre freely.

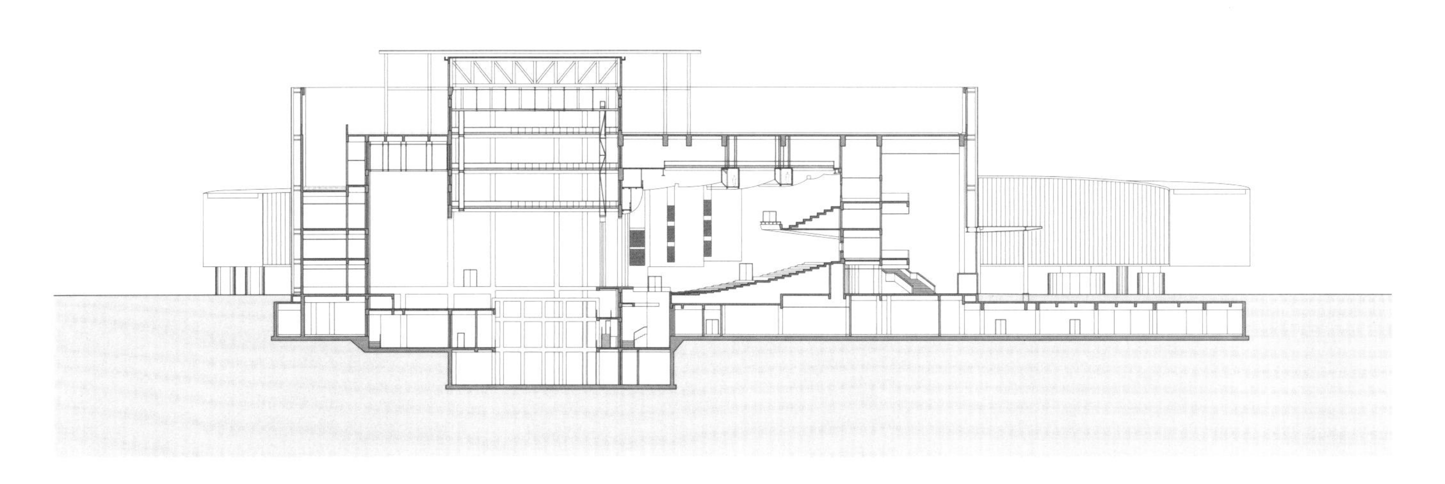
0 1 2 5 10m

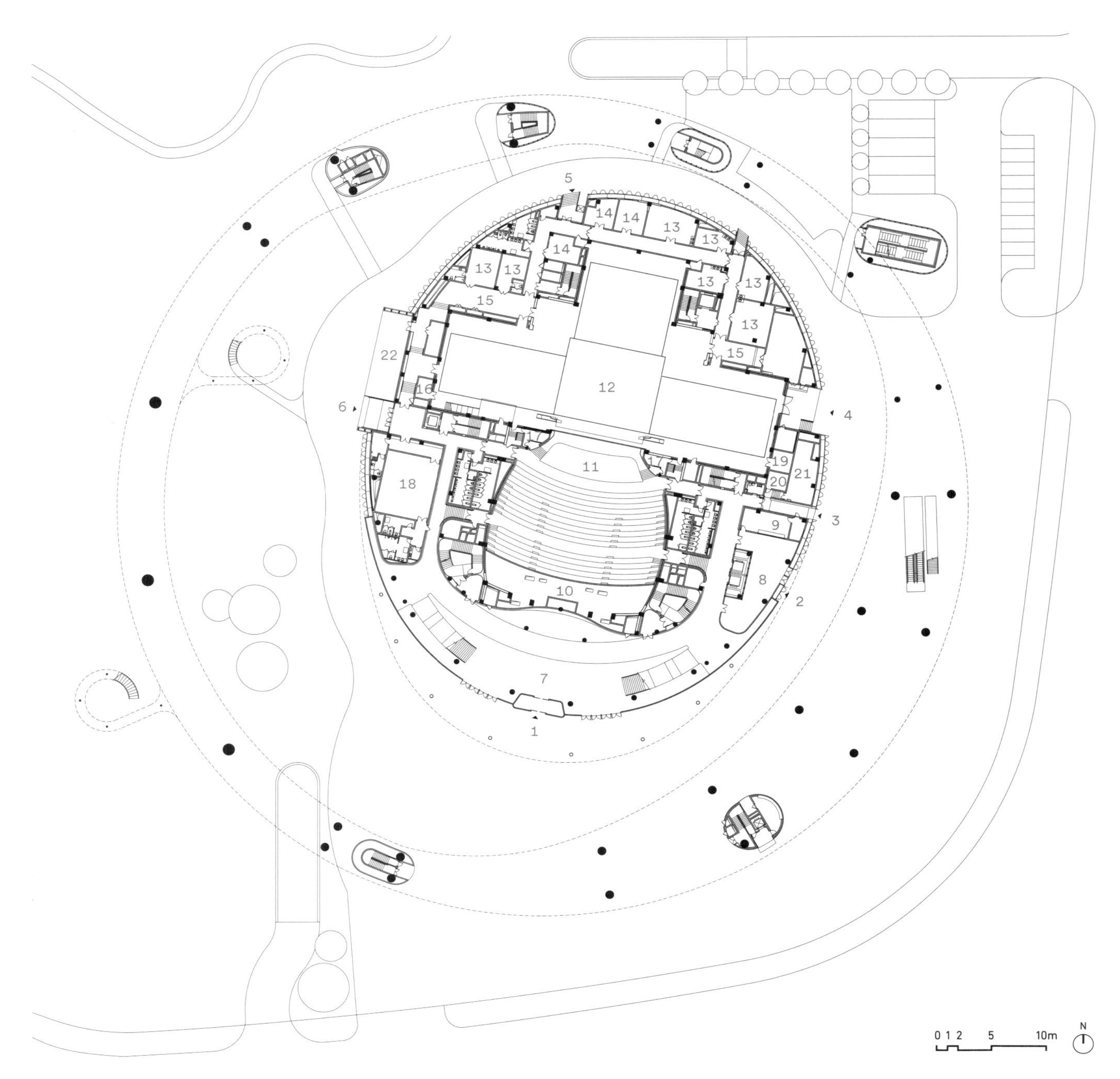

一层平面图

First floor plan

1	主入口	Main entrance
2	售票大厅主入口	Entrance of ticket hall
3	消防疏散口	Exit
4	货运入口	Freight entrance
5	演职人员入口	Casting entrance
6	贵宾出入口	VIP entrence
7	门厅	Lobby
8	售票大厅	Ticket box
9	票务中心	Ticketing center
10	设备	Device
11	乐池	Orchestra pit
12	舞台	Stage
13	化妆间	Makeup room
14	服装间	Costume room
15	候场区	Green room
16	舞台监控	Stage monitor
17	音响间	Acoustics
18	贵宾休息室	VIP room
19	琴房	Piano room
20	有线电视转播室	Cable TV transit
21	消防控制室	Fire control room
22	露天小剧场	Open-air theatre
23	门厅上空	Void
24	声、光控室	Sound and light control rooms
25	休息厅	Lounge
26	咖啡厅	Coffee
27	舞台上空	Void, part of stage
28	空调机房	Air-conditioning room
29	控制室	Control room
30	录音室	Recording room
31	配套用房	Supporting room
32	道具间	Prop room
33	休闲餐饮	Catering
34	备餐	Meal preparation
35	办公	Office
36	小卖部	Buffet
37	售票	Ticketing
38	电影厅	Theatre
39	储藏	Storage

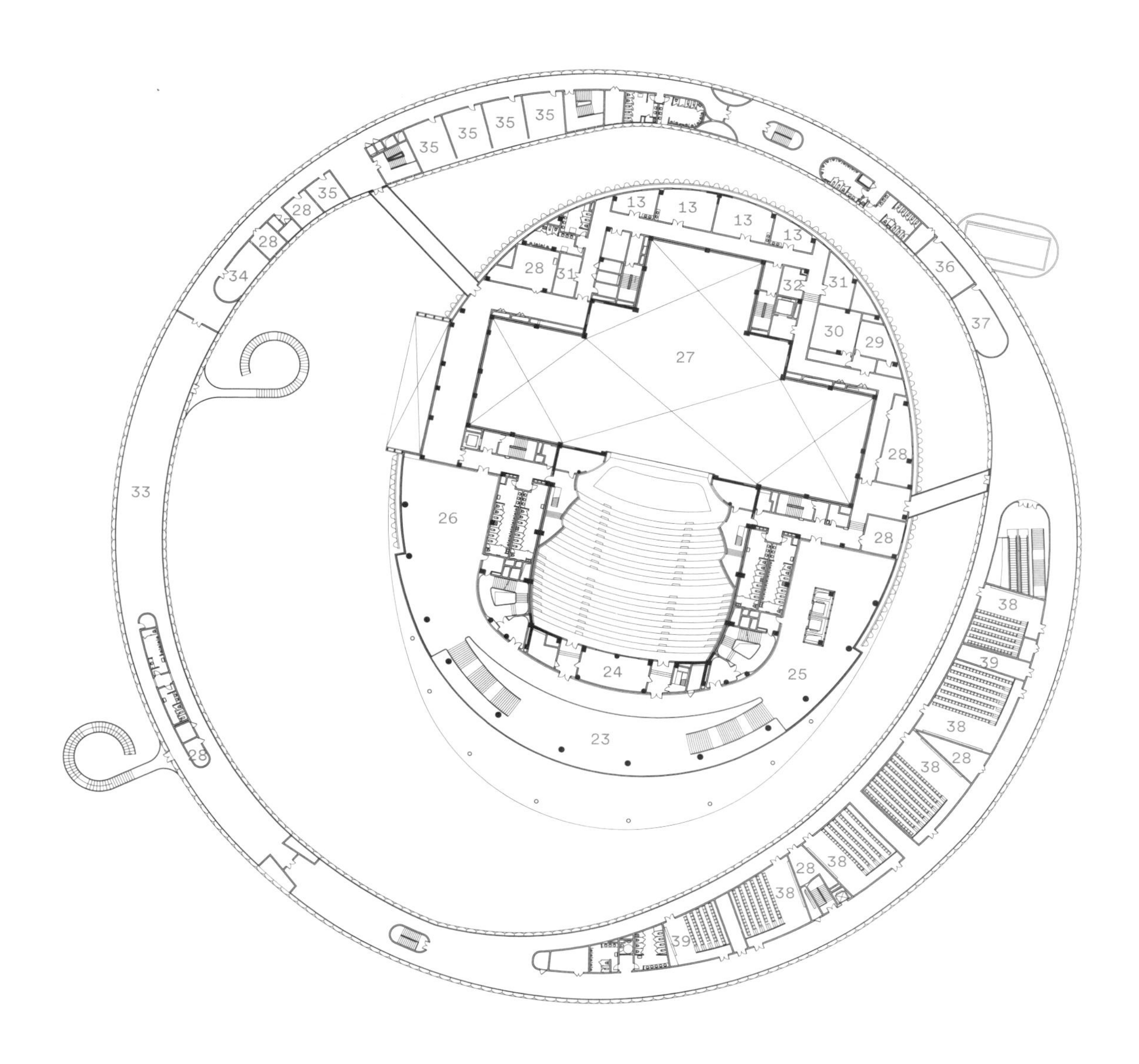

二层平面图
Second floor plan

安全通道
质量在我手中
平平安安回家
高高兴兴上班
川

安全警句千条万条

VERSACE

南京下关区永宁街地块
Yongning Street Plot, Xiaguan District, Nanjing

05

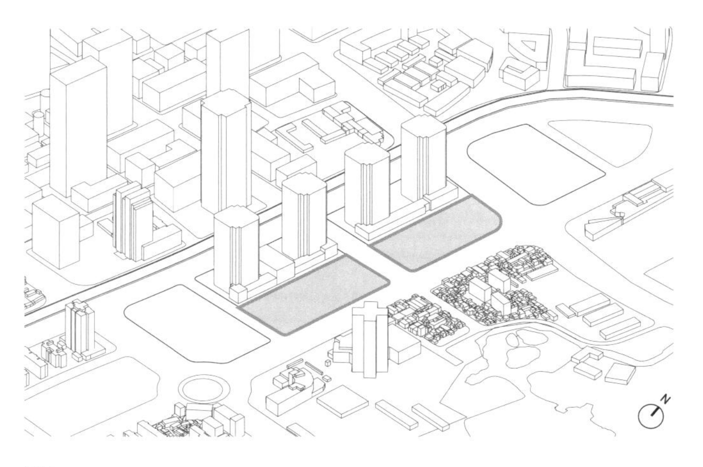

基地
Site

最终确定的布局
Final layout

东西向板式住宅

商品房的设计，大多跟着市场策划走，采用成熟的产品形式，尤其是住宅户型，一般不会轻易尝试有难度的新做法，以避免风险。于是，这种建筑类型似乎成了设计技巧基本无所作为的地方了。但永宁街住宅设计是一个特殊的案例，这个高层住宅在特定条件下形成了复杂的平面组织。

基地位于南京滨江，南北相接两块地，形状都南北狭长。地上建筑容积率 6，主力户型面积要求 80~90 平方米左右。江南人习惯居住南向为好，试了多种正常朝向的点式或板式的总体布局，均不理想，不是间距太拥挤压抑，就是会出现太多北向或东西向的户型。于是建筑师换一个角度去设想这个总体布局——一般在江南不太常用的东西向的板式会有利于布局吗？结果得到了两块基地上四栋东西向的板式与两头的点式组成的纵长的布置，这形成了南北贯通的区内中心景园，总体布局从空间感与容量解决上都优于其他。

交错对跃户型

接下来的问题就看户型是否可以达到要求——能为东西向的板楼设计一套符合江南生活朝向需求的住宅吗？最后，一个特殊的相互交错对跃的户型解决了这个问题。以一个 90° 直角转折的交错对跃的标准户型为主体，整体的平面犹如一挂鞭炮。从中间走廊入口，每户左边的底层跃到上层的右边，右边的则跃到上层左边。两户对称互跃，交织咬合在一起，依次重复形成长长的东西向板楼。一共四栋楼，两栋 90 米，两栋 60 米。每户都既有东南 45° 又有西南 45° 的朝向，上午与下午都有阳光日照。每户的视线都可以延伸到板楼的两侧户外。相对两栋楼房之间的视线又由于一同朝向东侧角度的关系，避免了相互近距离对视干扰。

入户走廊最终选取了一个空间曲折的方案。底层有两个出入口，一个主入口，一个厨房出入口，厨房入口应对消防规范，平时不用，室内布置活动操作桌面。每户底层有厨房、餐厅、客厅、书房与一个小卫生间，二层为主卧室、次卧室、卫生间与阳光房。这样的户型每户总面积不大，建筑面积 90 平方米，但功能完整，二房二厅二卫，同时上下再有大阳台与阳光房。

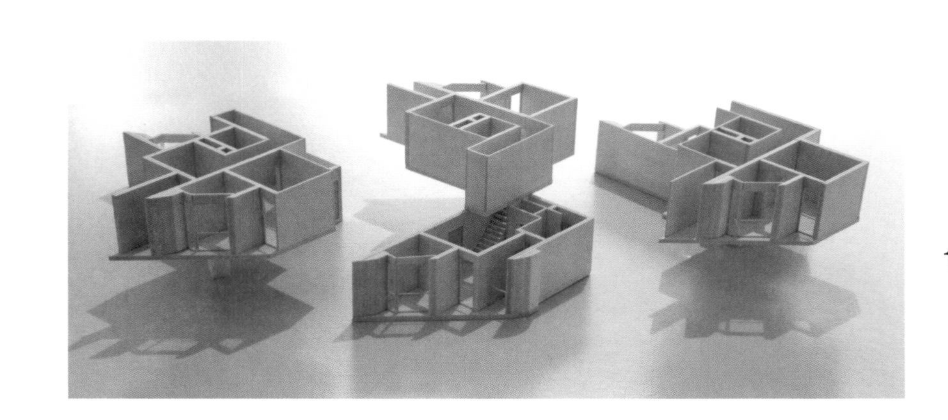

East-West Oriented Slab Apartments

In most cases, the design of commercial housing follows market trends and adopts mature prototypes, especially for residential units. To avoid market risks, new types are seldom proposed and design skill seems unnecessary in this field. However, the residential design of Yongning Street is a special case. This high-rise residence has a complicated layout under particular conditions.

Located in the riverside of Nanjing City, this site consists of two parts bordered with each other in the south-north direction, both narrow and long from south to north. The floor area ratio of aboveground buildings is 6 and the major type of apartment is required to be about 80–90m². People to the south of the Yangtze River prefer to live in a room facing the south. Many kinds of tower-like or slab-like layout with conventional orientations have been tried, but none of them is acceptable. Either the distance between buildings is too narrow and depressed or too many rooms are facing north or east-west. Consequently, the architect has refreshed his mind. Will the east-west oriented slab apartments that is not popular among the people to the south of the Yangtze River be helpful to the layout? Finally, the architect has designed on these two lands a longitudinal layout with four slab-type buildings of east-west orientation and two tower-like buildings at the two ends, a central garden extending from north to south. This layout proves to be better than any other schemes from the perspective of spatial perception and plot ratio.

Joint Duplex Unit

Next, can the apartment unit meet the requirements? Is it possible to get a residential apartment that can adapt to the habit of people to the south of the Yangtze River in an east-west oriented slab building? Eventually, a special type of apartment has been designed to address this problem. Entered from the middle corridor, the lower floor of each apartment on the left connects to the right side of the upper floor, and the lower floor on the right connects to the left side of the upper floor. The two apartments connect with each other symmetrically, and then this pattern is repeated, forming a long east-west slab-type building. There are four buildings in total, two of them 90m long and the other two 60m long. Each unit has both a southeast 45° orientation and a southwest 45° orientation, and can be exposed to sunshine both in the morning and afternoon. Each unit enjoys the outdoor view of two sides of the building. Since the windows of the neighbor buildings both face the same direction, to the east, eye contact is avoided.

The final plan of the access corridor is zigzagged. There are two entrances on the lower floor, one main entrance and one kitchen entrance, which is mainly designed to meet the fire control regulations instead of daily use. That's why we design a movable countertop here. The lower floor of each unit is designed with a kitchen, a dining room, a living room, a study room and a small toilet. The upper floor is arranged with the main bedroom, the second bedroom, a bathroom and a sunroom. With a floor area of 90m², this type of unit is not spacious but designed with complete functions. There are two bedrooms, one living room, one dining room and two toilets in each unit, with an extra big balcony and a sunshine room on both the upper floor and lower floor.

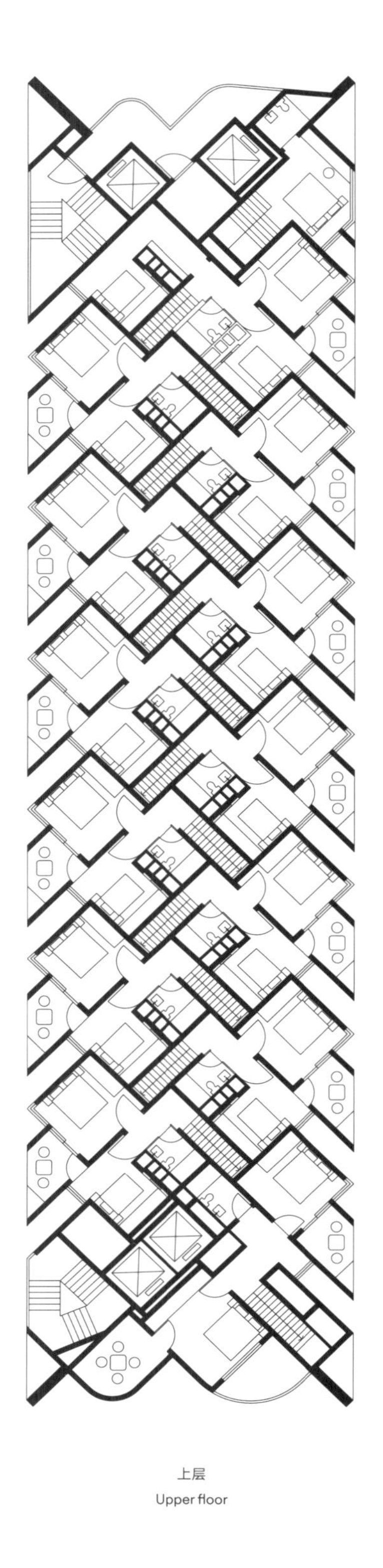

上层

Upper floor

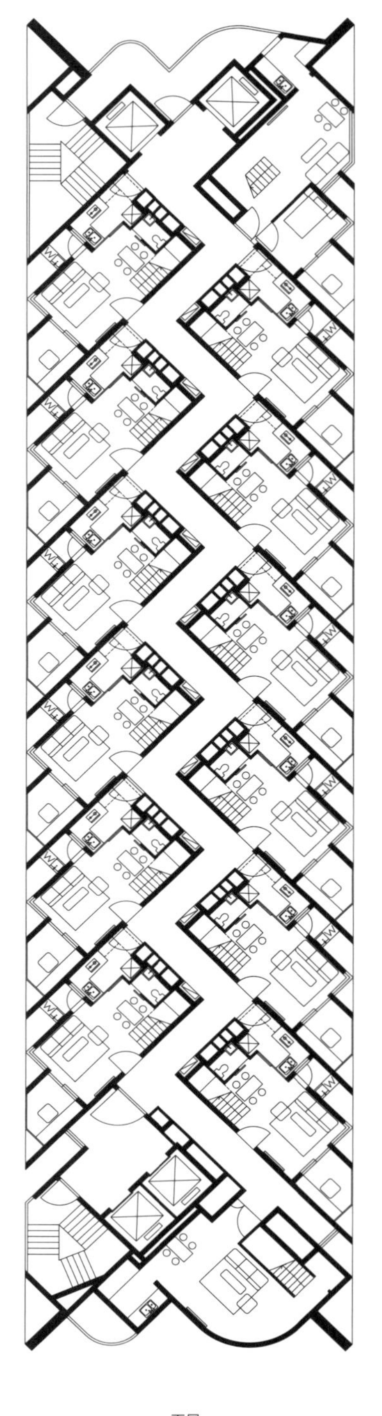

下层

Lower floor

标准层平面

Typical plan

0 1 2 5 10m

N

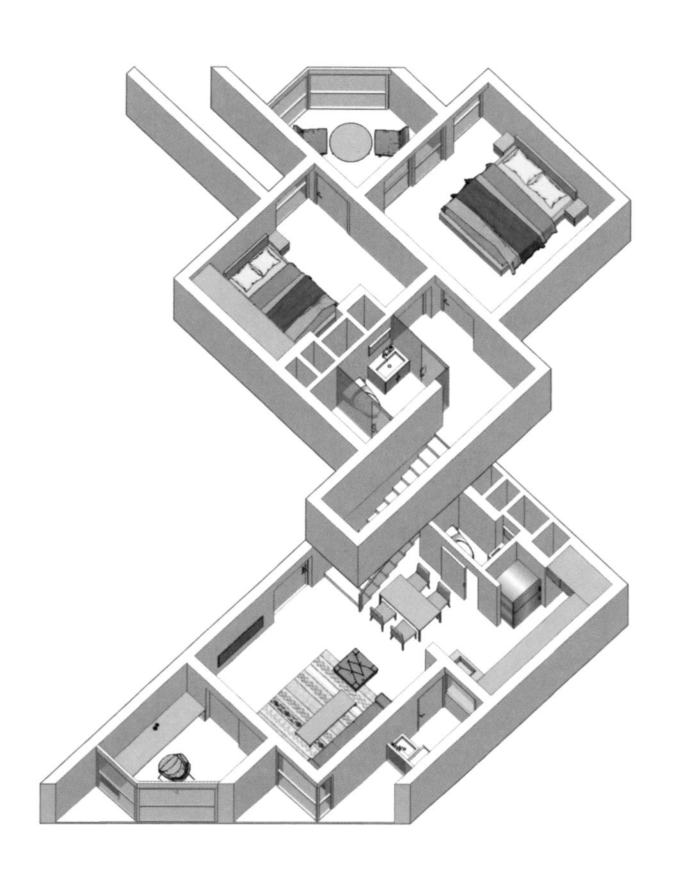

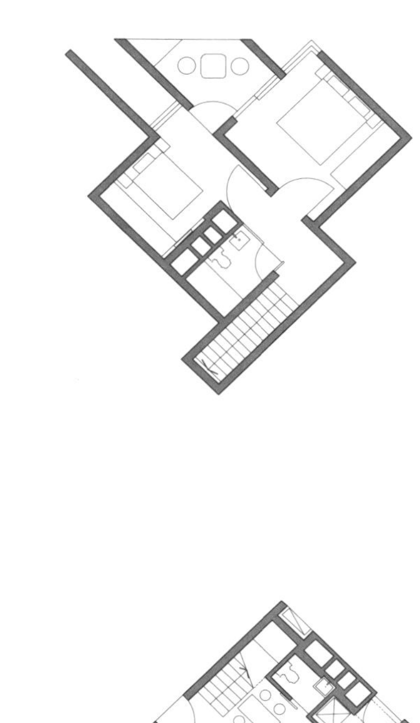

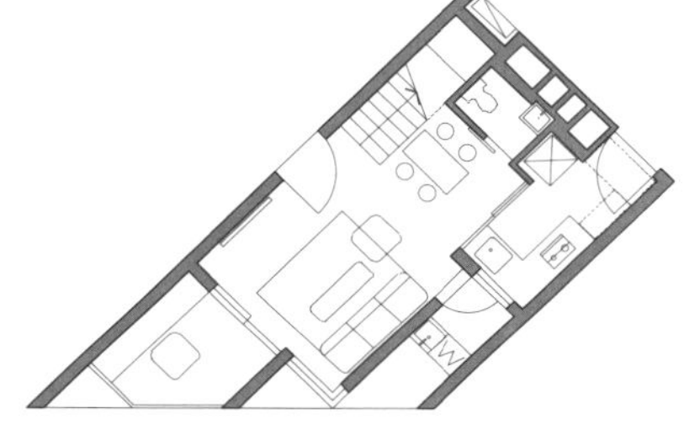

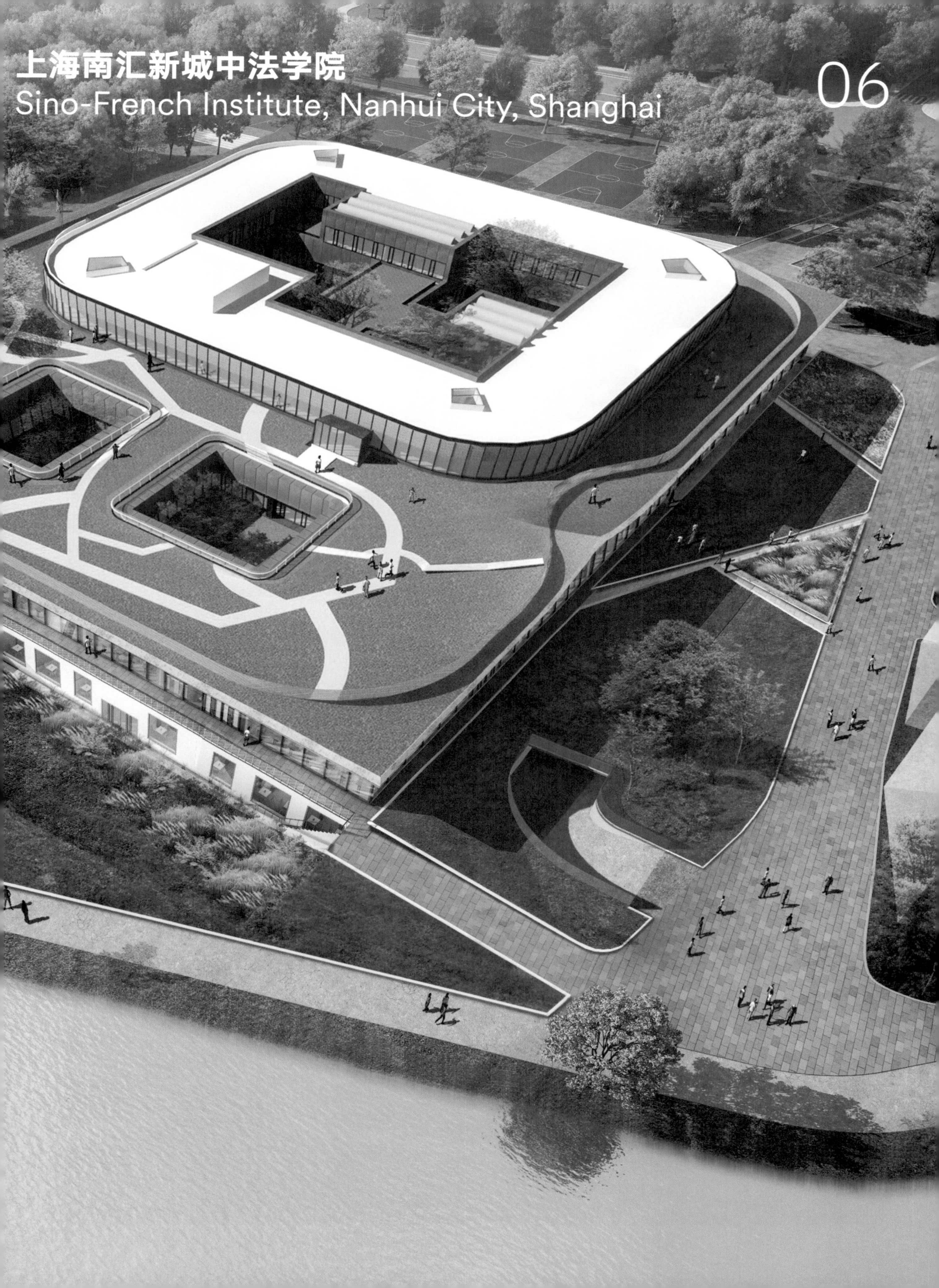

上海南汇新城中法学院
Sino-French Institute, Nanhui City, Shanghai

06

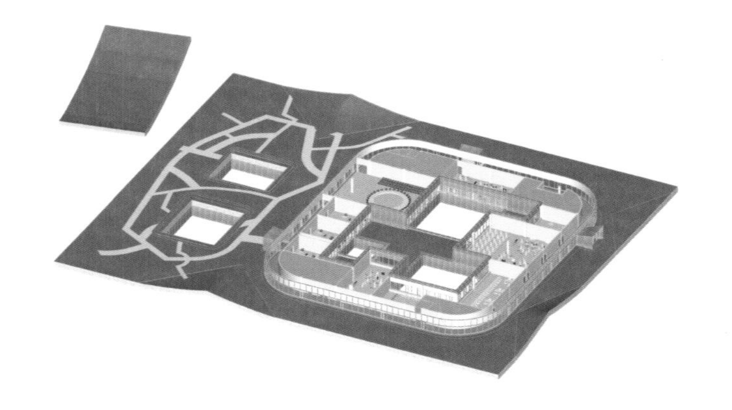

二层 部分活动室及所有办公室，屋顶绿化散步道以及屋顶小剧场

2F Activity rooms, all offices, roof garden walkway and small roof theater

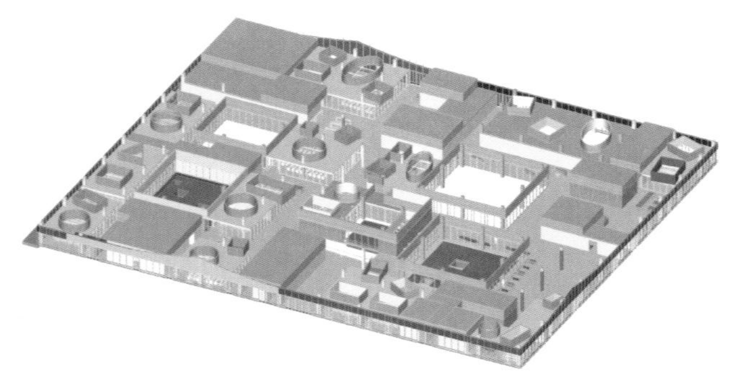

首层 主要文化活动场所和公共聚集场所

1F Main cultural events and public gathering places

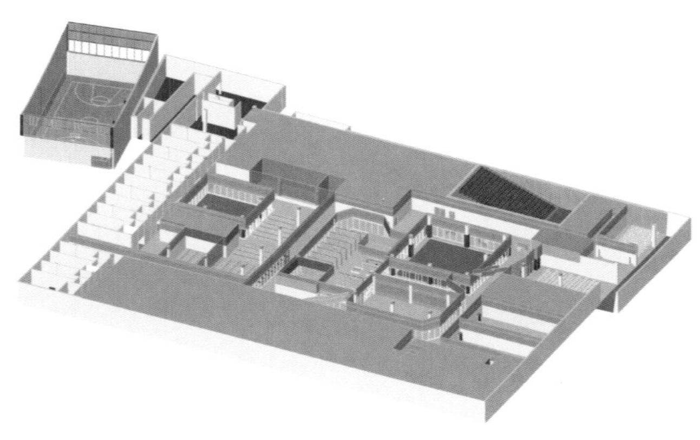

地下一层 设备、储藏、报告厅、游艺室，以及部分文化活动室

B1 Equipment, storage, lecture hall, game room, and cultural activity rooms

本项目是为中央美术学院设计的中法学院第二期，位于上海临港地区的春花秋色景观带中。由于使用方是艺术院校，学生课余的展览、交流活动都非常频繁，我们最终把设计目标定为如何为这个学校的老师与学生创造一个更加有活力的学院。

项目从外部与内部两方面着手体现概念：外部设计一个漂浮在景观带中的供学生老师公共活动的“绿毯”，既呼应整体景观又创造了屋面室外活动区。内部在首层创造购物中心式的空间模式，我们将首层内部在平面上最大化铺开，竖向高度尽可能做高（最终做到 5.4 米），在一个完整的大型内部空间中嵌入多个通透且拥有独立盖顶的“屋中屋”式教室。教室周边穿插各式室内公共交流空间。首层共设置 5 个室外庭院，保证良好采光通风的同时作为多重内部空间的进一步延伸。

The Sino-French Institute in Nanhui City is located alongside the “Spring Flower Autumn Color” landscape belt in Shanghai Lingang Area. The project is appointed by the Central Academy of Fine Arts as the second phase for Sino-French Institute. As an art school, different kinds of activities, such as exhibitions and communications will be frequently launched in the future, so we have decided to create a vibrant school both for the students and the faculties.

The strategy of the project consists of two parts—on the one hand, an exterior landscape “green carpet” floating in the air provides the students and faculties with spacious roof activity space. For the interior, we have organized the ground-floor space into a generic space like in a shopping mall. The area of the ground floor is expanded as big as possible whilst the floor height as high as possible (5.4m). On the other hand, several transparent classrooms with independent ceilings are plugged in the generic ground floor space, which defines various interior communication space, forming the “house within house” mode. For lighting and ventilation, five courtyards are set in the ground floor plan, which expand the interior space to the exterior, enriching the layers of space.

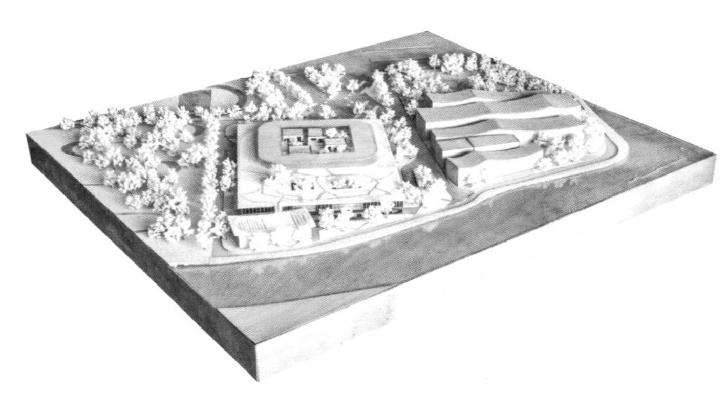

2017
塑說

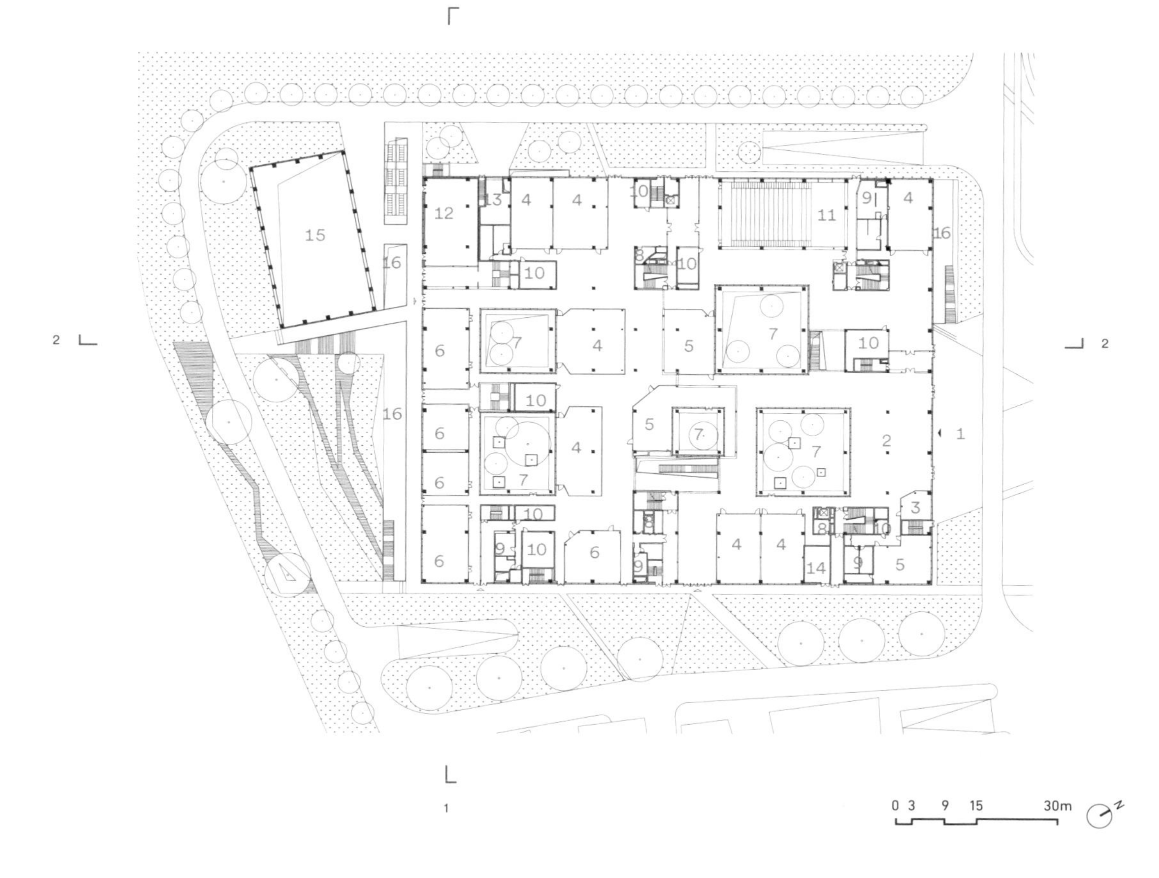

首层平面图

First floor plan

1	主入口	Main entrance
2	门厅	Lobby
3	门卫	Guard room
4	普通教室	Classroom
5	文化教室	Culture classroom
6	游艺活动室	Entertainment room
7	室外庭院	Courtyard
8	工作间	Studio
9	卫生间	Toilet
10	新风机房	Fresh-air room
11	报告厅	Auditorium
12	开关站	Switch station
13	垃圾房	Garbage
14	消防控制室	Fire control room
15	篮球馆	Basketball hall
16	下沉广场	Underground courtyard

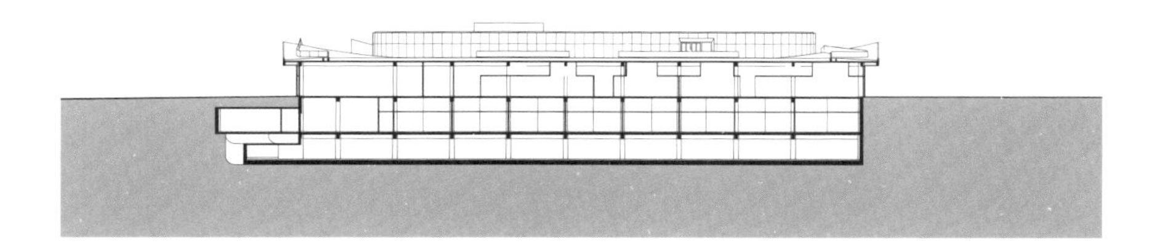

1–1 剖面图

1–1 section

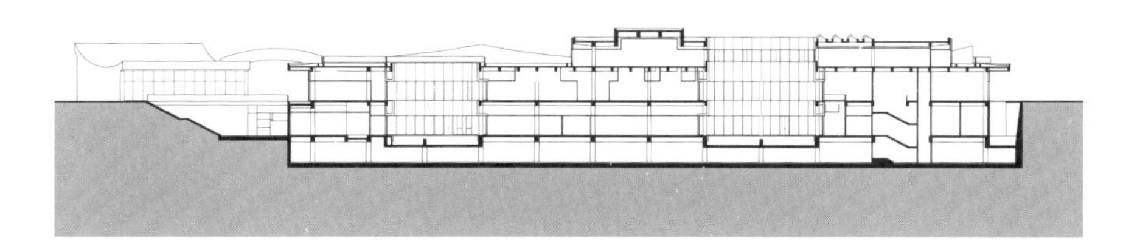

2–2 剖面图

2–2 section

双栖斋
Twin Trees Pavilion

07

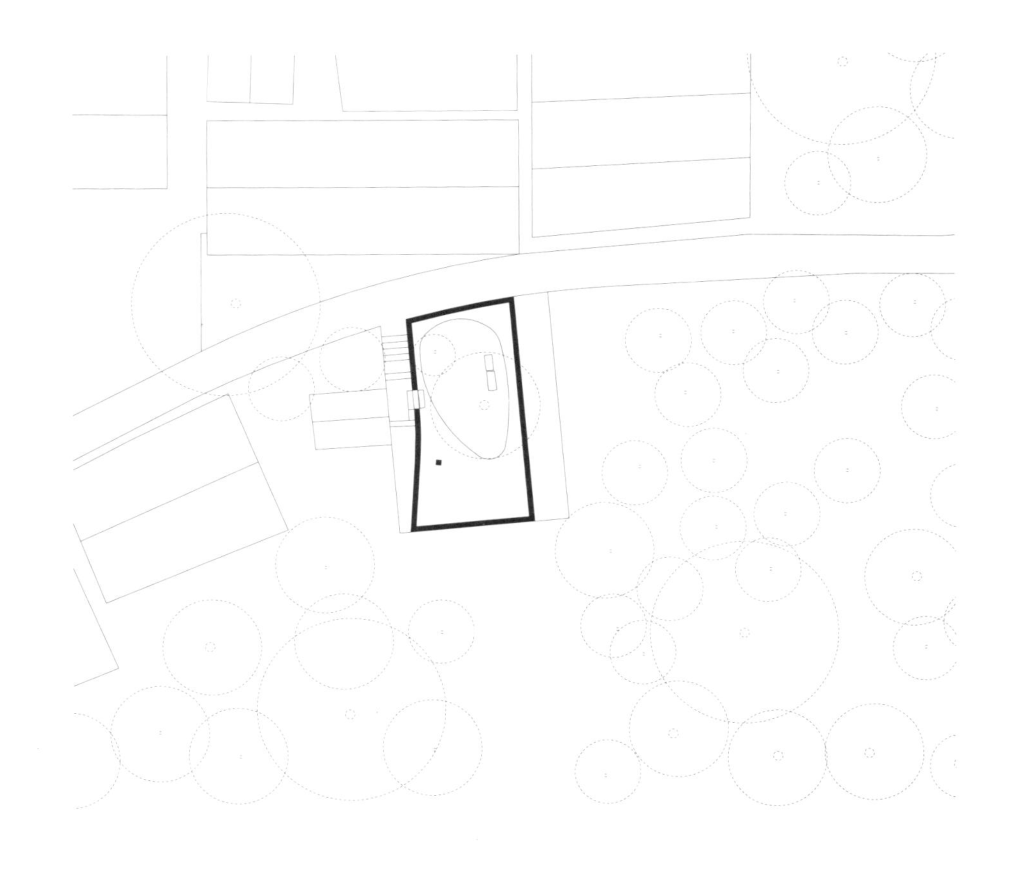

0 1 3 5 10m N

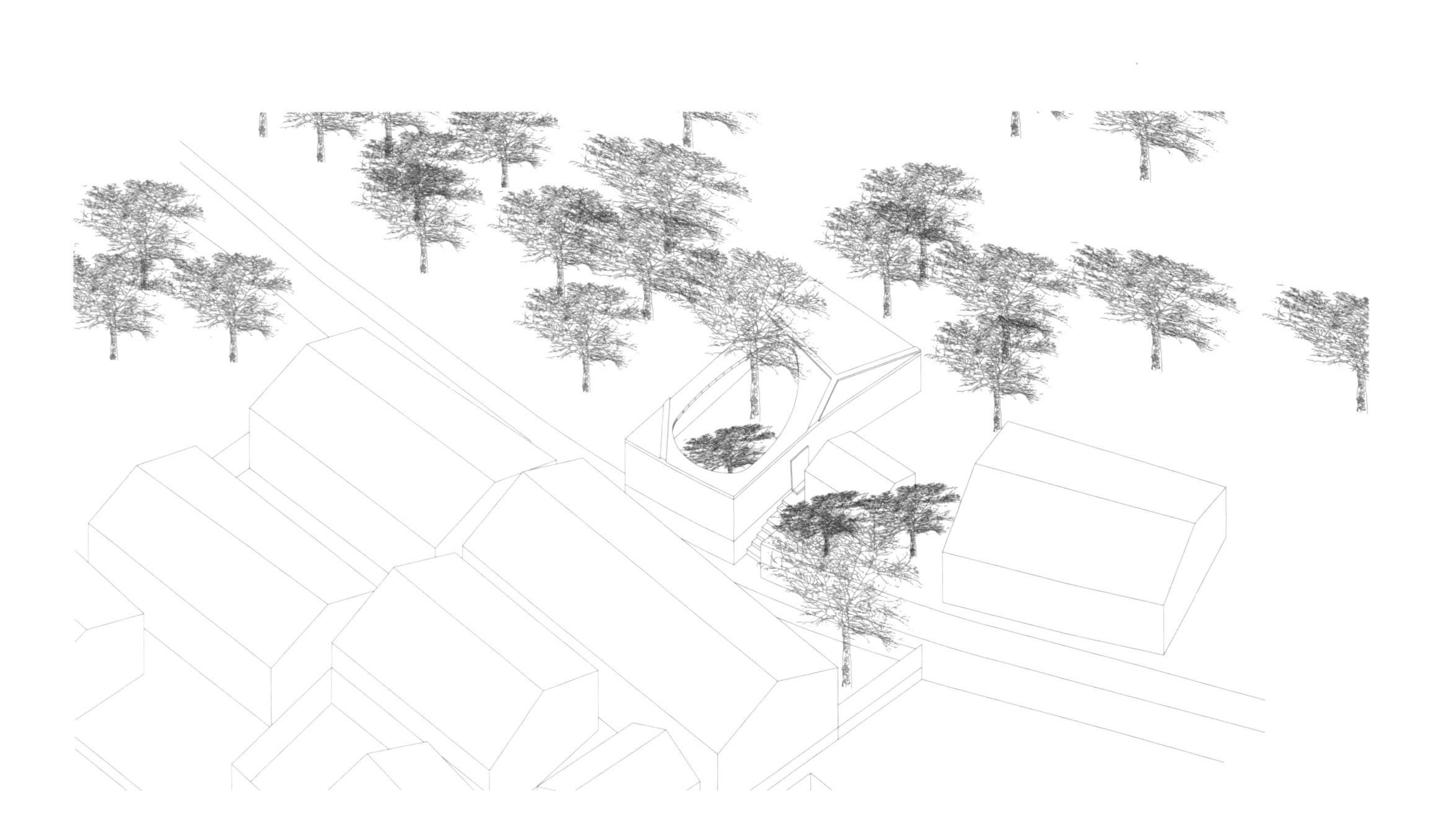

东山陆巷镇是中国太湖之畔一座传统江南风貌的小村镇。双栖斋位于陆巷镇的东面，毗邻种满桔子树的山坡，是主人传统风格主宅院旁的一个独立配房，用来进行聚会与休憩。

双栖斋的基地以前是一个 56 平方米的猪圈，基地上有一株 20 多年的楝树与一株 15 年的桔子树，基地从南至北有一定坡度。房屋采用了砖混结构，拆除的猪圈的老砖被全部用在新的承重墙上，略带倾斜的混凝土屋盖围绕着两株树木形成了一个天井，地面保留了原来坡度的自然形态，用砾石铺地。

建筑空间是开敞的厅堂模式，形成一种人工与自然混合的状态。室外根据村落的风貌要求使用了白色的涂料，室内保持了建筑材料的自然面貌。

Archmixing has changed a former pigpen into a cozy pavilion in a small traditional town located in the Yangtze River Delta Area. It is a supplement for two traditional style residences, providing an extra place for meeting and recreation, just nearby a hill planted with orange trees.

Two big trees, 20-year-old neem and 15-year-old orange dominate the site and lend inspiration to the design as well as the name. The hybrid structure is employed in this project. All bricks from the dismantled pigpen are recycled to build the new walls in various laying patterns. A moderate yard is created around the trees by leaving an oval void in the slightly sloping concrete roof paralleling the graveled ground in the original situation.

This is an open pavilion with a mixing artificial and natural atmosphere. The exterior is painted in white to meet the historical preservation requirement while the interior keeps the natural appearance of the building materials.

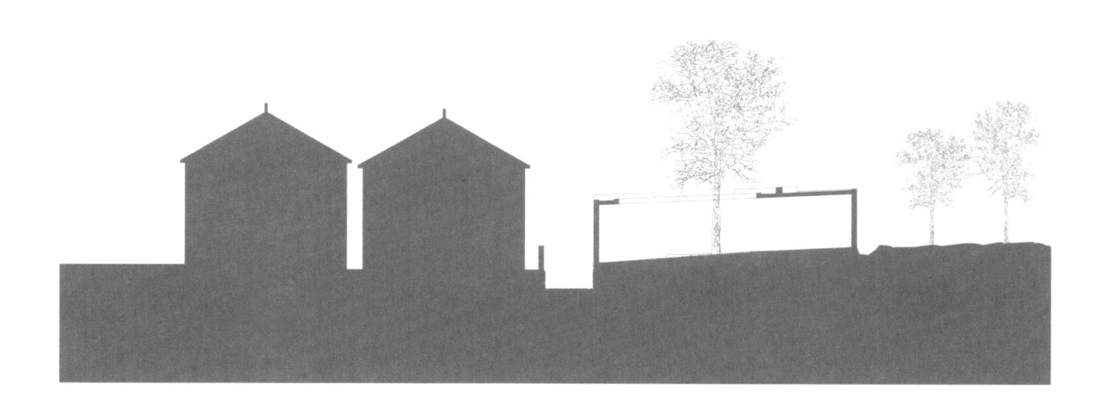

莫干山庾村文化市集蚕种场改造
Silkworm Hatchery Renovation, Yucun Culture Market, Mogan Mountain

08

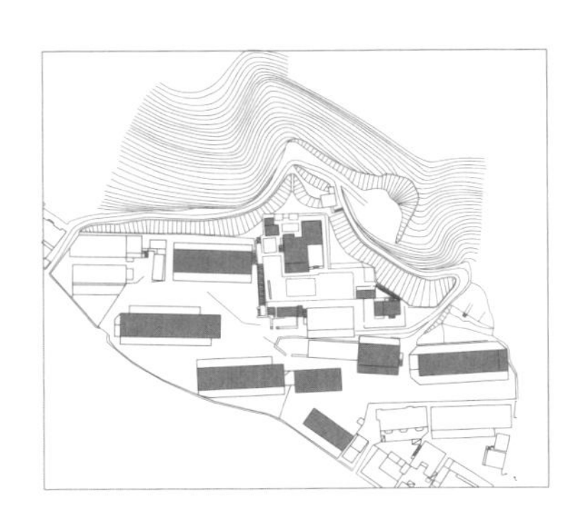

基地原状
Original site

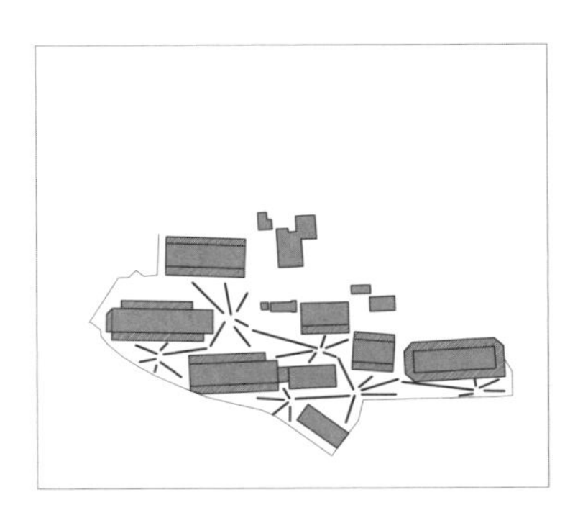

营造有中心感的外场空间
Creating an outdoor space as the center

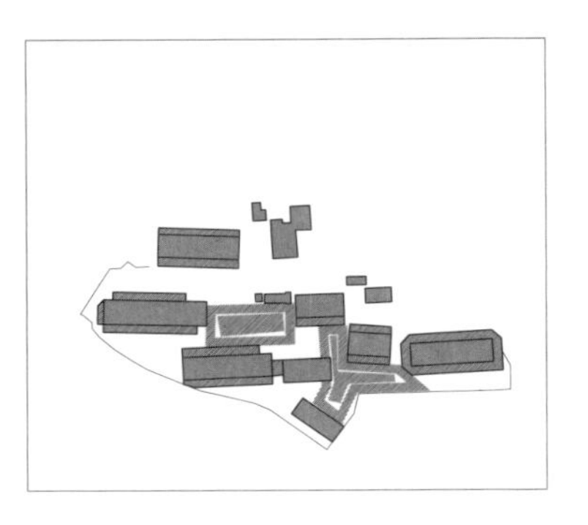

方案 1　实体覆盖作为中心
Plan 1　Covering as the center

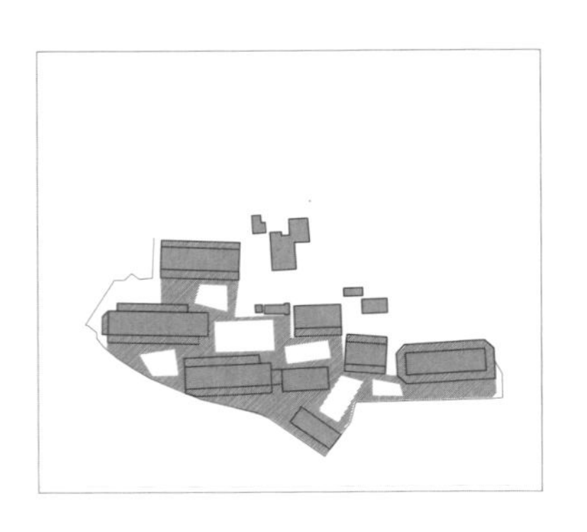

方案 2–1　空间作为中心
Plan 2–1　Void as the center

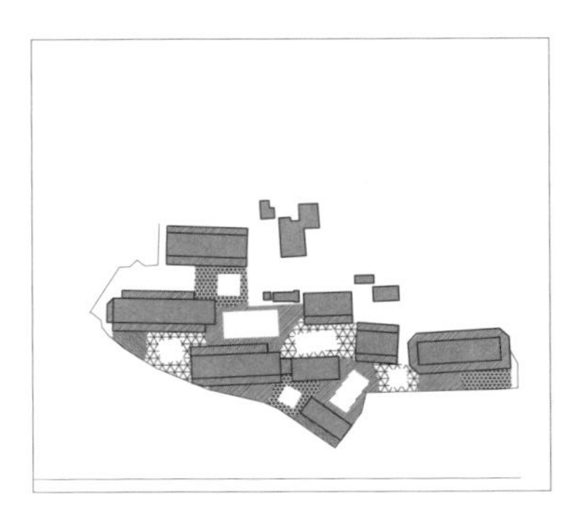

方案 2–2　每个都有不同的编织法
Plan 2–2　Different methods of weaving

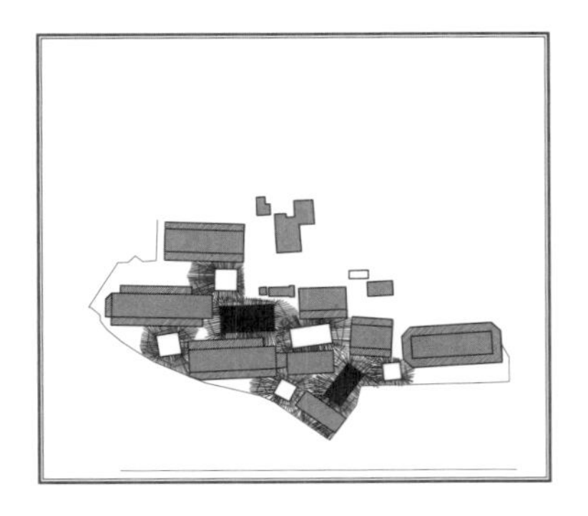

方案 2–3　确定与强调的内边界，自由的外边界
Plan 2–3　Identifying the inner frame, leaving the outer boundary free

搭竹棚

莫干山庾村蚕种场改造内容包括：搭竹棚，改造场地、青年旅社和V-Life绿色产品展示厅。其中占地3000平方米、使用期一年左右的搭竹棚工程是这个建造项目中的一项临时构筑。用当地盛产的竹子覆盖大面积室外场地旨在快速聚集人气，这些竹棚在蚕种场开启改造和初期招商阶段可以起到改进场地和暖场的作用，在培育成熟过程中慢慢退出使用，或被改造更新，或被彻底拆除。对阿科米星来说，这次实践的本意并非专门针对乡村，而主要是一次非正规建造的体验，是为城市改造工作积累经验的一次实践——在周期短，投资低的前提下，尝试用不同方式组织空间的整体设计和有效建造。

竹棚的定位

为了通过组织形成整体性，又能允许变化与现场调整，也为了便于施工实施、不怕误差，我们确定采用内边矩形、外边发散的独立竹棚形成一系列重复形式的中心，这一方面可加强明确的中心感与四向放射感以实现场地整体的效果；另一方面也解决了场地的不规则，以及竹棚之间彼此相连呼应的问题。竹子内边限定出中心空间，采用矩形投影的明确边界，竹子外边呈放射状，自由发散形成模糊消失的边界。内边支架构成完整的矩形，其余支架则用不连续的单杆随机支撑。这个设计中，连续转折且封闭的线只有一条，就是中心的那个矩形，而其余单杆支架两头都是出挑的。

打钉子

我们用直接“打钉子”的桩基和固定竹子的建构方式来实现快速建造和节约成本。单边支架的基础是将圆形钢片下焊接一根钢筋，形成一个“大钉子”，施工时将“大钉子”敲入地中，单边支架的小钢柱立于其上。使用“大钉子”基础一是不破坏现有混凝土地表，二是可适应各种高差变化，或狭小空间。

竹子的固定方式是先将螺钉打穿竹子较粗一端，然后将其固定到“中心矩形”的内边支架之上，在固定几根之后，请外侧的支架上的另外一位工人，根据角度、间距等需要进行旋转调整后，再分别固定。这样的先后固定的方式十分灵活。而竹子低廉的价格，也确保了即使在钻孔时有一定的损耗，也不会过多影响总体造价。

由于采用的是“不规则”的形式，传统的建筑施工图制图方式无法明确表达。我们制作了1:50的手工模型搬到现场作为施工和调整的依据。

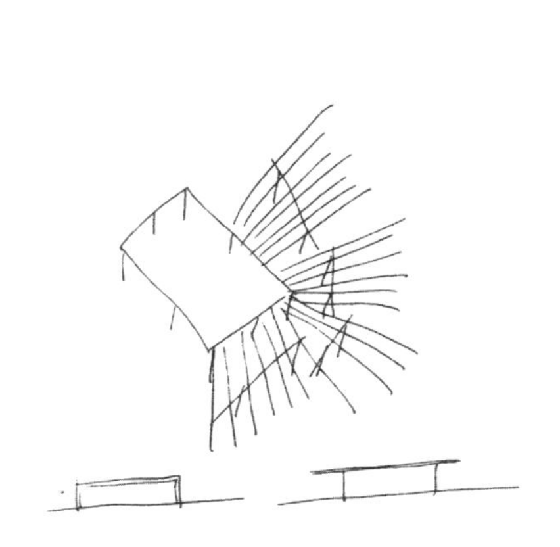

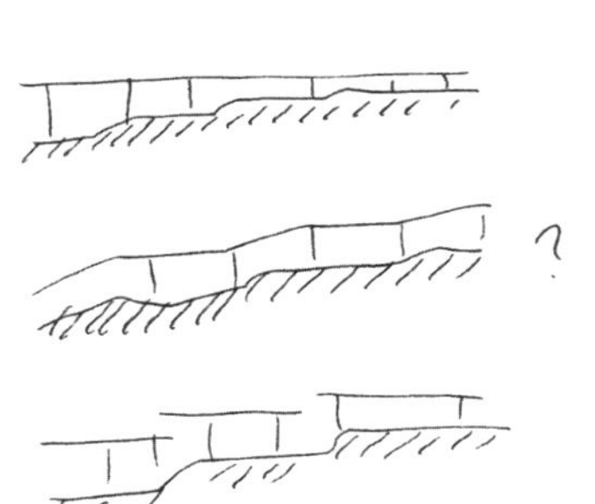

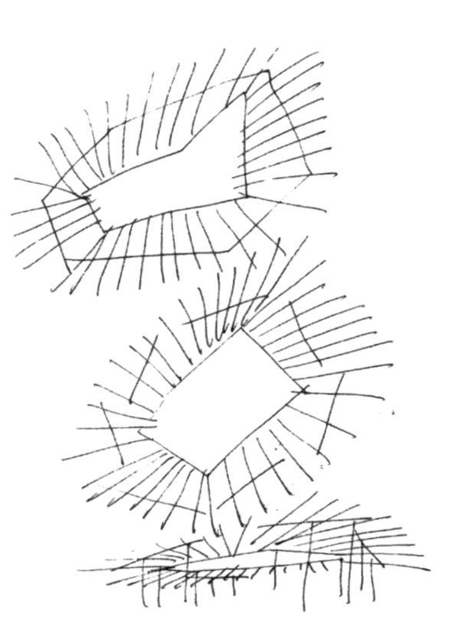

Building Bamboo Shed

The renovation of Mogan Mountain Yucun Culture Market includes bamboo shed construction, site renovation, hostels and V-Life green products display hall. Among these renovations, the bamboo shed, occupying 3,000m², is designed to be a temporary construction lasting about one year. These sheds serve as site improvements and warm-ups during the initial stage of renovations and investments and will gradually be out of use, either renovated or removed completely. For Atelier Archmixing, instead of rural construction, this informal renovation will inspire similar programs in the urban areas. How can we apply various methods to organize the overall spatial design and realize effective construction under limited time and budget?

Positioning the Bamboo Shed

In order to form a spatial unity through the organization, while also allow change, on-site adjustment and construction errors, we have designed bamboo sheds into simple and repeated forms. The clustering individual sheds can not only enhance the feeling of the center and the dispersion to build unity, but also adapt to the irregularity of the site and achieve continuity of all sheds. A central space is circumscribed on the inner rim of bamboos to form a complete rectangle, while a delineated boundary is projected from the center. The rest of the scaffolds are randomly supported by discontinuous single support. In this design, there is only one closed line, the rectangular center, while both ends of all other single supports are cantilevered.

Driving Nails

Simple tectonics are applied to achieve quick and economical building. A round steel sheet is welded with a U-shape reinforcing bar to form a "big nail" as the foundation of the free-standing steel frame. The "big nail" is directly hammered into the ground to support the steel pole for the frame. This big nail strategy has two advantages. On the one hand, the existed concrete ground is less destroyed, and on the other hand, it can adjust to uneven ground and narrow space.

After erecting the steel structures, the wide end of each bamboo is nailed into the center rectangular supporting frame first, and then the slim end is fixed with nails on the periphery steel frames after reorientation, rotation and adjustment. It's rather flexible to fix the two ends consequently with nails. Since Moganshan is famous for bamboo production, some mistakes are also economically acceptable. Traditional construction drawings have difficulties in describing this irregular design clearly, therefore we made a 1:50 large model and brought it to the site to guide the construction and adjustment.

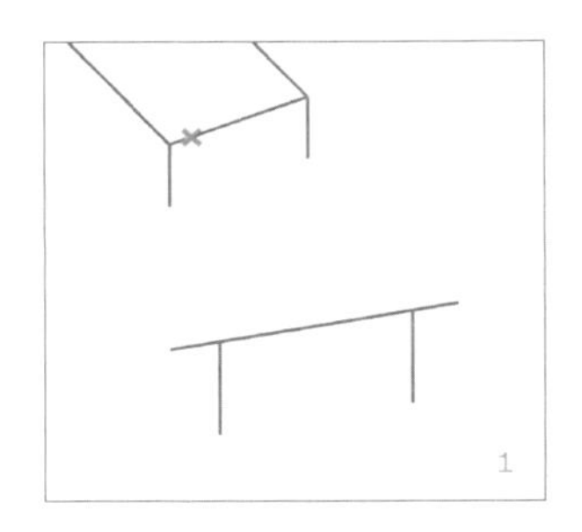

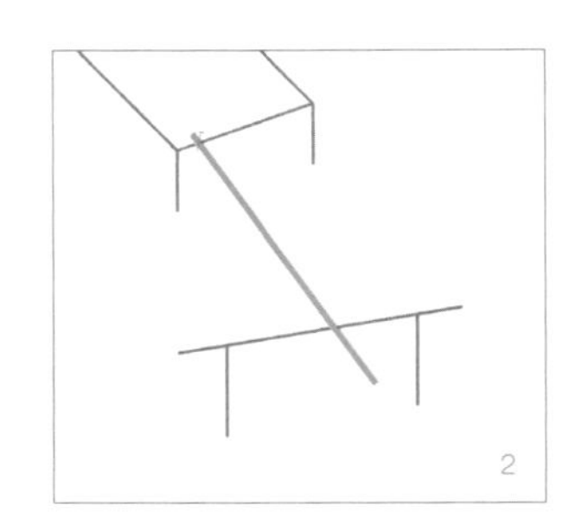

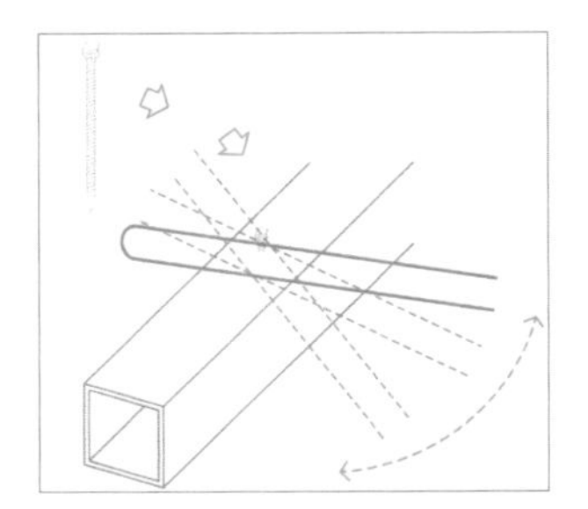

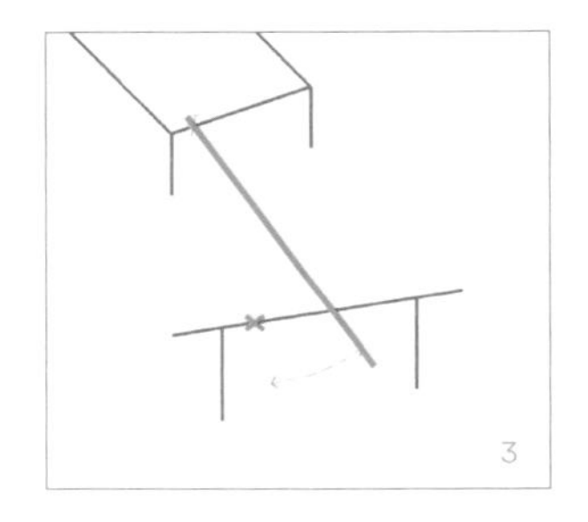

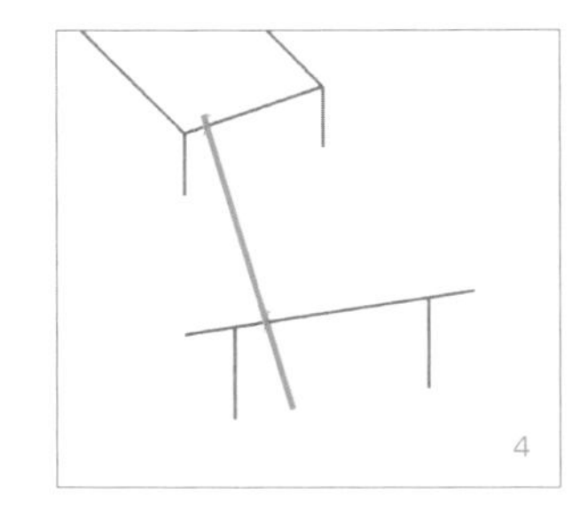

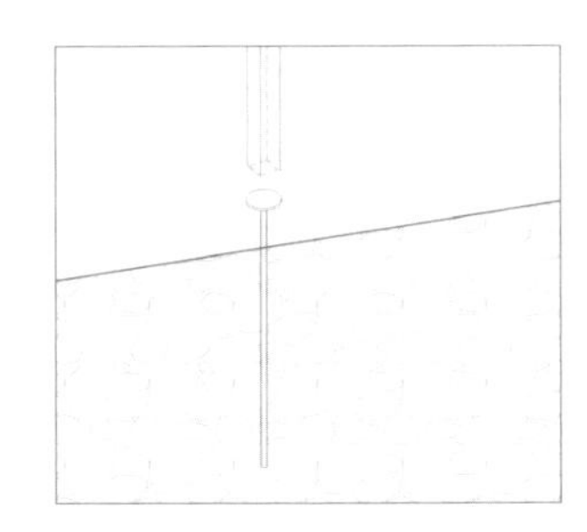

竹子与钢结构节点分析图
Construction diagram of bamboo and steel structure

钢结构与地面节点分析图
Construction diagram of steel structure and ground

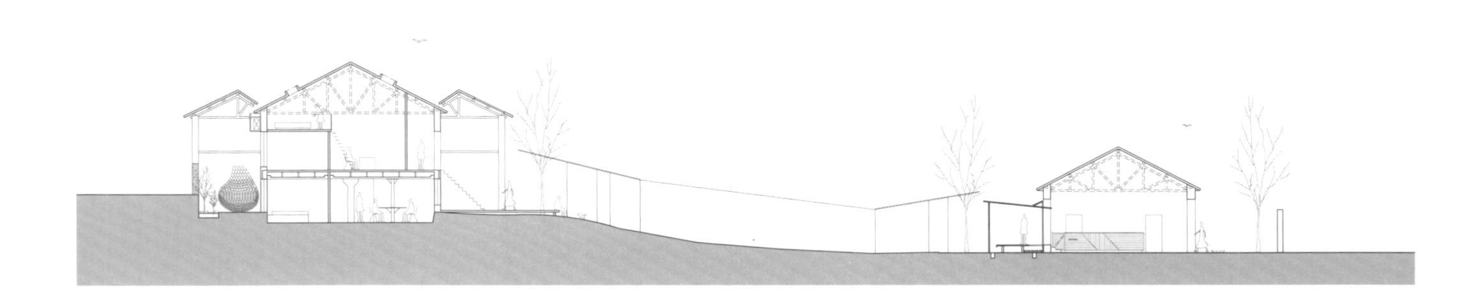

富春俱舍走马楼
Fuchun Kosa Zoumalou

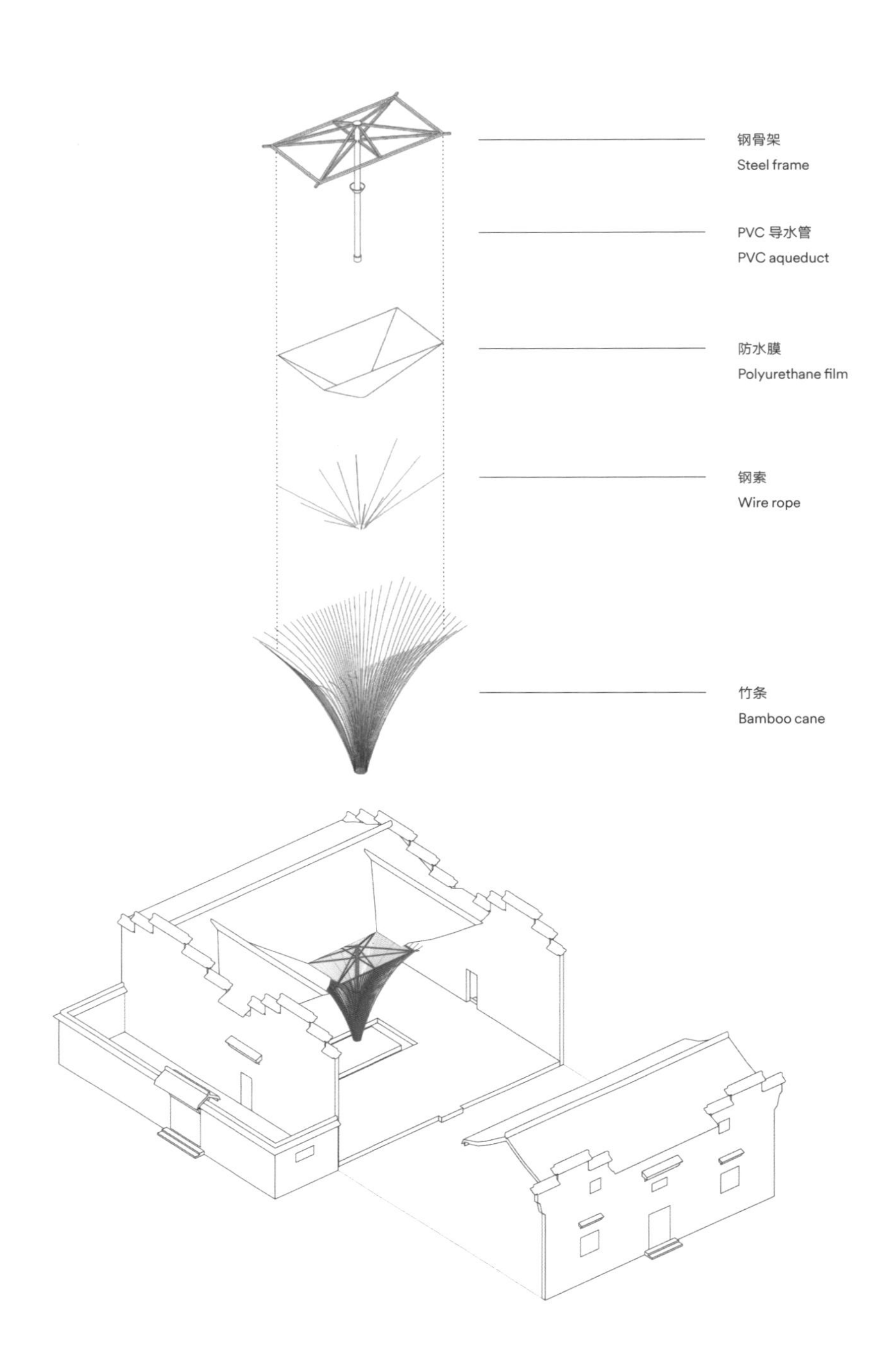
钢骨架
Steel frame
PVC 导水管
PVC aqueduct
防水膜
Polyurethane film
钢索
Wire rope
竹条
Bamboo cane

富春俱舍位于中国顶级水上黄金旅游线"富春江—新安江—千岛湖—黄山"的中点，坐落于国务院首批核准的国家级风景名胜区内，是富春江、新安江、胥江汇流之处的唯一一片依山傍水的平地，并为富春江国家森林公园所环抱。

"走马楼"位于富春江的一处码头附近，是一个传统徽派四合院建筑，业主想将其旧有的住宅改造成一个艺术酒吧和接待厅。整个老宅的占地面积非常小，且中间天井占了不少空间。方案在天井中设计了一个局部可以开启的"伞形"独立装置，来回应空调密闭性、雨水收集、采光通风等一系列问题。而"伞形"装置下方开放出来的空间用于摆放沙发座椅。同时老宅特有的"雕梁画栋"亦可清晰地展示。

我们希望这样的老建筑尽量能被使用，而不是单纯作为被"供奉"的对象。在使用过程中，建筑的价值被充分体现，更为重要的是建筑和人的互动产生出更多的情感交流。这和以前人们会经常修补陶器，一直把玩的道理是一样的。

Fuchun Kosa is located at the midpoint of China's top water golden tour connecting Fuchun River, Xin'an River, Thousand Islands Lake and the Yellow Mountain, which is also among the first state-level scenic spots approved by the State Council. This area is the only flat land on the hillside, surrounded by the Fuchun River National Forest Park.

Zoumalou is a typical Chinese Huizhou courtyard house sitting near a dock, and the client wants to turn this old vernacular residence into a reception hall equipped with a gallery bar.

The two-story house is of modest size, while the enclosed courtyard is rather spacious and decent, surrounded by richly carved wood beams. To make the best use of this spatial and decoration feature, a fantastic transparent umbrella structure is employed to shelter the center yard, where people can now relax on sofas without worrying about the changing weather.

This is an independent bamboo structure covered with Polyurethane films, part of which can be opened to receive the sunlight and fresh air. This technical intervention successfully solved all challenges including air conditioning, rain collecting, daylighting and ventilation.

Through this project, we highlight our design philosophy that historical buildings should be put into everyday use instead of being adored as a shrine. Only in this way, they can achieve their full value by building emotional connections with contemporary users.

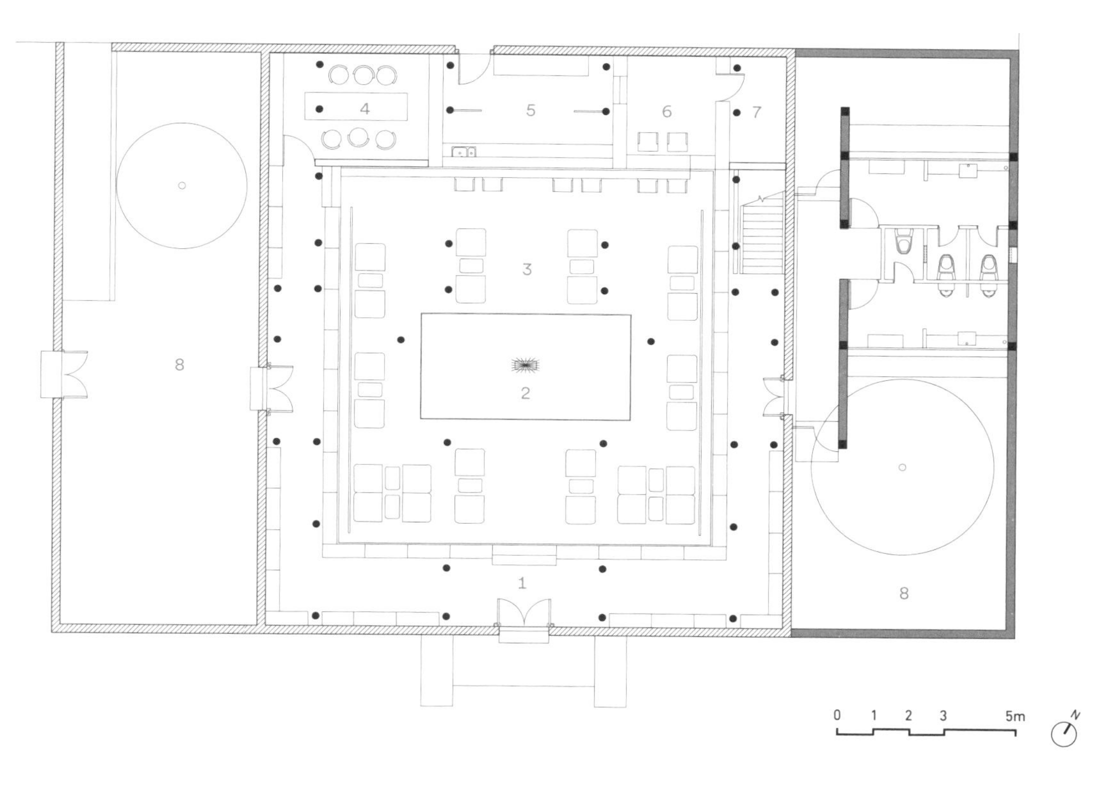

首层平面图

First floor plan

1	走廊	Corridor	5	吧台区	Bar
2	天井	Atrium	6	接待区	Reception
3	公共区	Teahouse	7	储藏	Storage
4	VIP 包间	VIP room	8	庭院	Courtyard

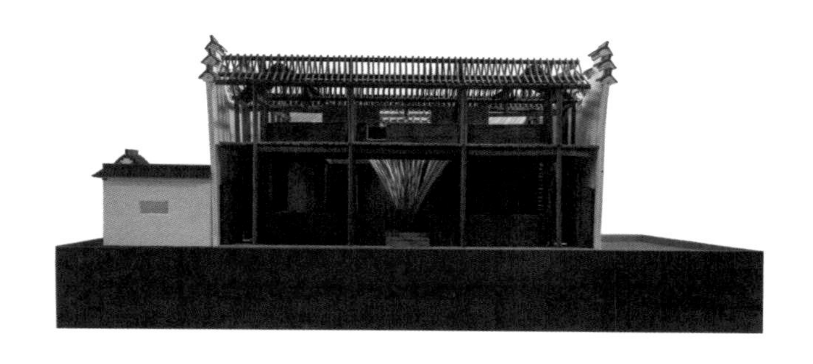

10

安龙森林公园东部码头小镇商业建筑

Commercial Building at East Dock Town in Anlong Forest Park

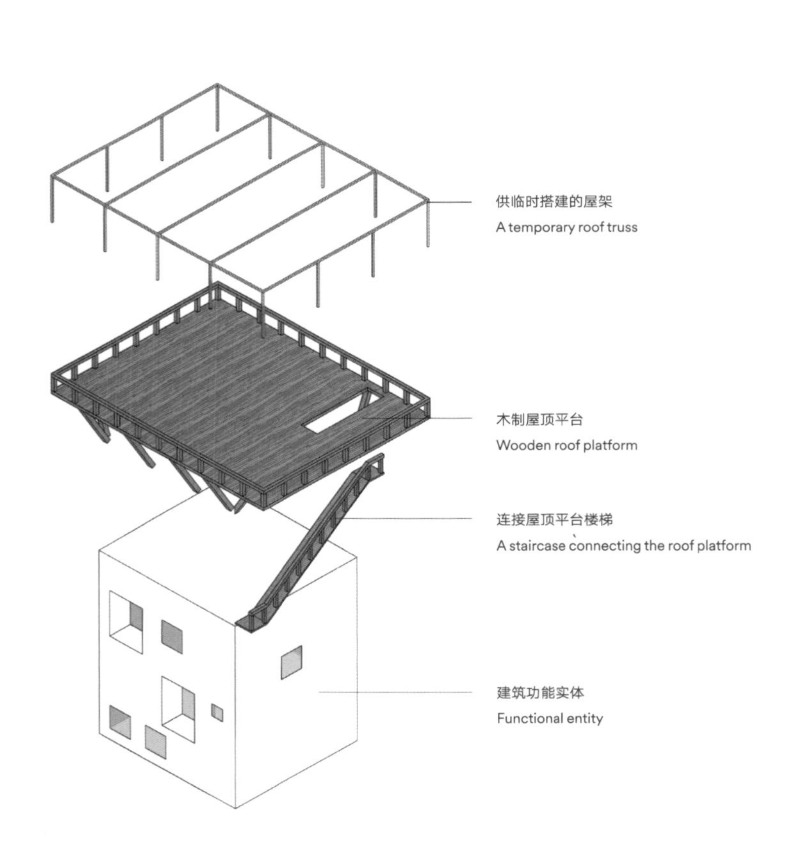
供临时搭建的屋架
A temporary roof truss
木制屋顶平台
Wooden roof platform
连接屋顶平台楼梯
A staircase connecting the roof platform
建筑功能实体
Functional entity

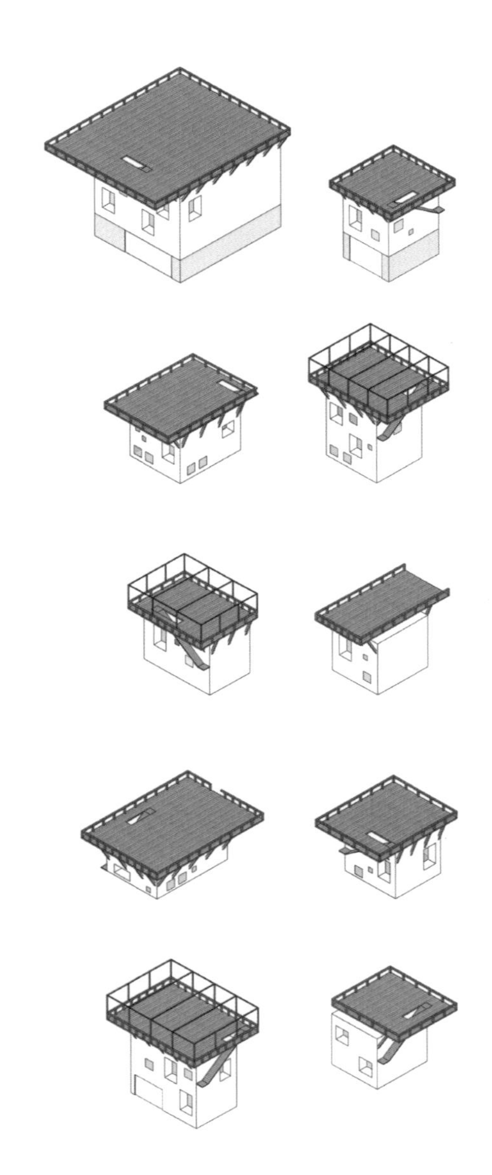

功能的不确定性

安龙森林公园东部码头小镇是一个规划项目，由分散在不同区域的一组组商业、文化、服务类建筑群落组成。6 号单体是一组商业餐饮与小客房的混合业态形成的建筑。小镇区域主要的小路在组团基地面前经过，组团坐落在山坡地上，面向千岛湖的水面。

这个设计要面对的问题是功能的不确定性。在开始设计的时候，并没有明确的任务书指出建筑在未来完成、开业的时候具体会如何运营，只是模糊地有一些分类，简单罗列了面积大小、使用性质与设备的要求。未来商业的经营具有不确定性，商家如何更换更是无从知晓，这也是商业业态的常态。然而对于设计的另外一种常态要求却是需要设计一个保持特色的建筑，这两种不同的常态之间的关系成为了一个难题。也就是说，设计必须对于不确定与确定、临时与永久、改变与不变这样的问题进行思考并提出解决方案。也正因此，这个最终未得以建成的设计才具有了某种典型的意义。

永久的平台与改变的单体

设计的解决方法是：设计一个建筑群落，由一簇相同形态组成基因的大小建筑单体组成。每个单体好似一个带着帽子的卡通人物，下面的身体是一个方方正正的立方体，由钢筋混凝土结构加砌体围护墙体组成，墙面上开着大小不同的窗洞与门，这个空间用来容纳未来不同的商业服务业态，上面的帽子是一个钢木结构的木平台，浮起挑出在身子之上，每个平台都可以从下面的房屋通过专门的室外楼梯上来，在这里可以舒适地观赏到千岛湖的湖光山色。这样的一个个单体被放置在山坡上，高低错开，每处平台均拥有良好的视线。在这样的构思里，单体下部的商业部分允许未来的变化与改造。窗户门洞、指示招牌、表面材料、装修的颜色，全都允许改变，这个看似结构形态结实稳固的部分实际被设定为未来最具有变化的部分。相反，钢木结构的露台看似结构柔弱、形态临时，但实际始终承担露台的功能，不需要改变，因而是稳定永久的。这个稳定永久的部分在设计中被刻意在形式上强调出来，承担了这个组团建筑未来长期保持的特色。

这样的设计还有另外一层形式思辨上的考虑。单体的两个承担不同使用功能、形式各异的部分还有一种微妙的关系：钢木结构的部分显得更为轻质，仿佛依附在结实粗犷、像是从山地上长出来的体量之上。但轻质的钢木结构的平台又仿佛是对于下面实用空间的一种修正，一种改造调整，弥补它缺乏的稳定性，也弥补它在功能使用上的不足，毕竟，在我们看到的大部分千岛湖水岸边的商业建筑中，亲近湖面景色、在户外坐望美景是每个人的渴望。

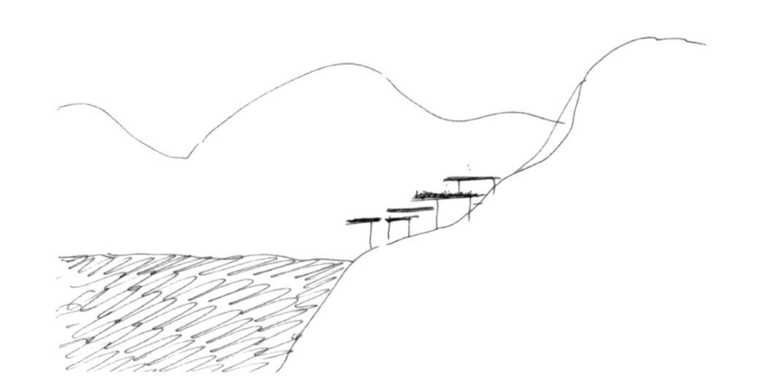

Functional Uncertainty

The East Dock Town in Anlong Forest Park is a planning project, which contains groups of commercial or cultural service buildings that scatter in different regions. No.6 Building is of mixed functions, including several commercial caterings and hotel rooms. It sits on a slope behind the main street in the town, facing the view of Thousand Islands Lake.

The biggest design challenge is the uncertainty of its function. There were no specific task assignments to show how to operate the business when it opens in the future, except for vague usages, floor areas and equipment requirements. Designers know nothing about how the future business would run and whether the operators would change, which is a rather common situation for commercial projects. What makes the task even harder is another common requirement, to design a building with notable characteristics. In other words, the design should result from balancing uncertainty and certainty, temporariness and permanence, change and non-change. Thanks to the above consideration, this unbuilt project has gained prototypical meanings.

Permanent Platform and Changeable Structure

We have designed a community composed of single buildings of similar appearances and traits of different sizes. Every single building looks like a cartoon character with a big hat. Its body is a square cube made of reinforced concrete and masonry walls, with windows and doors of different sizes open for various commercial and services in the future. The hat above is a platform of timber-steel structure, cantilevering over the body. Each platform can be reached via the corresponding outdoor stairways from the building below, where people can relax and enjoy the water and mountain views of the Thousand Islands Lake. All independent buildings are constructed on the slope at different altitudes so people can get nice scenery on each platform. In this design, the commercial parts of the building are open for future change and adaptation. These changes might happen to windows, commercial signs, colors, materials and decorations on the surface. This part, even though with a sturdy and solid structure and form is actually designed to be subject to changes in the future. On the contrary, the soft and temporary platforms of timber-steel structure will always be kept unchanged and permanent. This stable and permanent part is emphasized on purpose to present the characteristics that this group of buildings will always maintain.

Such design also includes contemplation of architectural form, a subtle relationship between the two parts of a single building with different functions and forms. The lighter timber-steel structure seemingly attaches to the sturdy and rough volume grown from the ground, while it also constitutes a correction to the practical space below, an adaptation to make up for its unstability and functional deficiency. After all, among most commercial buildings we see along the bank of Thousand Islands Lake, a building which enables visitors to stay close to the lake view and sit outside to appreciate the scenery is a universal desire.

EXPRESS

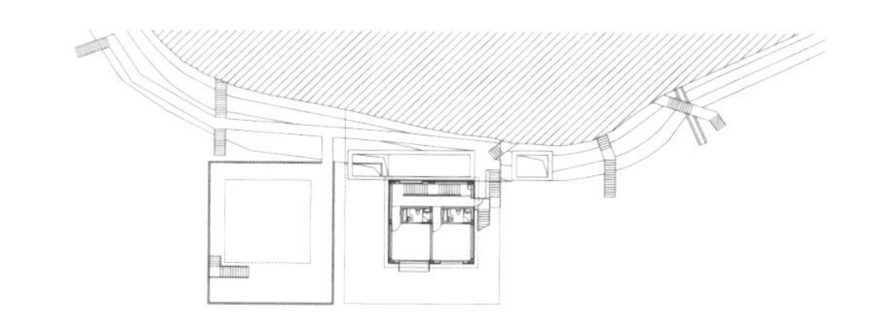

六层平面图
Sixth floor plan

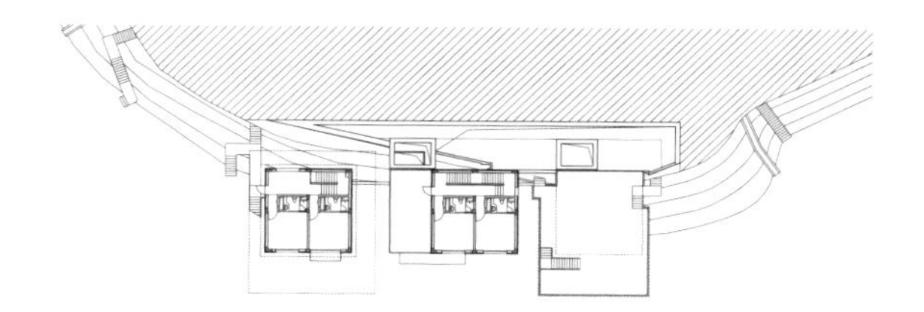

五层平面图
Fifth floor plan

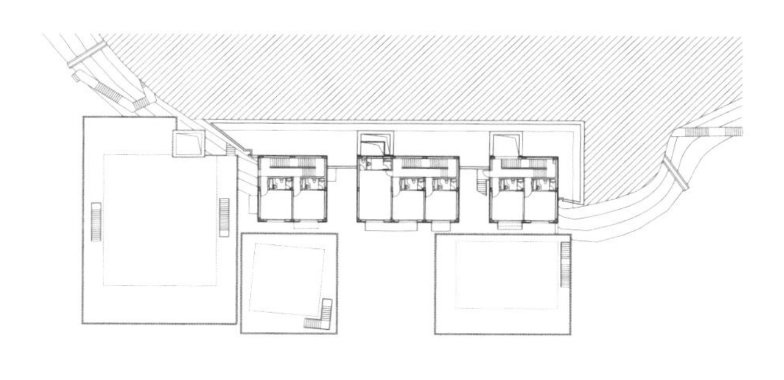

四层平面图
Fourth floor plan

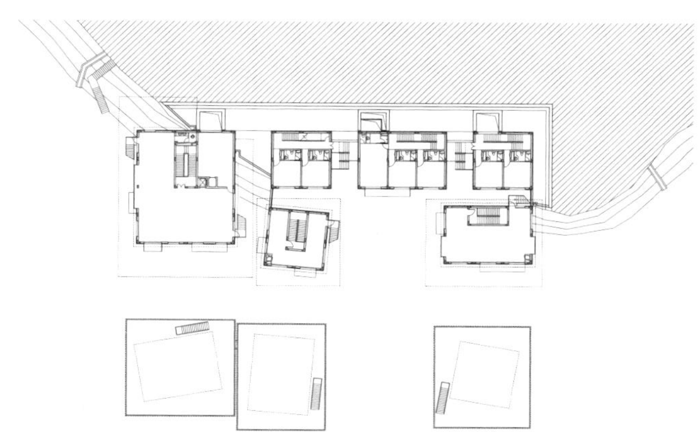

三层平面图
Third floor plan

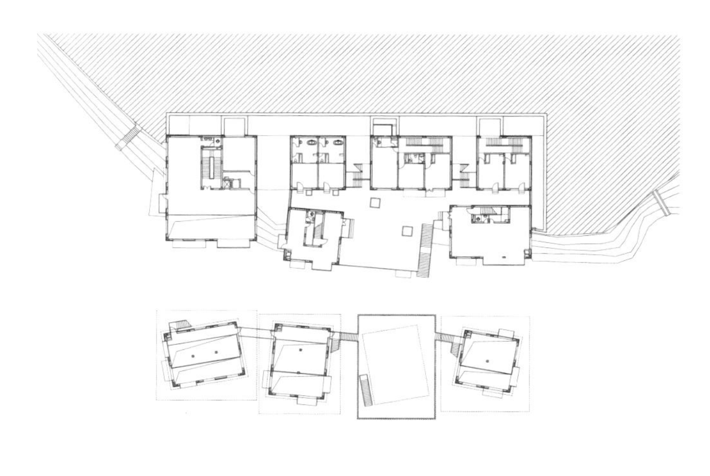

二层平面图
Second floor plan

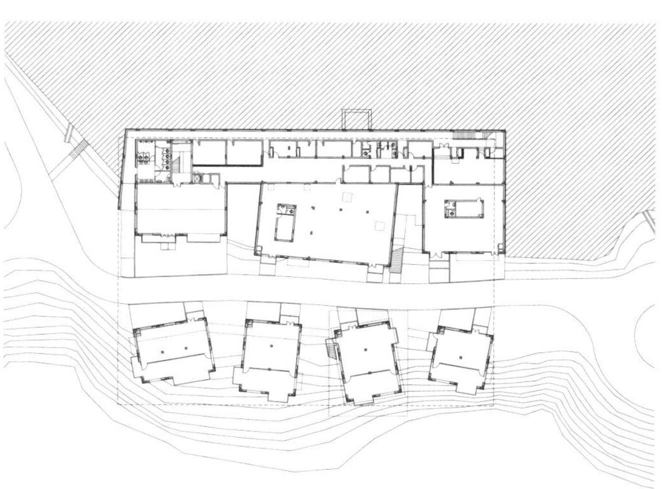

一层平面图
First floor plan

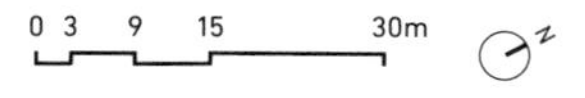

11

阿那亚金山岭艺术中心

Art Center of Aranya, Jinshanling

从场地而来的建筑

山中的艺术中心坐落在一条山脊线上，位于自山脚蜿蜒而上的山道一侧，它承接着北方特有的苍茫山岭，也连接起从山脊线铺向另一侧山谷的社区。社区居民归来或访客前来，沿山路逶迤而上，会首先看到这个艺术中心。基地东侧和南侧面向山路，居高临下，可远望连绵群山；北侧俯瞰沿山坡铺陈而下的别墅群；西侧则紧邻社区边界——收束社区的街道。这特殊的位置让它成为社区入口标志与枢纽。

小小的建筑如何成为面对大自然的一个不卑不亢的地标，又能够与社区街道的近人尺度协调融合，这是建筑场地设计要实现的目标。一片外观体量厚实的多边形单坡屋顶，一个与挡土墙结合包绕基地的基座，使建筑物面向群山有一个纪念性的尺度。同时，屋顶高度与倾斜方式的设计又能够使其在面向社区一侧收束成低矮立面，与社区别墅的尺度相近。

沿着道路由挡土墙包绕的基座和建筑成为整体，并通过它的形式和材料处理调节着路上行人的尺度感受。基座三角台地部位的尽端为高 2.1 米的梭形收头，其与毗邻的别墅基座挡墙之间布置人行台阶，让行人自然地拾级而上，靠近艺术中心或折而进入社区内部街道。基座在建筑面向群山一侧，形成一小片广场，铺地与建筑内部的连续性使建筑难分内外。这里被设计为这个方寸小的基地上的空间尽端，将视线引向山岭与天空。

砖亭之亭

设计另一个要面对的问题是建筑与自然的关系，从山脚到这里的旅途一路兜转，人们已经饱览山色，如何在这个建筑里创造新的体验？

15 个各具功能的砖砌“柱亭”，共同组织出一个被覆盖的“大砖亭”。柱亭之间，风景在这里成为一种“邂逅偶遇”。沿途熟悉的风景和光线从柱体之间渗透进来，风景的画面被柱亭分割，会再次在人们的想象中联缀起来，成为一个新的体验。建筑空间以这样的方式来塑造定义周围群山的风景的同时，也塑造了建筑的特征。

不同尺度的空间被组合在一起——既提供两三人交流的空间，又提供集会活动的场所。“柱亭”本身与它构成的空间既是日常的，又是具有永恒意味的。

高厚的钢格栅结构屋顶，以及与挡土墙、室外铺地一致的砖材的使用，为覆盖下的空间营造出“室外感”。室内不再有直射光，袒露在柔和均匀的自然光的氛围之中。底部百页式的设计隐匿了光源，同时让整个室内随着时间变化而发生微妙的亮度和气氛变化，使人们对时间的感受变得模糊，产生新的感知维度。深色的屋顶亦不争抢人们对于空间中主角“风景”的注意力。

The Building Adapted to Local Condition

The art center in the mountain is located on a ridge line. It stands beside a winding path that starts from the foot of the mountain. Its location connects the typical boundless mountains in northern China and also the community in the valley on the other side of the ridge line. When residents or visitors go up along the winding path, they will first be greeted by this art center. Its east and south sides face the mountain road below its foot and continuous mountains can be seen from these directions. Its north side is overseeing villas spreading down along the hillside. Its west side is adjacent to the community and acting as an end of the street. Because of this special location, it becomes the entrance landmark and a hub for the community.

For site strategy, this design aims to make the small building a proper landmark in the nature that is neither too humble nor too arrogant, at the same time it is supposed to be in harmony with the community street. A thick polygonal roof and a base that wraps around in combination with the retaining wall bring the building a monumental scale facing the mountains. In addition, the design of the height and the slope of the roof gives the building a low facade on the side facing the community, similar to the scale of the villa.

The base made up of retaining walls surrounding the site along the road is integrated with the building and adjusts the scale feeling of pedestrians through its form and materials. The end of the triangular platform is 2.1m high. Pedestrian steps are designed between this end and the retaining wall of the villa. Visitors can walk up step by step to the art center or turn into the internal street of the community. The base forms a small square on the side of the building facing the mountains. The seamless paving between the outside and the interior makes it difficult to distinguish the inside and outside of the building. This place is designed as a spatial ending for this site, drawing people's eye sight up to the mountain and sky.

The Pavillion of Brick Pavilions

Another challenge of this design is to tackle the relationship between building and nature. Visitors coming from the foot of the mountain have already enjoyed a wonderful scenery, so how can we create a new experience in this building?

Spaces of different sizes are combined, and spaces for two or three people's communication or spaces for gathering are both offered. The "column pavilion" and the space it constitutes are both everyday and eternal.

The high and thick roof of steel grid structure, and the use of bricks same as the retaining walls and outdoor paving, create an outdoor feeling for the covered space. The indoor space is not exposed to direct sunshine. It is illuminated by soft natural light. The shutter design at the bottom of the roof conceals light source and makes the entire indoor space go through a subtle change of brightness and atmosphere as time passes by. In this way, people will have a vague perception of time and a new dimension of perception. The dark roof does not overshadow the leading role of "landscape" in the space.

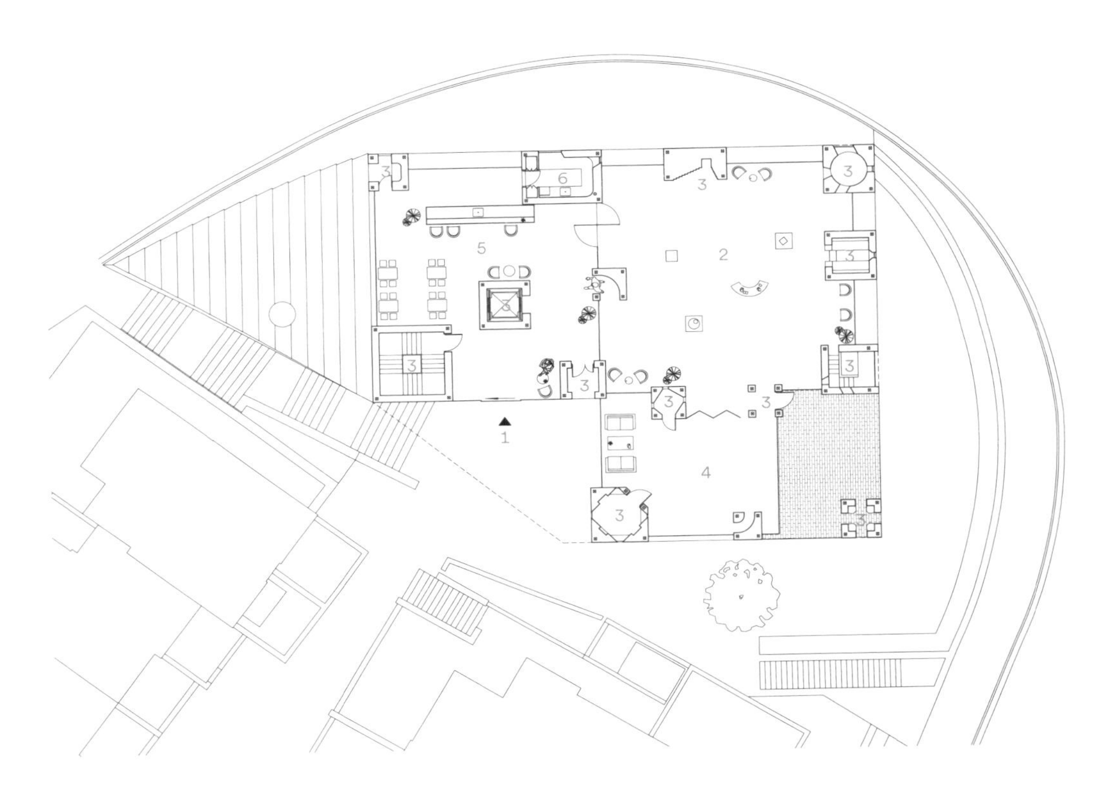

一层平面图
First floor plan

1	入口	Entrance
2	多义空间	Multi-functional space
3	亭	Pavilion
4	附属空间	Affiliated area
5	咖啡区	Cafe
6	准备间	Service room

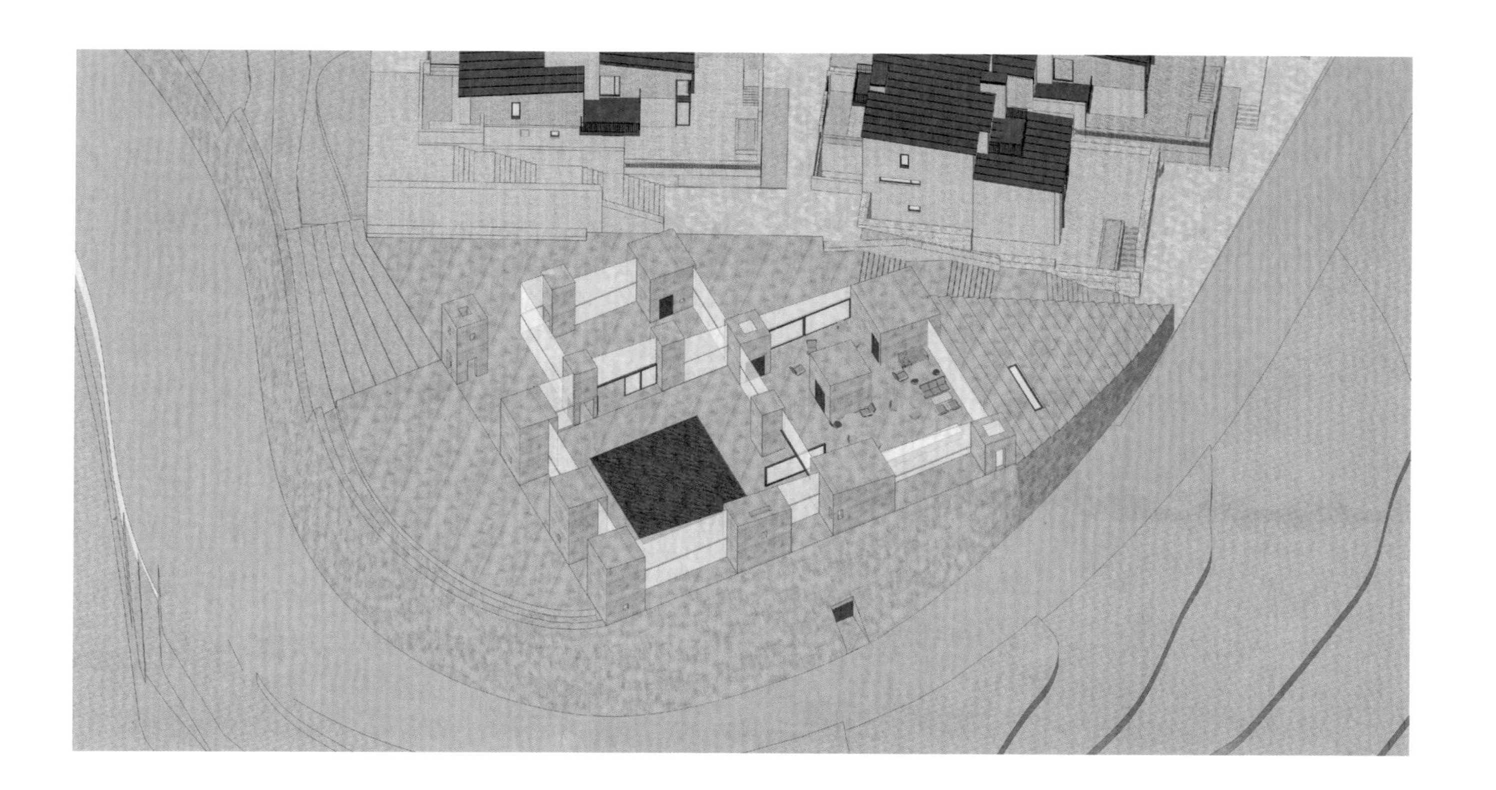

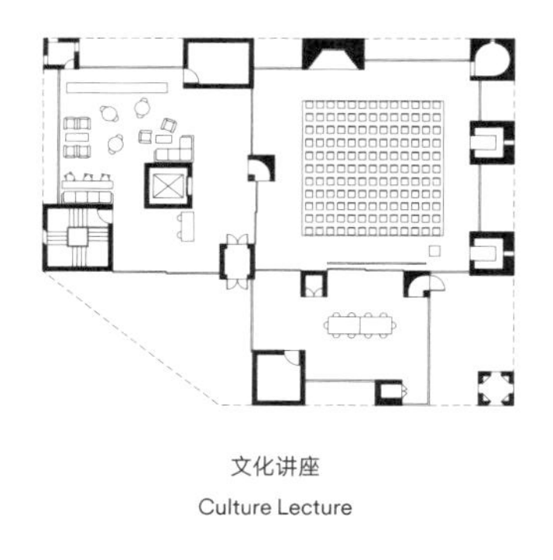

文化讲座
Culture Lecture

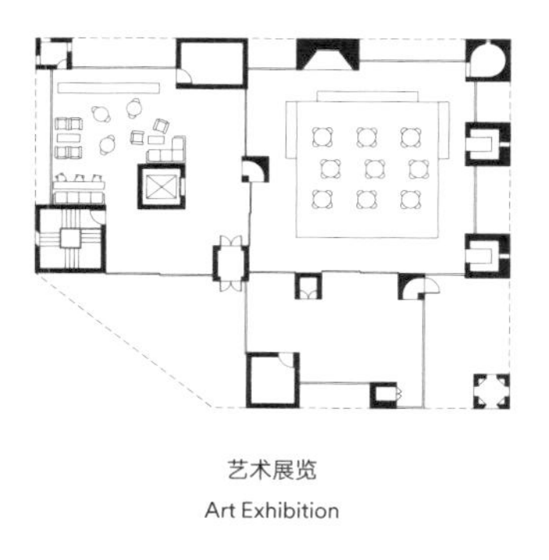

艺术展览
Art Exhibition

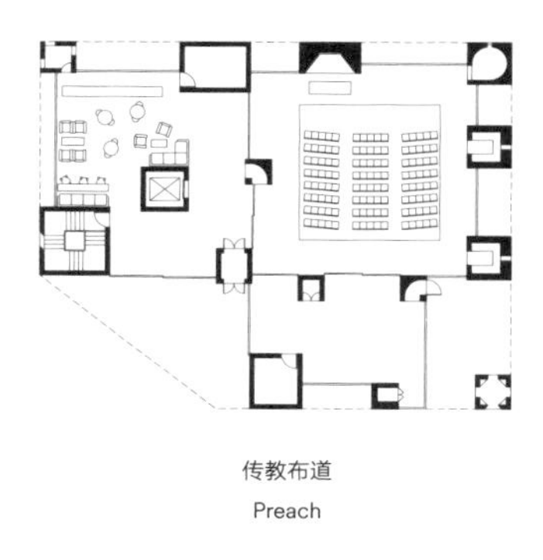

传教布道
Preach

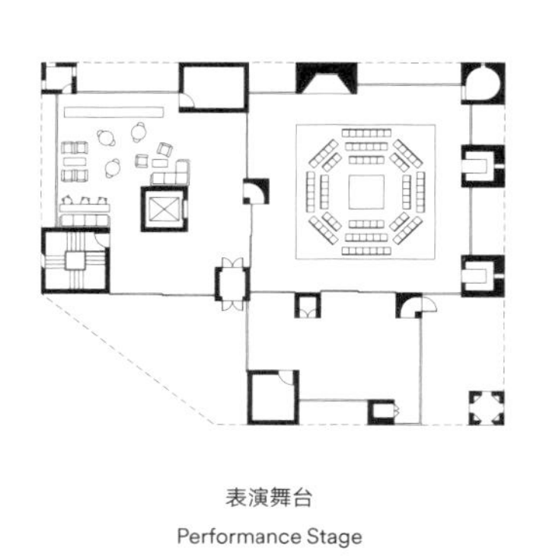

表演舞台
Performance Stage

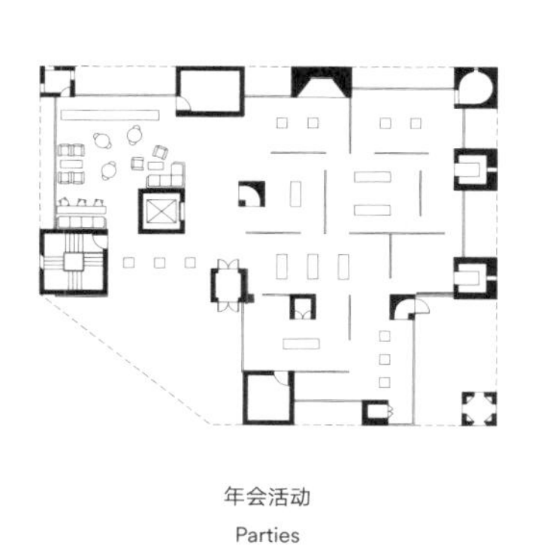

年会活动
Parties

幼儿教育
Preschool Education

不同家具摆放对应不同使用模式
New function new arrangement

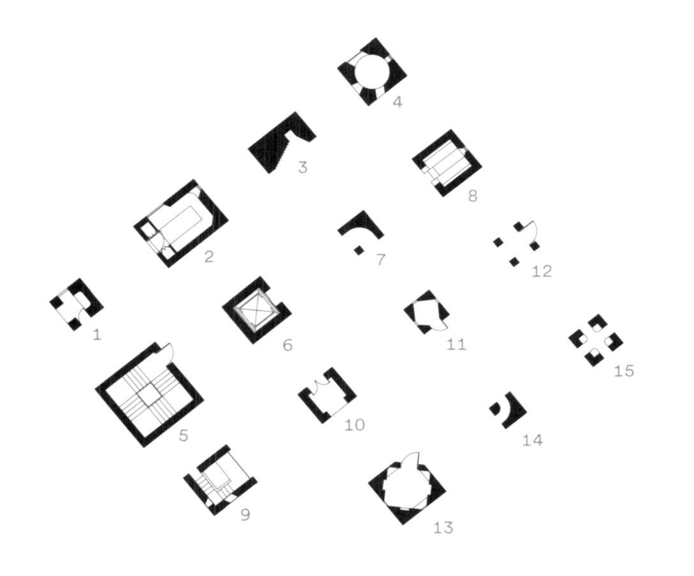

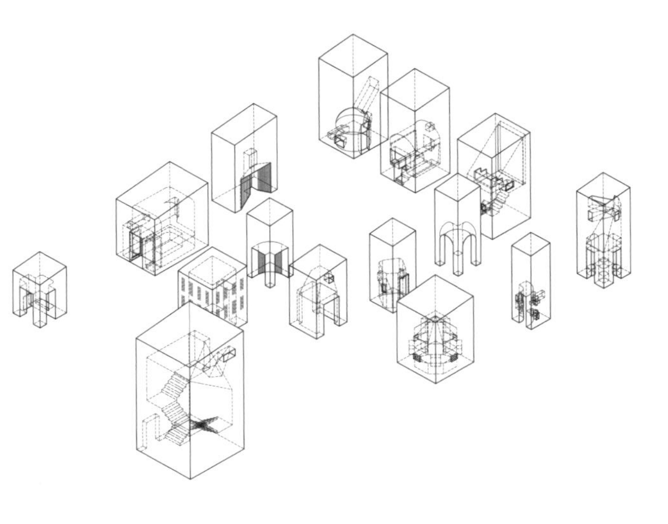

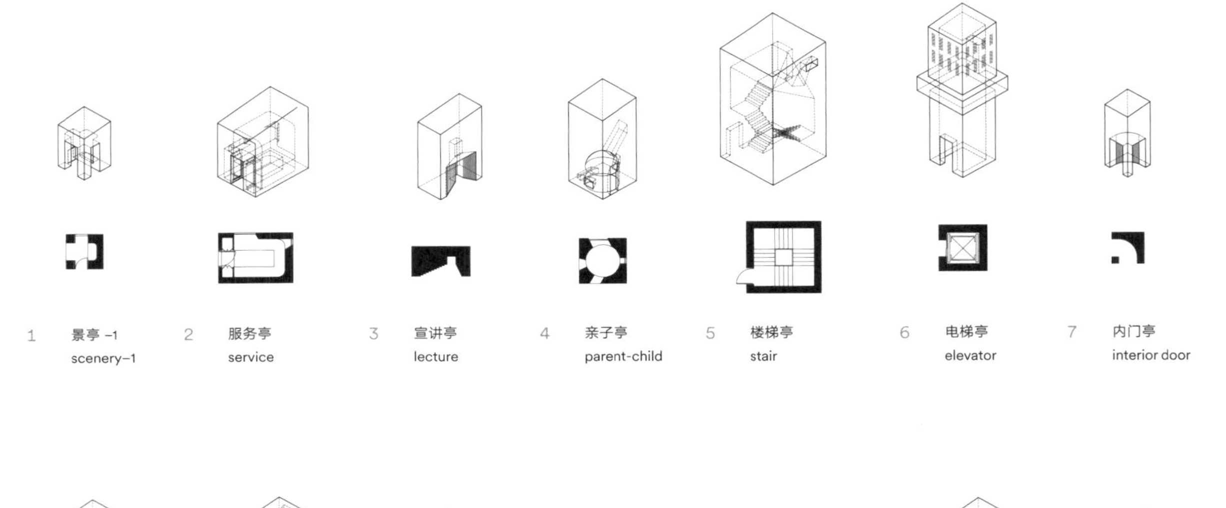

1 景亭 -1 scenery-1
2 服务亭 service
3 宣讲亭 lecture
4 亲子亭 parent-child
5 楼梯亭 stair
6 电梯亭 elevator
7 内门亭 interior door

8 读书亭 -1 reading-1
9 读书亭 -2 reading-2
10 玄关亭 entry
11 仓亭 -1 storage-1
12 后门亭 exit
13 管理亭 management
14 仓亭 -2 storage-2
15 景亭 -2 scenery-2

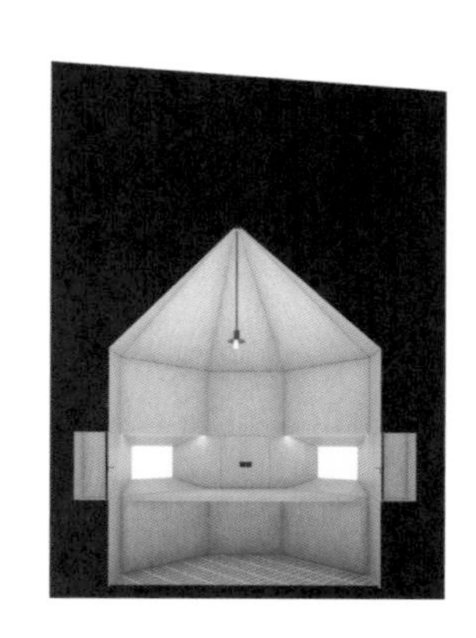

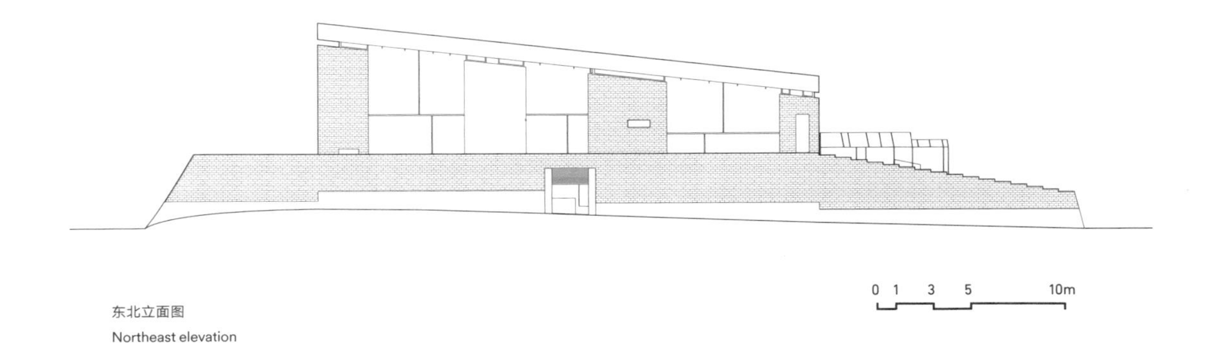

东北立面图
Northeast elevation

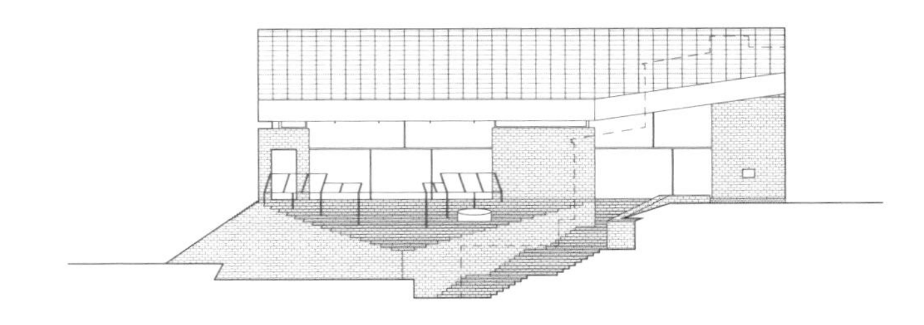

西北立面图
Northwest elevation

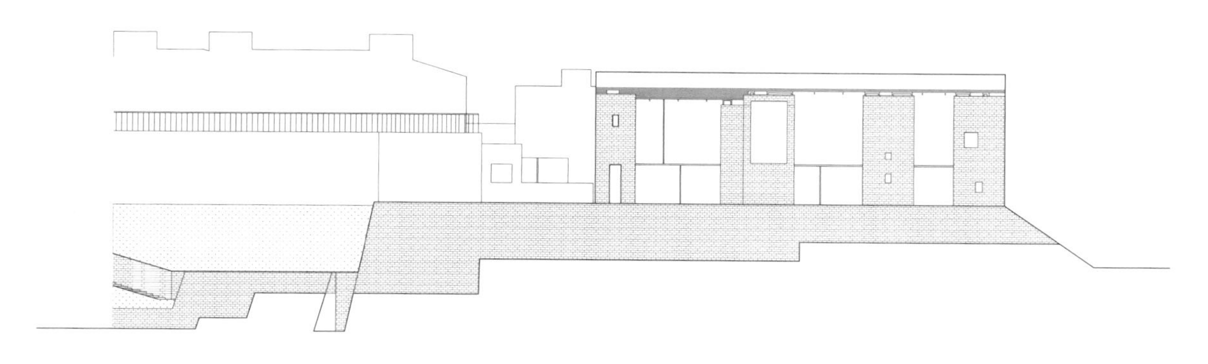

东南立面图

Southeast elevation

黎里
LILI
12

区域总平面，从小镇进入基地

Regional site plan

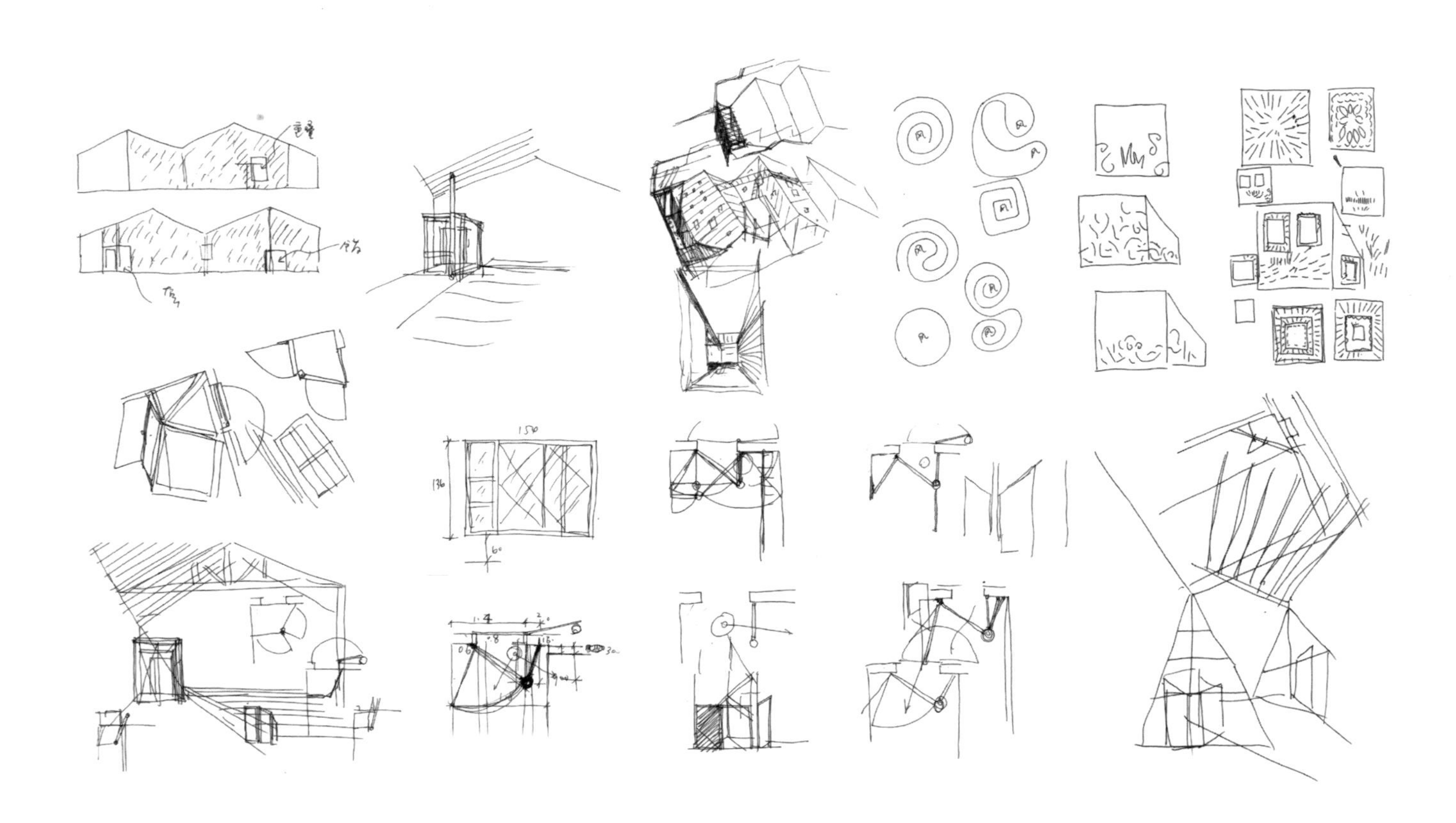

这个叫“黎里”的房子是一个改扩建项目。原房屋是一座一层小车间，有连续两个双坡顶，位于黎里古镇，周围被坡顶民居包围着，毗邻柳亚子故居，从临河的街上必须穿过一条细窄幽暗的背弄才能进入。房子及其庭院被重新规划成混合使用的两种功能：一种是利用原车间改造而成的多功能艺术文化活动空间，供公众使用；另一种是在北院新建两个房间作为居住空间。两种空间相隔一个双折的坡屋顶：坡顶的下面是公共空间，而上面高低转折的坡顶则作为居住者的室外活动空间。

两种空间自然地形成不同的氛围。公共空间内有 40 个传统民居天窗，引入的斑驳日光使得这里一改原来因为四周被民居包围而形成的阴暗感，天然的室内材料在日光下形成沉郁的质感。对住户而言，于坡顶之上独有一方天地理解小镇：小镇的坡顶尽收眼底。

建筑师在整个建造过程中尝试扮演了导演的角色。他指挥民间的施工队伍，全部利用民间习惯的施工方法，用低造价和短时间完成了这一项目。

The house named “LILI” is a project of renovation and extension. Located in the ancient town Lili, Jiangsu Province and surrounded by sloping roof dwellings, this former single-story workshop with two continuous gable roofs is adjacent to the former residence of Liu Yazi, a famous Chinese poet born in this area. It can be reached only by crossing a narrow and dim back alley from the main street along the canal. The building and its courtyards are expected to accommodate two mixed-use new programs. The workshop will be changed into a multi-functional public space holding art and cultural activities. A new residential space with two rooms will be added in the north yard, between which stands the double folding roof. The area beneath serves the public, while the above is a private outdoor space owned by the host.

Naturally, these two spaces generate different atmospheres. In the public area, patchy sunlight through 40 traditional style roof windows drives out the gloominess caused by the dense surroundings and adds a thick feeling to the natural interior materials. For the residences, standing above the roof lends a specific perspective to understand the town because of the panoramic view of the town roofs.

During the whole construction process, the architect tries to play as a director. By supervising the local craftsmen and adopting their folk building methods, he accomplishes the project with a low cost and short period.

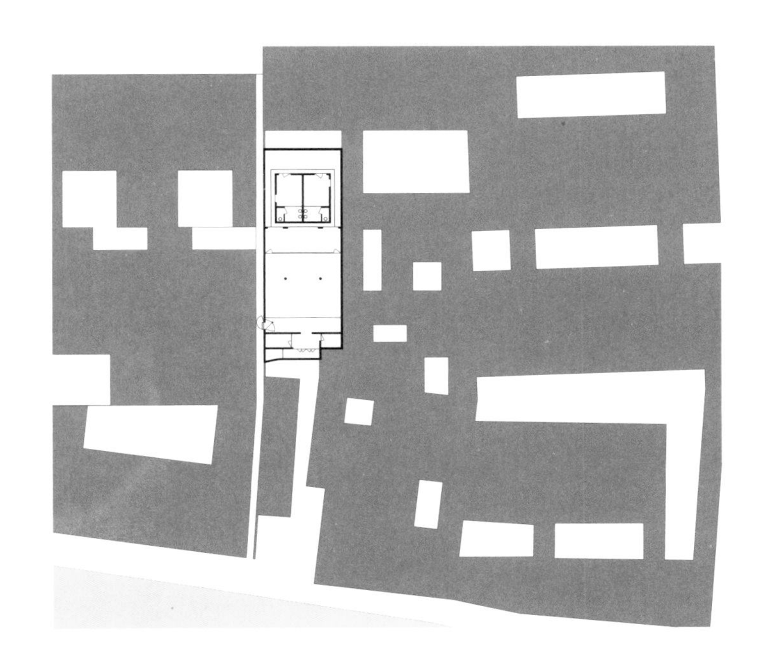

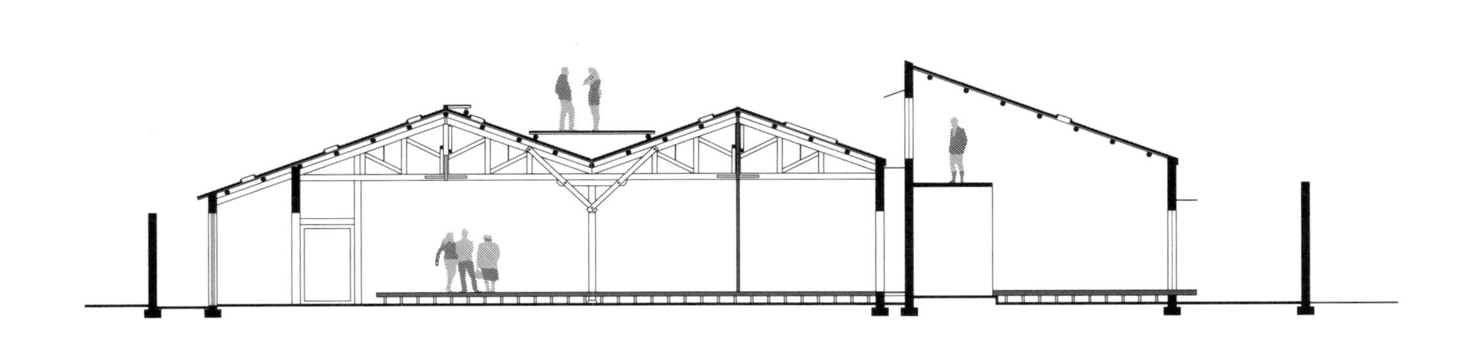

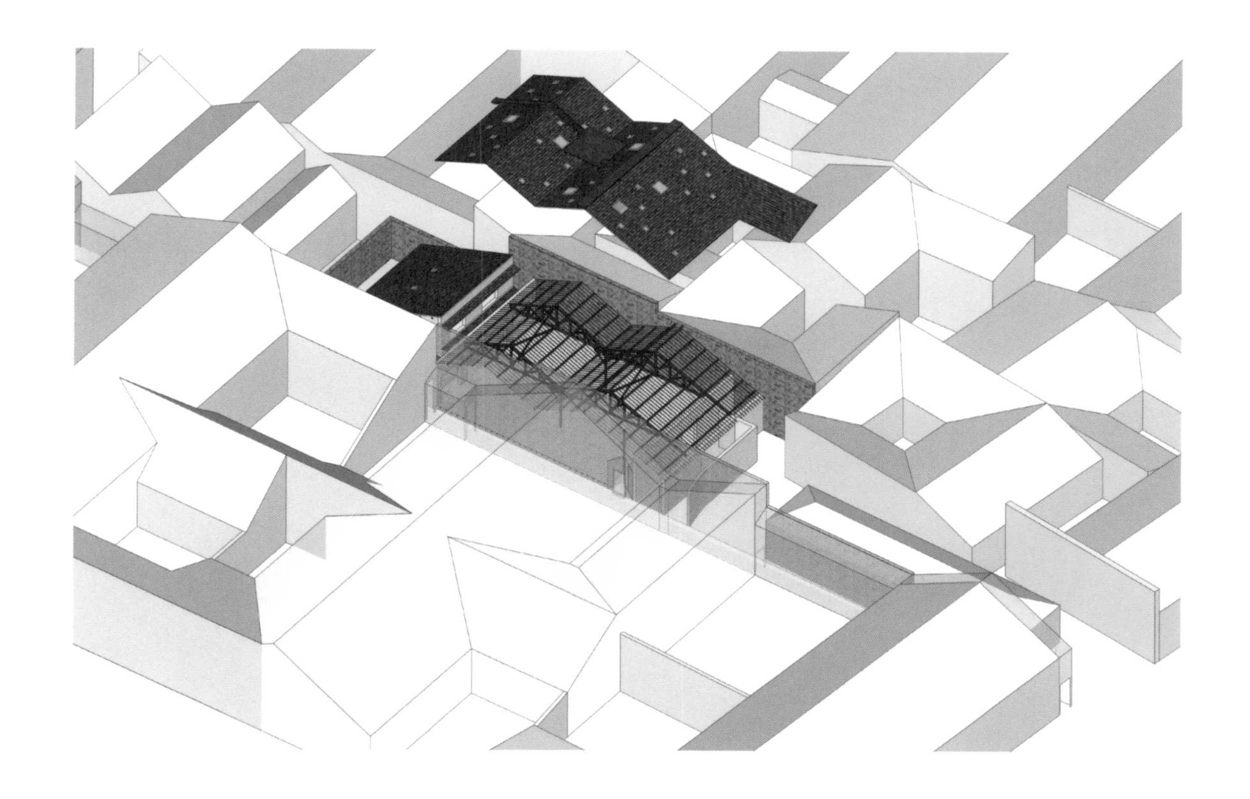

衡山路 890 弄（衡山和集）8 号楼外立面改造
Facade Renovation for No.8 Building, Lane 890, Hengshan Road

13

徐家汇公园
Xujiahui Park
华山路 Huashan Rd.
衡山路 Hengshan Rd.
肇嘉浜路 Zhaojiabang Rd.
徐家汇中心
Xujiahui Center
虹桥路 Hongqiao Rd.
漕溪北路 N. Caoxi Rd.
N

变化中的变化

衡山坊历史保护区由建于 1934 年的里弄住宅和建于 1948 年的花园洋房构成，拟改造成一个精品商业区，衡山坊 8 号楼的立面改造就是这个大城市更新中的一部分。基地所在的徐家汇商圈是上海最繁华的商业中心之一，其经济和城市的活力来源于不断的变化。因此这个项目对于阿科米星来说，最大的挑战是，如何在不牺牲历史保护街区统一性的情况下，实现新旧建筑之间的差异以及商业上的活力。

这幢老洋房坐落于整个街区的中心，一角朝向商业区的主入口，外面就是衡山路天平路的十字路口，还有广受欢迎的徐家汇公园。为了平衡差异性和统一性，建筑师为这座变身为精品店的三层老洋房赋予了一层可变的肌肤——白天优雅安静，夜晚魅力四射。这种戏剧性的效果由一种原创的发光砖产品而非传统的幕墙或照明灯实现，这种独特的发光砖与传统青砖尺寸相同，并和青砖混合形成表皮。白天，它就是纯粹的清水砖效果，很容易融入环境；当夜幕降临时，墙面会突然闪亮起来，像是一盏质感丰富、略显神奇的灯，这不仅使建筑从周围环境中脱颖而出，也吸引了路上的行人。然而，使这座建筑从中心商业圈内脱颖而出的不是寻常的霓虹灯效果，而是建筑本身在白昼之间不断变幻的戏剧感。这种从白天的“静谧”向夜晚的“璀璨”过渡的精妙设计也传达出一种“上海性格”：隐藏时的板正与放开时的漂亮都透着点矜持。

发光砖

发光体组件由五个不同的部分组成：主体结构是与传统砖一样大小的不锈钢框，其上插入由三个层次构成的照明设施。其中拓彩岩透光板这种新型材料在创造精致纹理时起到了至关重要的作用。发光砖和青砖按照传统的砌筑方式有序地铺砌，并形成了大小和密度各不相同的方形发光区。四个立面构成了统一又简单的连续表皮。

发光的建筑表皮是由相互叠合的方形发光图案组成，这些发光图案大小不同，密度各异，并与方形的门窗框相交叠。由于 LED 砖与其他青砖的形状大小都一样，所以它们可以一起构成不同肌理的方形发光区域。方形也是这座建筑门窗的形状，它在表皮中成为了一种基本元素。为了突出这种表皮覆盖的效果并生成不同亮度的肌理，在大的发光区域内还隐藏着一些小尺度高密度、亮度特别高的发光区。

入口和橱窗突出于建筑外墙，以满足功能和审美的需要。入口全部由不锈钢板构成，同样的材料也应用于独特的凸窗外壁，窗户内壁则覆盖了铜板。不锈钢使邻近的发光砖产生炫丽的幻影，亮铜色则为建筑增添了迷人的温暖感和繁华感。

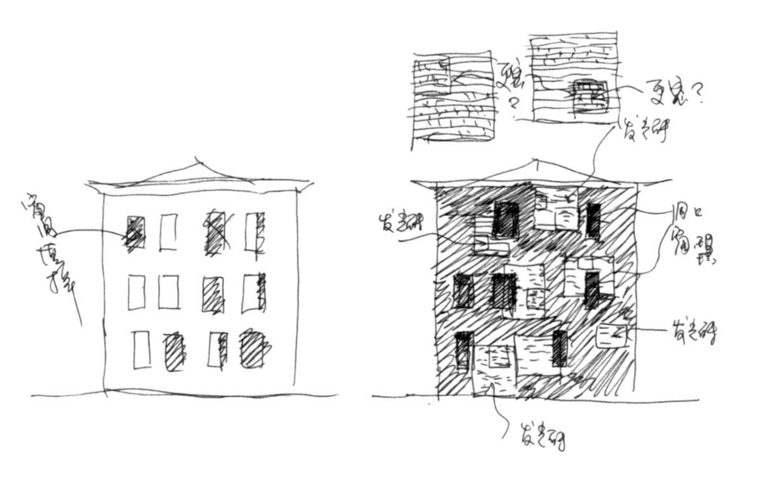

Change in Change

This facade renovation project is a small part of a big urban regeneration program, Hengshanfang, which is composed of a group of historical Lilong houses built in 1934 and garden villas built in 1948. This area is planned to be transformed into a boutique commercial area. The site locates in Xujiahui District, one of the most prosperous commercial centers in Shanghai, where constant changing plays a crucial role in guaranteeing economic growth and urban vitality. Thus, for Atelier Archmixing, the biggest challenge lies in how to achieve iconic difference as well as commercial vitality without sacrificing the coherence of a historical preservation area.

The old villa stands in the center of the whole block, with a corner exposing to the main entrance. Outside the entrance, a cross of two leafy streets and a popular park are close at hand. In order to balance the integration and distinction, the architect has decided to cover the three-story structure with a changeable skin—elegant and quiet at daytime, charming and attractive at night. Instead of conventional curtain walls or floodlighting, a luminous brick, an originally designed and customized product is selected for achieving this dramatic effect. These special luminous bricks are the same size as traditional black bricks. These bricks are mixed in the surface of the building. During the daytime, it looks exactly like a ganged brick wall that fits well in the surrounding environment. When night falls, the wall starts flashing like a magic lantern. Standing out from the surrounding environment, the attractive building gives passers-by a pleasant surprise. However, instead of the neon light, which makes the building distinctive in this downtown commercial area is its ever-changing dramatic fact during day and night. Quiet at daytime, resplendent at night, this ingenious design expresses a typical “Shanghai style”: being traditional or fashionable, it has always kept restrained elegance.

Luminous Brick

The luminous brick is composed of five different components: its major structure is a stainless steel frame which is as big as traditional bricks. The frame is inserted with three lighting facilities, among which a new material called Photosensitive Mineral Resin Sheet is a critical component in creating a delicate appearance. By laying these bricks together with the black bricks in a well-organized way, the wall is decorated with square light areas with different size and intensity. These four facades constitute a simple, unified and continuous architectural surface.

The luminous facade is composed of overlapping rectangular lighting patterns. Varied in size and intensity, these patterns are installed intersected with window frames. Since LED bricks and traditional ones are of the same size, they can form rectangular lighting areas of different texture. The rectangular shape is exactly the shape of windows and doors. It is used as a basic form on the surface. To highlight this covering effect and create different brightness, the extremely denser lighting blocks of small size is hidden in those large light areas.

For the functional and aesthetic purpose the entranceway and display windows are projected from the exterior walls. The entranceway is totally made of stainless steel plate. The exterior frame of the bay windows is built with the same material, while their interior surface is covered with copper plates. When the night is lighted up, those lighting bricks adjacent to the stainless steel produce gorgeous reflections, and the bright copper adds a charming warmness and prosperity to the building.

入口
Entrance

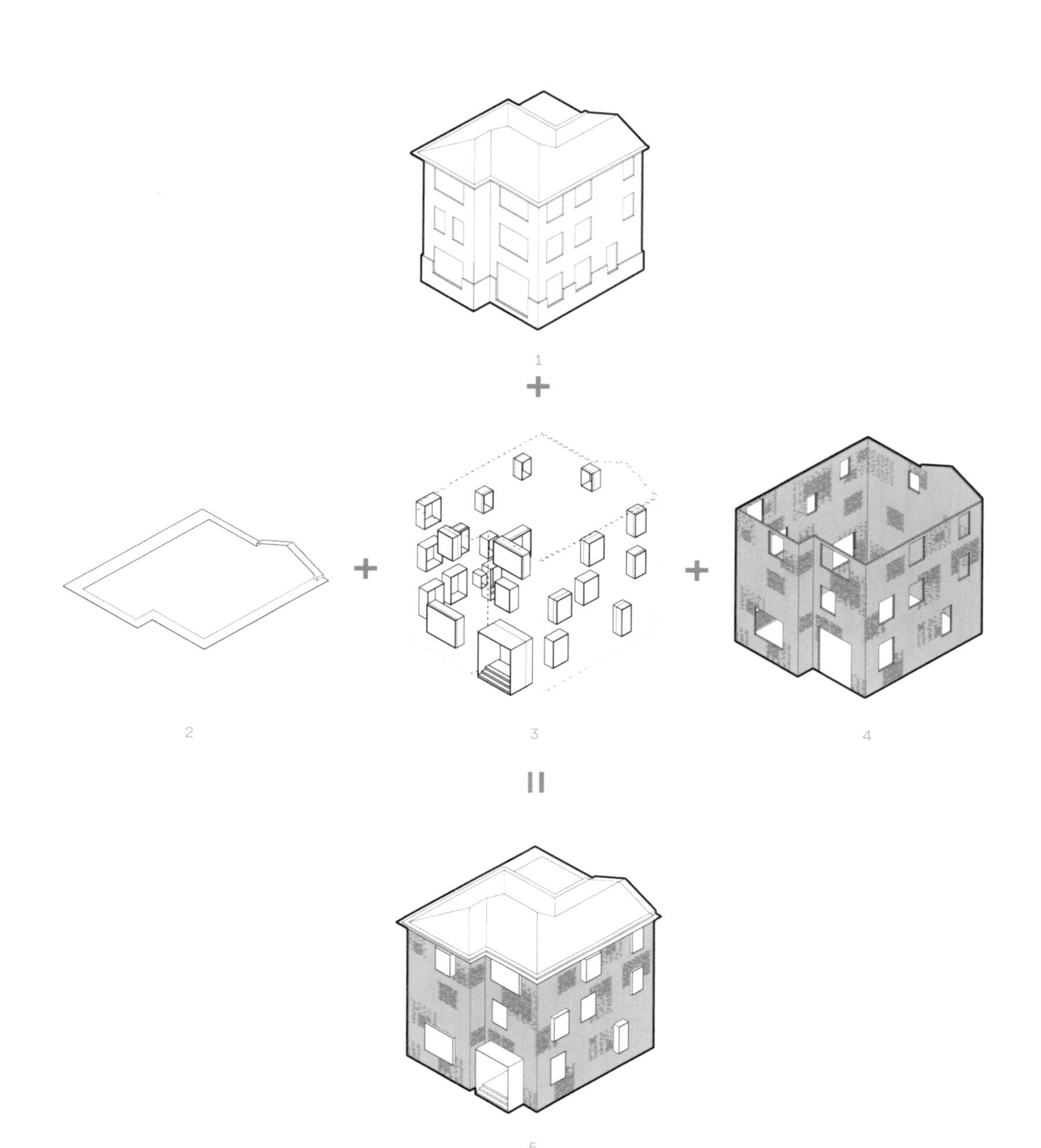

1	改造前建筑	Before renovation
2	金属收檐	Steel cornice
3	门窗改造	Renovated windows and doorway
4	砖墙表皮	Brick skin wall
5	改造后建筑	After renovation

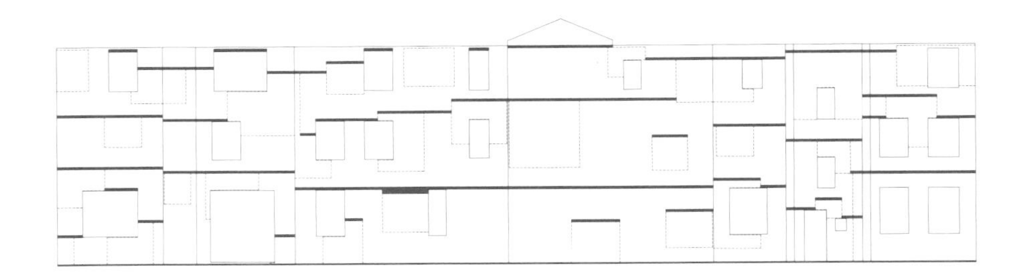

外砖表皮后洞口及构造暗挑梁立面展开图

Unfold elevation of openings behind exterior brick facade and hidden cantilevered beam

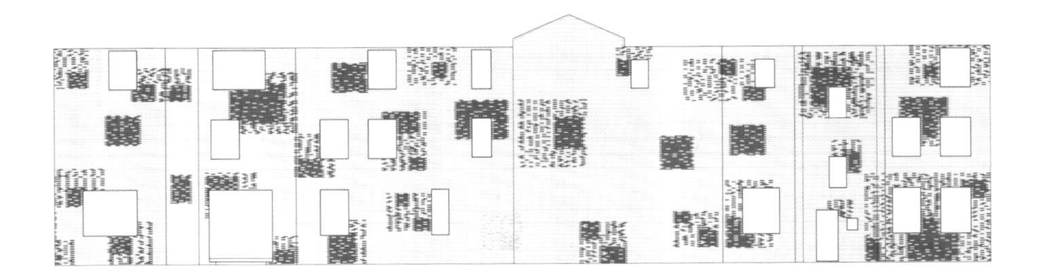

外表皮青砖及发光砖立面排版展开图

Unfold layout of the dark brick and luminous brick

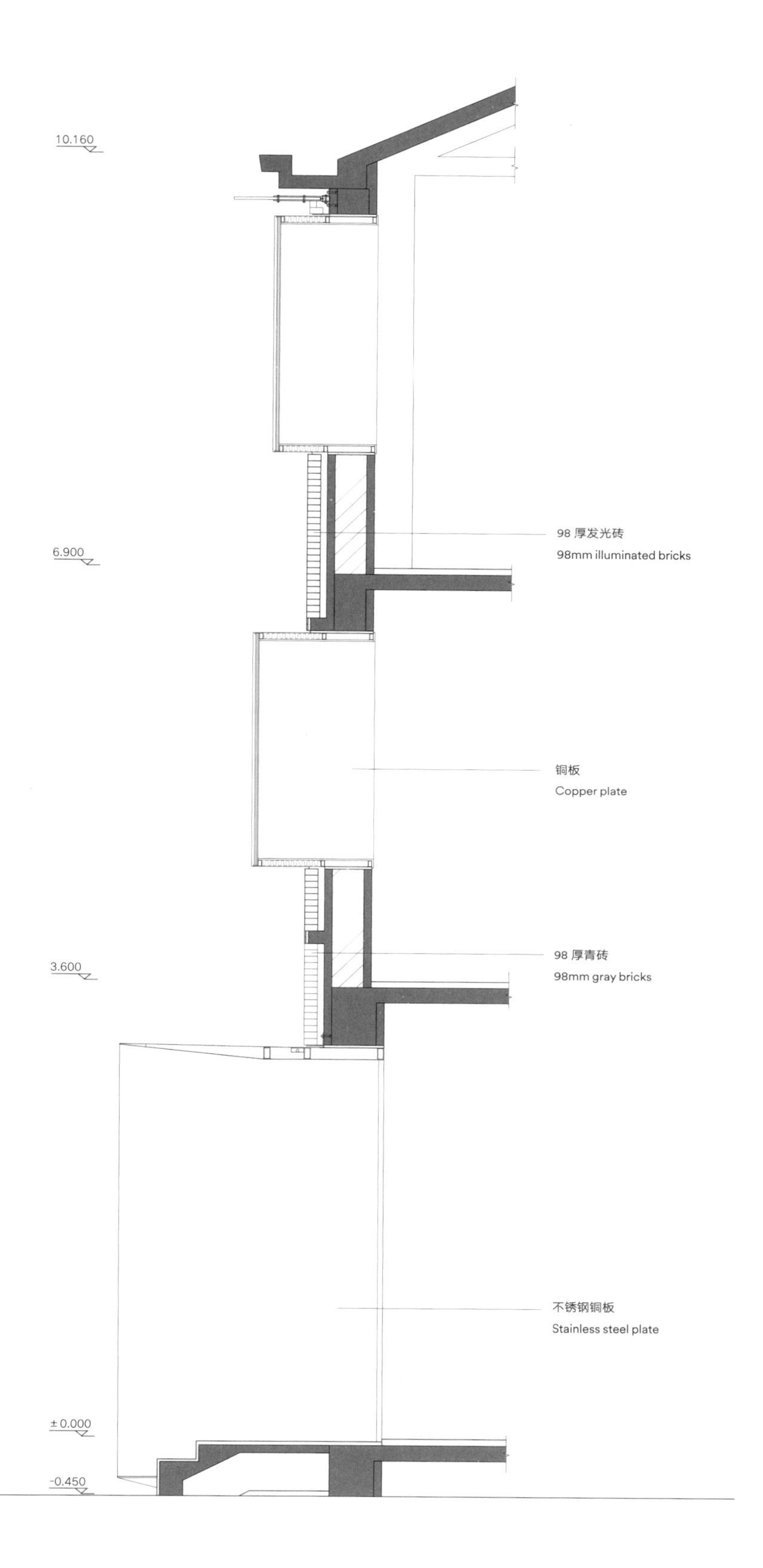
10.160
6.900
98 厚发光砖
98mm illuminated bricks
铜板
Copper plate
3.600
98 厚青砖
98mm gray bricks
不锈钢铜板
Stainless steel plate
± 0.000
-0.450

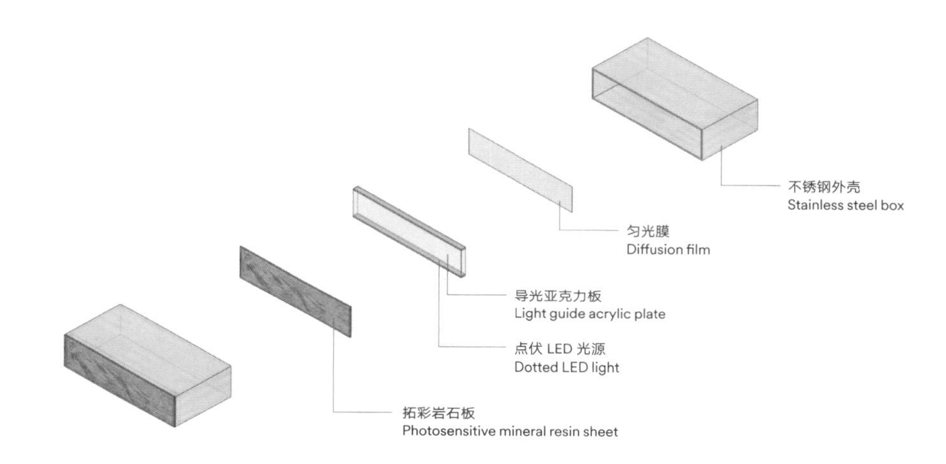

单块发光砖分解轴测
Composition of luminous brick

徐汇区龙华街道敬老院立面改造
Facade Renovation for Longhua Street Elder Care Center

原方案空间模式

Original spatial composition

改造后的空间模式

Spatial composition after renovation

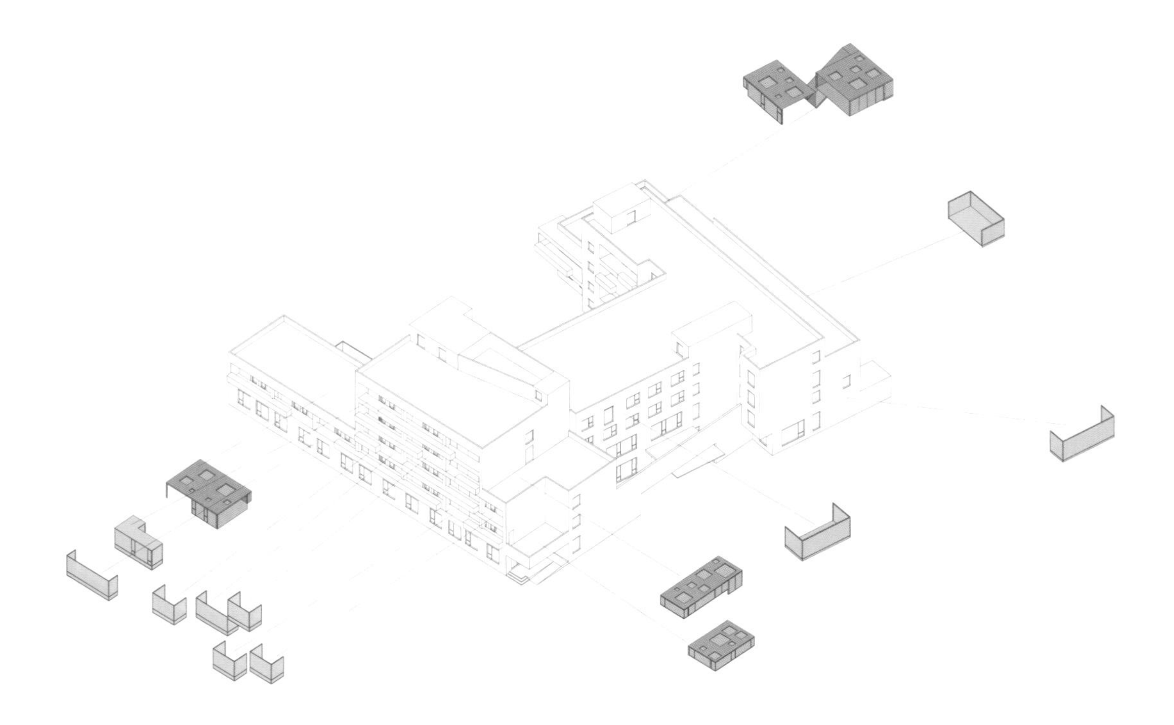

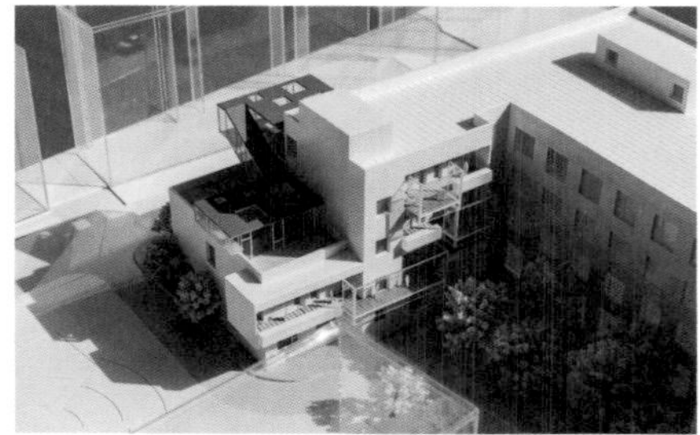

从立面改造到功能调整

龙华街道敬老院是一个普通的立面美化工程。在着手设计时，建筑师发现，原来的平面布局与其说是一座养老院，不如说是一家医院。一条中间走廊将各种房间功能化地串联在一起，北面是辅助用房，南面是老人居室，居室外的阳台一个个是孤立的，每一个还被空调室外机占去了很多面积。医院中病人住院的时间一般不会太久，所以中间晦暗的功能性走廊是可以接受的，但老人院的老人通常需要在这里长期生活，一条无法交流、照不到阳光的暗走廊是难以忍受的，而彼此隔离、空间狭小的居室阳台也无法承载有质量的公共交流活动，这样的设计对一座公共功能配套并不充裕的社区老人院来说，实在不够人性化。于是，设计者成功说服业主，结合立面改造对老人院的使用功能做出调整，增加公共活动的空间。

建筑立面成为生活功能外溢的痕迹

具体做法包括：第一，把立面上原本孤立的每间老人住房的南向阳台横向打通并连接起来，局部空间还有所外扩，以利于驻留。这样就为很多居室在功能性的走廊之外，创造了一个新的公共端口，可以在这里沐浴着阳光相互交流。原来阳台上占据空间的分体空调室外机被集中放置在外立面新设置的设备平台上，用百叶遮挡，不再对生活阳台造成干扰。第二，在每层中间走廊局部段落增设外挑的空中庭院，使走廊不仅可以用于行走和连通，也可以坐下来聚会和享受太阳。第三，利用原来跌落形成的大小不同的闲置的屋顶空间，为每层的老人增设一个公共活动室，用于看电视、打牌、打麻将或聊天等公共休闲娱乐活动。

因为该项目结构早已确定，不可能重新做桩基，荷载增加不宜过大；另一方面，建筑师认为，新增的区域在保温节能等方面要求也不必太苛刻，因此在满足强制性规范的原则下，改造尽量采用了轻质的结构、材料和相对简单的构造方式，以减少对原来主体结构的影响。它们就像在日常城市中随处可见的非正规搭建那样，在主体结构之上用最经济、简单、有效的方式增添必要的生活设施。在日常生活中，居民自主搭建的可能是晾衣架、花架、空调机位、阳台、鸽棚，乃至厨房、卫生间等，而在这个老人院中，建筑师在主体建筑表面用轻钢结构外挑、叠加的则是不同功能、形状和大小的公共活动空间。两者的共同之处就是：建筑立面成为生活功能外溢的痕迹。

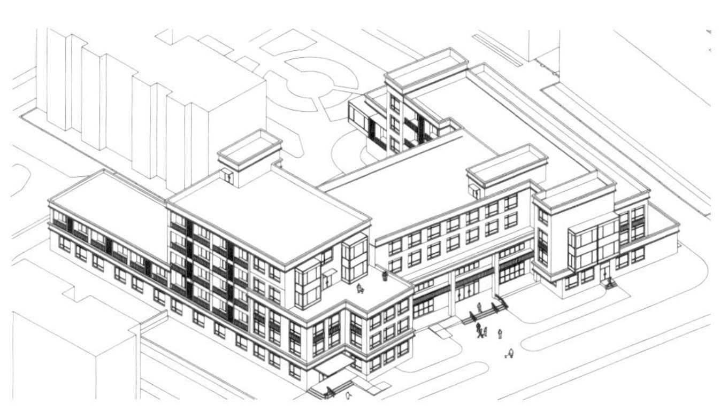

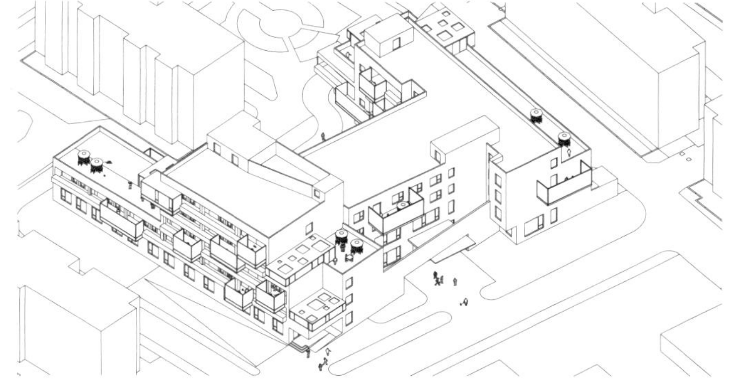

改造前后对比
Before and after renovation

From Facade Beautification to Functional Adjustment

The project of Longhua Street Elder Care Center was initially about beautifying its facade. But the architects found that the original layout was similar to that of a hospital, and various kinds of rooms were functionally connected by a corridor with auxiliary rooms on the north and old people's rooms on the south. All the balconies outside were isolated, and a large part of each balcony was occupied by the external units of air conditioner. A dark and gloomy functional corridor may be acceptable for a hospital because the patients would not stay long. However, in a daycare center, old people usually live there much longer. It is intolerable that people can't communicate and enjoy the sunshine in this dark corridor. The isolated small balconies cannot support high-quality public communications. Due to the lack of public functional facilities, the previous design for this elder care center was not user-friendly. Finally, the designer successfully persuaded the client to improve its functions in combination with the facade renovation, mainly through providing more public spaces.

Facade Recording the Outflow of Living Functions

Firstly, we have connected all isolated south balconies on the facade, some of which are expanded outward so that people can stay there comfortably. In addition to the interior corridor, many rooms enjoy new access to this public area, where people are able to talk in the sunshine. All the external units of air conditioners are placed on the terraces newly designed on the facade. They are sheltered by louvers, and will no longer dehumanize the balconies. Secondly, out-extended hanging courtyards are added to the central corridor on each floor. In this way, the corridors not only serve as channels for walking and circulation, but also are suitable for gathering and sunbathing. Thirdly, public activity rooms are set on each floor, taking advantage of various vacant roofs. People can watch TV, play cards, play mahjong, chat or be engaged in other public leisure activities here.

Since the structure has already been determined, it is impossible to redo the pile foundation or add extra loads on the whole structure. The architect assumes there is no need to make strict requirements on thermal insulation and energy efficiency. Therefore, complying with the mandatory norms, light structures, materials and relatively simple construction methods are selected as much as possible to reduce the impact on the major structure. Like those ubiquitous informal constructions in the everyday city, the architect has decided to add necessary living facilities on the major structure in a most economical, simple and effective way.

In everyday life, the residents usually set up on the residential surface laundry racks, flower shelves, air conditioner units, balconies, dovecotes, kitchens and toilets, etc. In the renovation for this elder's care center, the architect has created public spaces of various functions, shapes and sizes. These spaces are out-extended or superposed by a light steel structure on the surface of the major building. For both the everyday construction and new renovation program, the facade has recorded the outflow of the living functions.

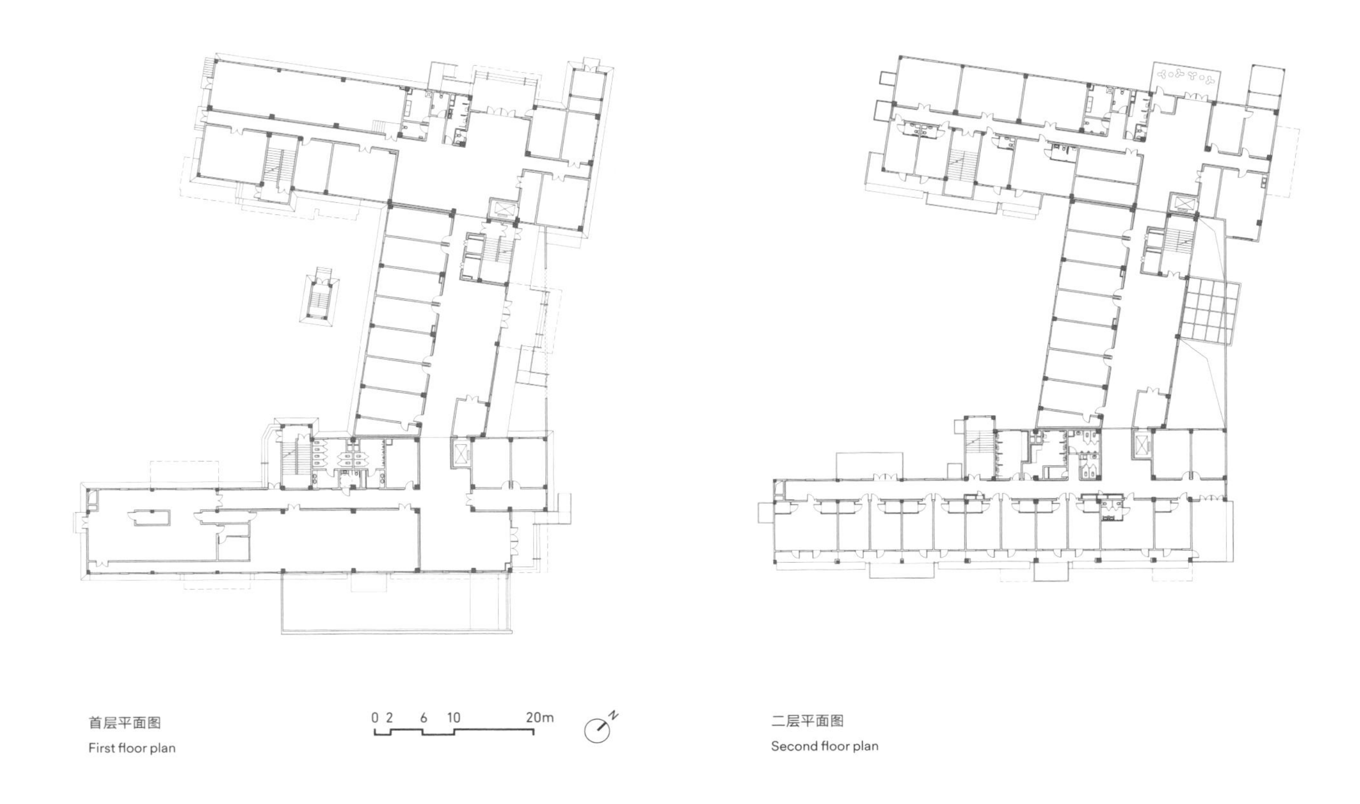

首层平面图
First floor plan

二层平面图
Second floor plan

三层平面图
Third floor plan

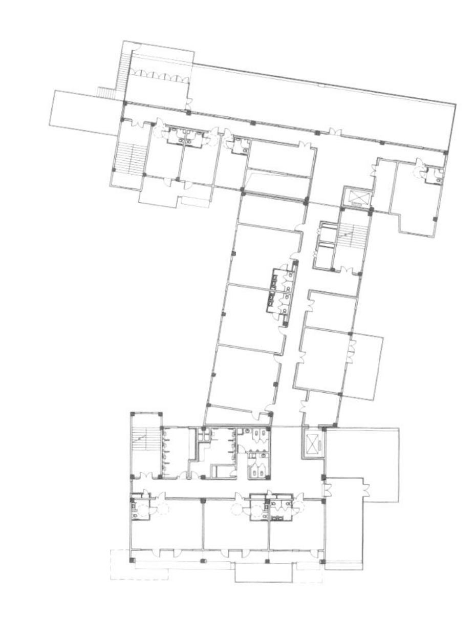

四层平面图
Fourth floor plan

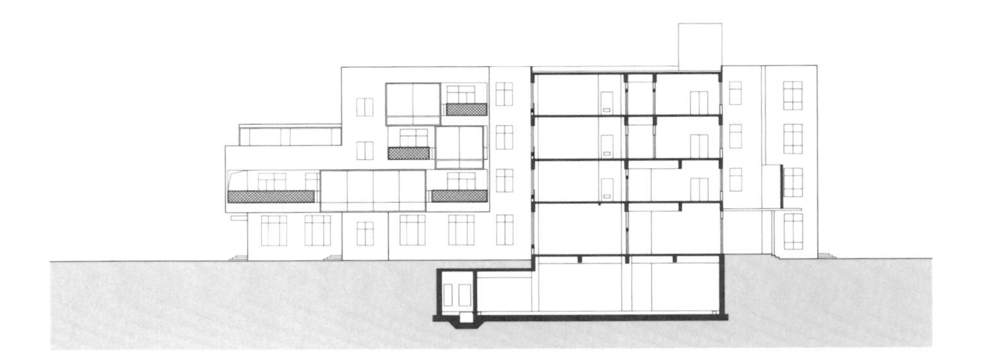

项目环境
environment

桦墅乡村工作室
Huashu Rural Studio

区域总平面图

Site plan

1	荷花池	Lotus pond
2	水库	Tank

基地原貌全景图

Panoramic view of the original site

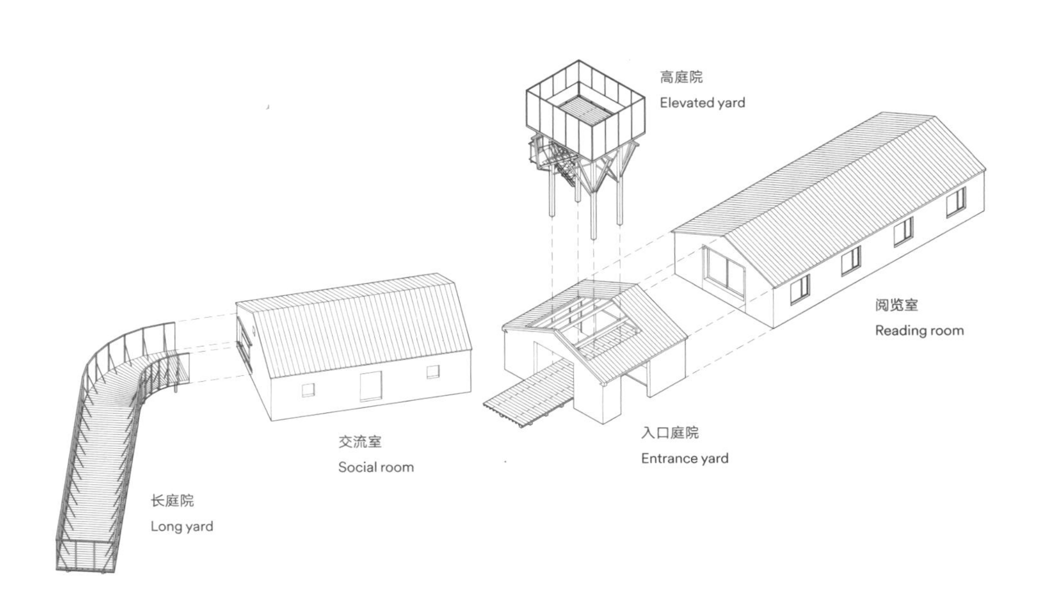

依附

改造设计若是始于"发现"，就可能意味着对于建筑物自然属性的人格化：建筑物不再只被看作附属于我们使用的身外之物，而可能被当作一个具有个性特征的鲜活生命。

桦墅的两座村屋，位于南京栖霞区桦墅周冲村南，周冲水库的堤坝下。开始想改造成一组工作室，最后被用作一处书院。从前，这里一座是碾米仓库，一座是农舍。碾米仓库外墙用石块砌筑，农舍外墙是砖混砌筑，表面水泥砂浆抹面。近旁的村落，远望的厂房，高处的水库，四周的农田池塘，环绕的远山近林，组成的景色恬静秀丽。

就室内环境与整体气质而言，农舍明亮一些，碾米仓库幽暗一些。两个房子彼此靠近，一起依附在水库的堤坝脚下。房子北侧为一片空场地，其东北角连接着一个荷花池。水库堤岸略高于房屋的屋顶，挡住了房子的视线，一旦登上堤坝，水库与村庄则具收眼底。

依靠依附的状态，两个房子的气质差异，高处的水库，略远的荷花池，这些是原有的建筑物给我们的第一印象。这就是我们所说的"发现"，也是我们设计的方法。一方面要拂掉杂质，让空间气质更好地显露出来，也就是让房子的个性被发现。另一方面是建立起房子与周围景观与环境的独特关系，也就是让房子去发现别人。

大眼睛和长鼻子

改造的一个重点是整理。光线明亮的农舍经过整理，空间变得明净而有仪式感，可用作阅览室；光线幽暗的仓库经过整理，氛围变得低沉而有质地感，可用作交流室。

改造的另一个重点是扩展与联系。明间在入口处扩展了一个入口辅助庭院与一个架空的空中庭院，在这里登高可以看到水库、群山、村庄的美景，所以架空的庭院也被我们称为"大眼睛"；暗间在房子的东侧扩展了一个弯弯长长的庭院，一直伸到北侧的荷花池边，在这里可以室外聚会，我们戏称这个庭院为"长鼻子"。这两处室外空间都采用了细小的钢结构或者轻巧的钢木结构，它们相对于各自房屋具有轻巧的状态，因此形式上也可以看作是另外一种依附的状态。

Attachment

If the renovation design begins with discovering, the nature of building may be personified. The building will be treated as not only a material possession, but also a featured fresh life.

The two village houses in Huashu are seated under the dam of Zhouchong Reservoir which is located in the south of Zhouchong Village, Huashu, Xixia District, Nanjing. They were initially planned to be renovated as a group of studios, but finally they were designed as an academy. These two houses used to be a rice milling warehouse and a farmhouse. The exterior wall of the former is laid with stones and the exterior wall of the later is masonry-concrete structure covered with cement plaster. Looking around, there are nearby villages, distant plants, reservoirs on the upper side, neighboring farm ponds, surrounding mountains and forests. All these elements constitute a tranquil and beautiful scenery.

Respect to the indoor environment and style, the farmhouse is brighter and the rice milling warehouse is darker. They are close to each other and nestle at the foot of the dam. In the north of the houses lies an open space, which adjoins a lotus pond at its northeast corner. The embankment of the reservoir is slightly higher than the roof. Thus the view in house is blocked. But once on the dam, people can enjoy a panoramic view of the reservoir and the village.

The first impression we have had on these two original buildings is their geographical location, different styles, the reservoir on upper side and a distant lotus pond. To discover, as we said, is exactly our philosophy towards architectural design. On the one hand, the impurities shall be removed to better reveal the space's temperament. In other words, the personalities of the house shall be discovered. On the other hand, a unique relationship between the house and the surrounding landscape and environment shall be established. That is to say, the house shall be made to discover other things.

Big Eye and Long Nose

To adjust is a key work for renovation. With adjustment, the bright farmhouse has been converted to a clean place with the sense of ritual, and it can serve as a reading room. The dark warehouse has been given a repressed atmosphere and textured feeling, which can be used for communication.

Another key issue for renovation is to extend and to connect. The entrance of the brighter room has been extended with an additional entrance courtyard and a hanging courtyard. This hanging courtyard is called "big eye", since people can hold a panoramic view there. The darker room has been extended to the east with a long and winding courtyard which directly leads to the north lotus pond. It is an ideal place for public communication. Because of its unique shape, this place was nicknamed as "Long Nose". These two outdoor spaces are constructed with small steel structure or light steel-timber structure. Compared with the original houses, they are lighter in structure and dependent in terms of architectural form.

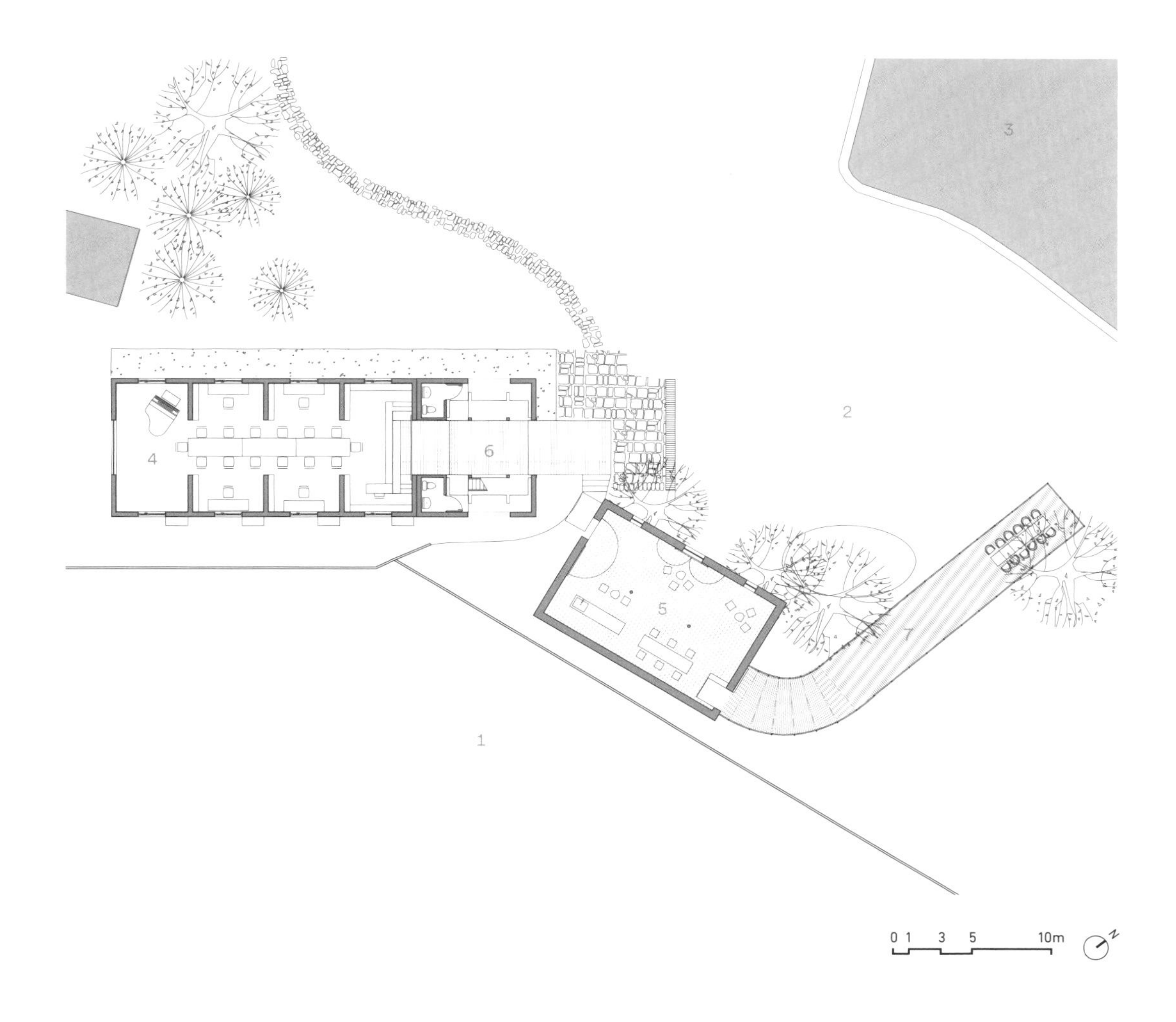

1	堤坝	Dam	5	交流室	Social room
2	菜圃	Vegetable garden	6	入口庭院	The entrance yard
3	荷花池	Lotus pond	7	长庭院	The long yard
4	阅览室	Reading room			

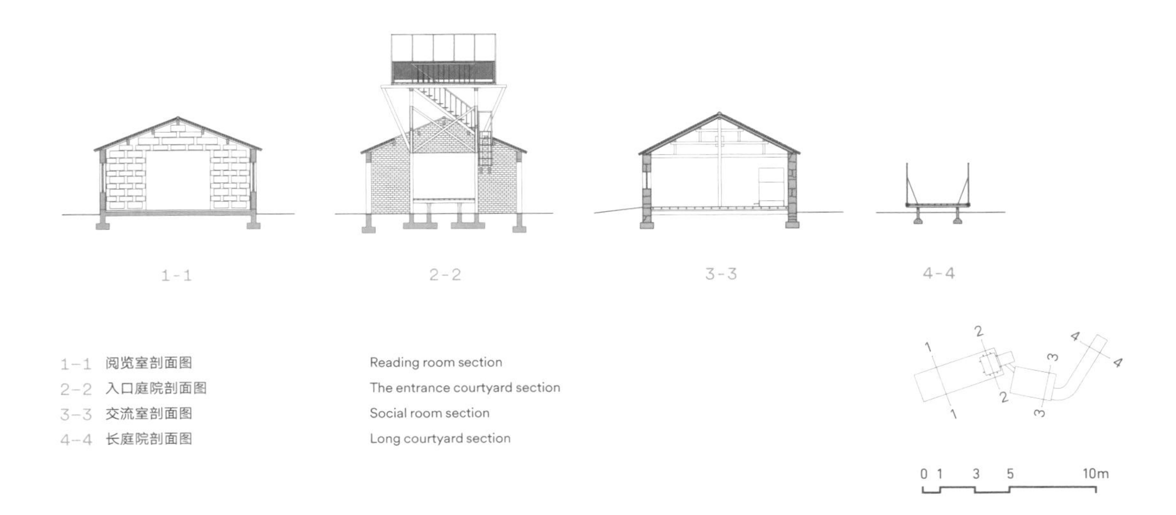

1-1
2-2
3-3
4-4
1-1 阅览室剖面图 Reading room section
2-2 入口庭院剖面图 The entrance courtyard section
3-3 交流室剖面图 Social room section
4-4 长庭院剖面图 Long courtyard section
0 1 3 5 10m

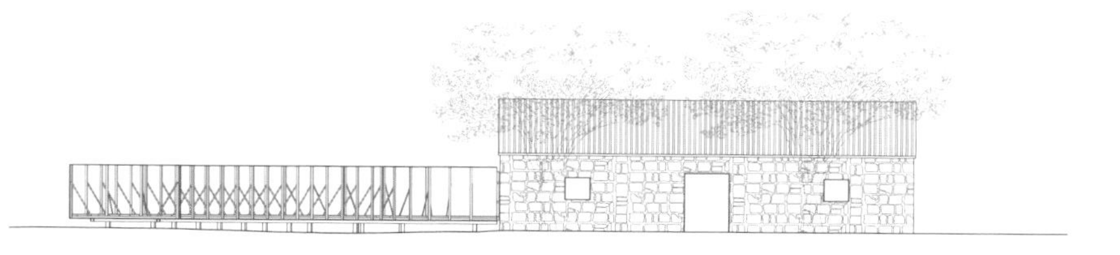

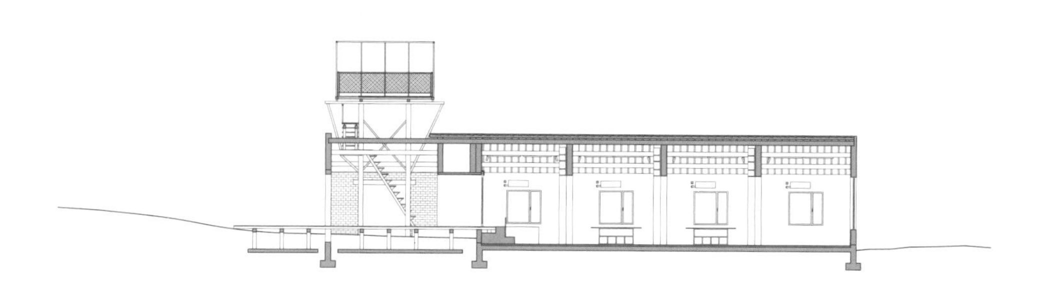

“永远关不上”的门

A door that can never close

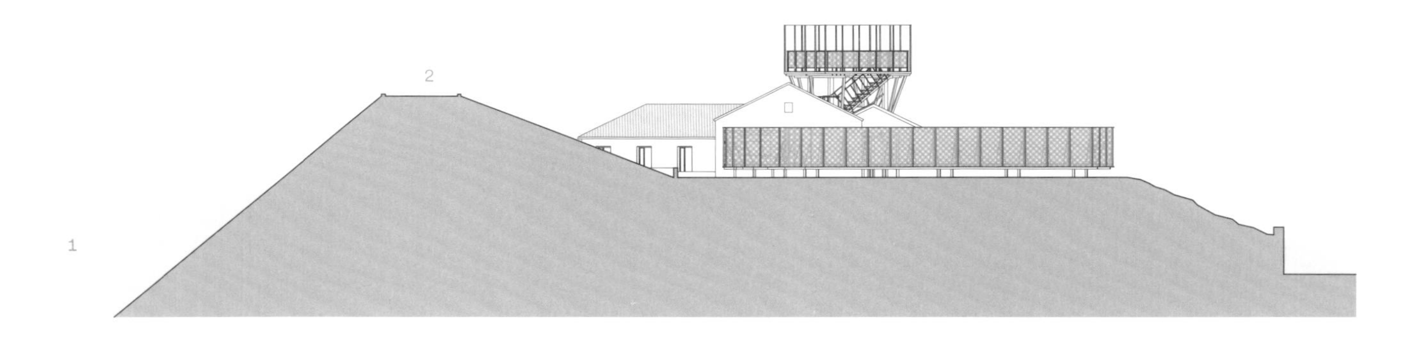

1 水库 Reservoir
2 堤坝 Dam

申威达 2 号楼改造
Renovation of No.2 Building of Shenweida Workshop Space

16

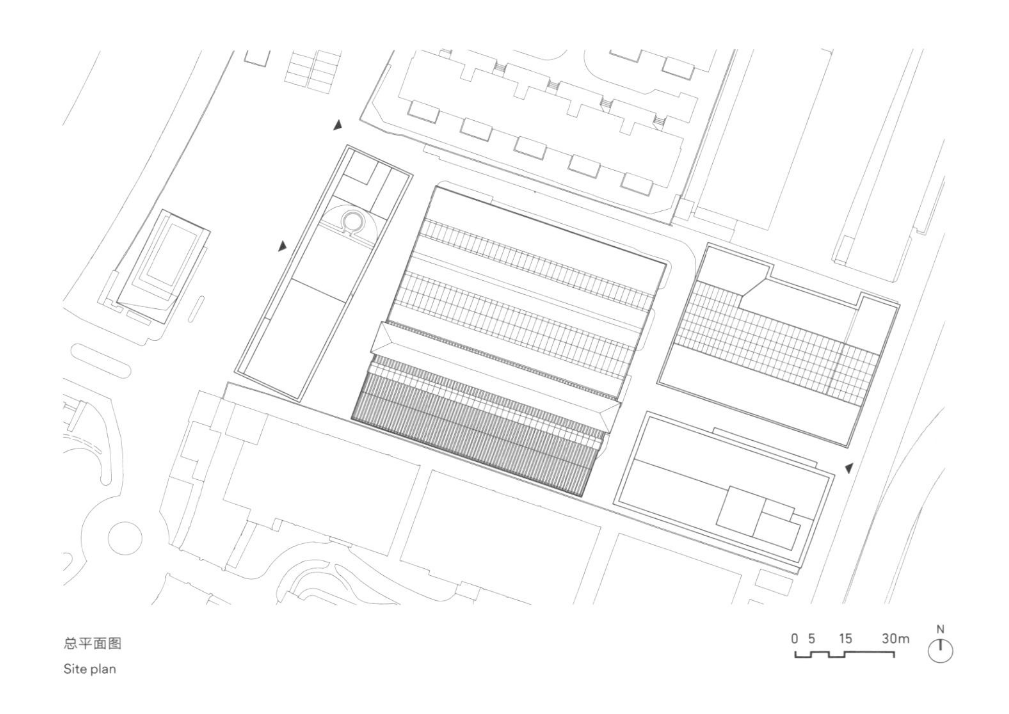

总平面图
Site plan

屋中屋

该项目的前身是一家刀具工厂，位于上海浦东新区。厂区内建筑大部分建成于 20 世纪七、八十年代。业主买下整个厂区后，拟保留原有建筑物，并将其改造成以办公功能为主，融合商业、酒店等多业态的园区。园区中共有五栋建筑，2 号楼是连续排架结构的单层多跨大厂房，曾经是厂区内最重要的生产车间。

2 号楼厂房进深方向长 63 米，横向共有四跨，其中三跨是混凝土排架结构，最南侧一跨为加建的钢结构（后期改造中将其拆除，新建一栋与原有外轮廓一致的 2 层钢结构建筑物），总跨度 60 米。在保留的三跨混凝土排架屋盖下方，设计置入了三条高度低于排架的办公构建物。办公构筑物由二层轻钢结构加中空玻璃围护组成，与原有的厂房结构完全脱开，成为独立的“屋中屋”，底层作为展示或者小商业配套空间，二层功能为办公，类似传统的“前店后坊”的模式。每一条办公构筑物由三个相同的模块化单元错动组合而成，每个单元均有独立的楼梯和水吧区。

整个厂区内原有的建筑密度较大，除去必要的厂区道路之外几乎没有空余的绿地和广场，严重缺少促进聚集交流的公共空间。设计拆除了 2 号厂房的部分屋面板，替换成轻质的透明阳光板，形成两条宽为 3.5 米和 7 米的采光天窗，同时将厂房原有的玻璃外门窗全部拆除，只保留窗框，将原有建筑改造成为一个四周通透、采光良好的外围护覆盖。

街道与广场

新置入的办公构筑物通过体量的错动组合，形成了在大覆盖之下进入办公单体之前的一系列“街道”和“广场”空间。将来围绕这些空间的底层，引入小型商业或展示等配套功能，会创造出一系列半户外的公共空间。这里可以作为底层商业的外摆区，也可以是办公展示空间的延伸，人们可以在这里聚集交流，一边休闲一边工作。这些在原有建筑大覆盖下的多义空间，成为了整个园区最有活力的公共场所，连接起整个园区的南北两区。上海多雨，气候冬冷夏热，真正能舒适地使用纯户外空间的天数并不多，而有着大覆盖的半户外空间，能提高使用效率，使得该办公空间在不同的气候条件下均能保持活力。

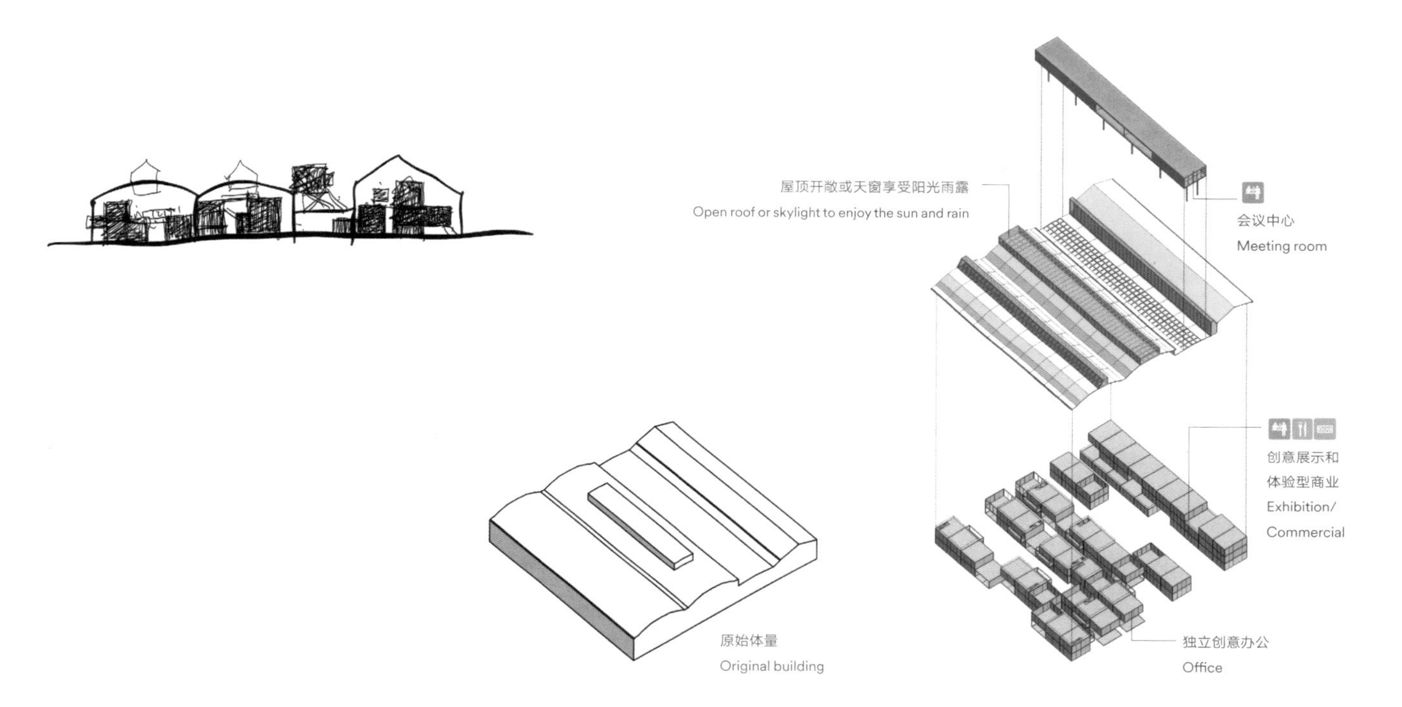

House Within House

This project, formerly a cutting tool factory, is located in the Pudong New Area of Shanghai. Most buildings in the factory were built in the 1970s and 1980s. After taking over the whole factory, the client intends to keep the original buildings and transform them into a mixed-use campus, mainly offices, also including hotels and commercials. There are five buildings in total. Building No.2 is a single-story multi-span large workshop with a continuous bent structure, which used to be the most important workshop in the factory. Building No.2 is 63m long in depth and has four spans in transverse direction, among which, three spans are built with concrete bent structure, and the last span 60m long at the southern side is an added steel structure that will be dismantled in later renovation and a two level high steel structure will be built in a profile consistent with the original building.

Under the reserved three span concrete bent roof, we have inserted three office structures with a height lower than the bent frames. The office structure is composed of a two level high light steel structure enclosed with hollow glass, detached from the original workshop structure as an independent “house in house”. Exhibitions or small businesses are on the ground floor, and the office area is on the second floor. This arrangement is similar to the conventional Chinese business space model of “store in the front and workshop in the rear”. Each office structure consists of three identical modular units combined in a staggered way. Each unit is equipped with an independent stair and a water bar area.

The original factory has a high density. Apart from the necessary plant roads, there is almost no spare green space and square, neither public space for gathering and communication. In this design, part of the roof boards of Building No.2 has been removed and replaced with light transparent sunlight boards, which forms two skylights with the width of 3.5m and 7m respectively. In addition, all the glass windows and glass doors of the original factory are removed. Only window frames are retained. The building is transformed into a well-lighted enclosure which can be seen through from all sides.

Streets and Squares

Under this large covering, through staggered and dynamic combinations, a series of "streets" and "squares" are formed in front of the office units. In the future, supporting functions such as small businesses and exhibitions will be introduced into the ground floor to turn this area into a series of semi-outdoor public spaces. This area can be used as an extension area for the businesses or exhibitions on the ground floor, where people can gather and chat, enjoy their work and leisure time. This multi-functional space under the original large covering will become the most dynamic public space in the campus, connecting the south and the north part. Thanks to the rainy days, cold winter and hot summer in Shanghai, there are few opportunities for people to make full use of outdoor space. Thus the semi-outdoor space with a large covering can be better used and help keep the vibrancy for the office space under all climatic conditions.

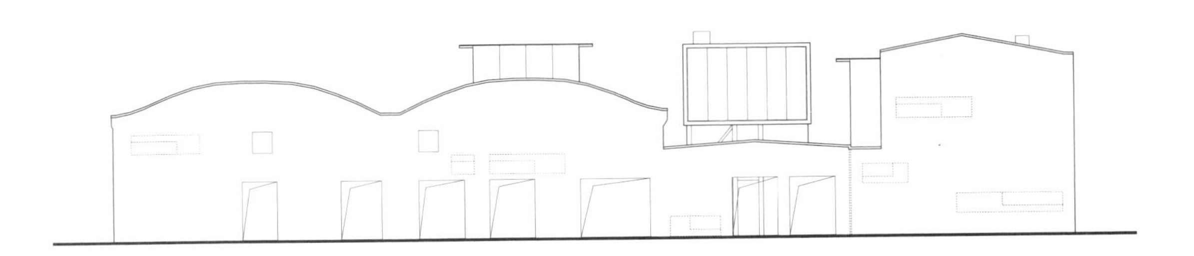

东南立面图

Southeast elevation

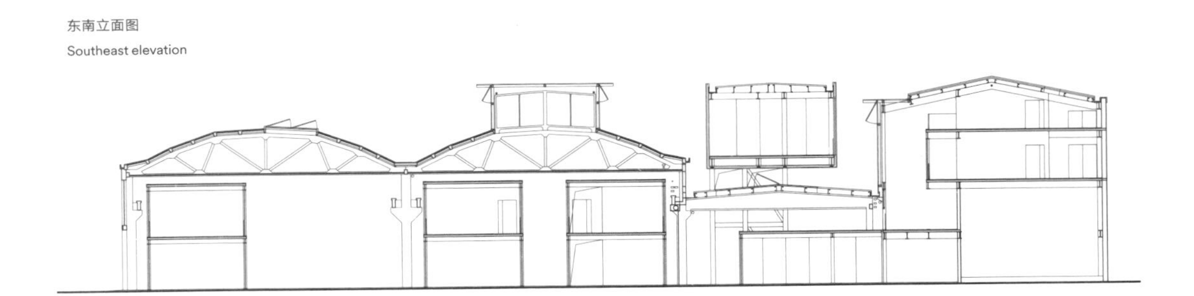

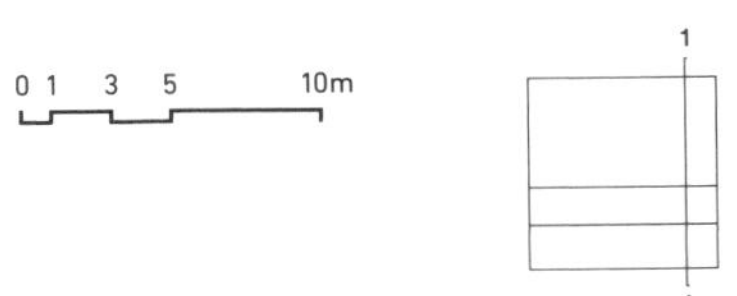

1–1剖面图

1–1 section

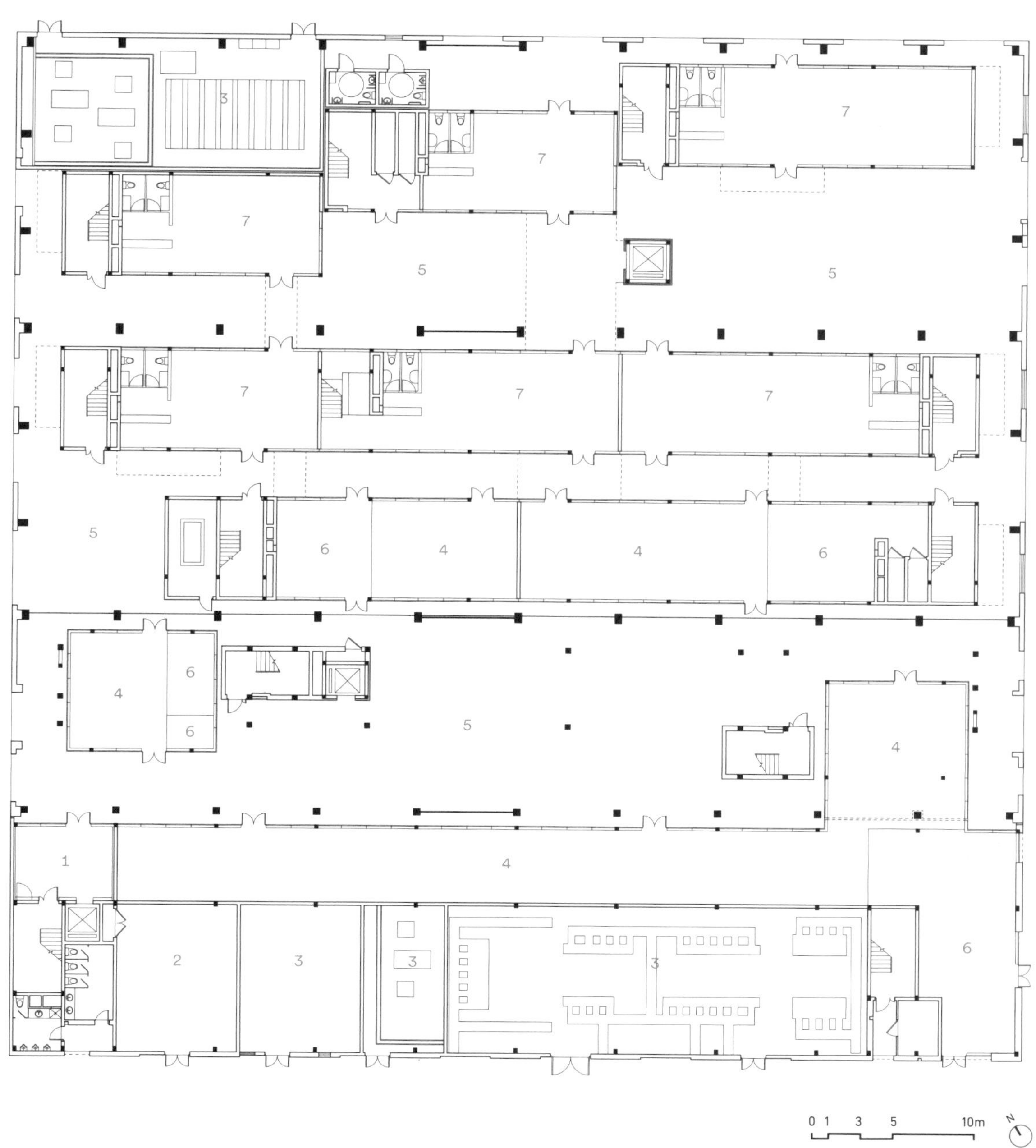

一层平面图

First floor plan

1	办公门厅	Office lobby
2	安保消防中心	Security center
3	设备房	Equipment room
4	餐饮	Catering
5	半室外广场	Semi outdoor plaza
6	厨房	Kitchen
7	办公	Office

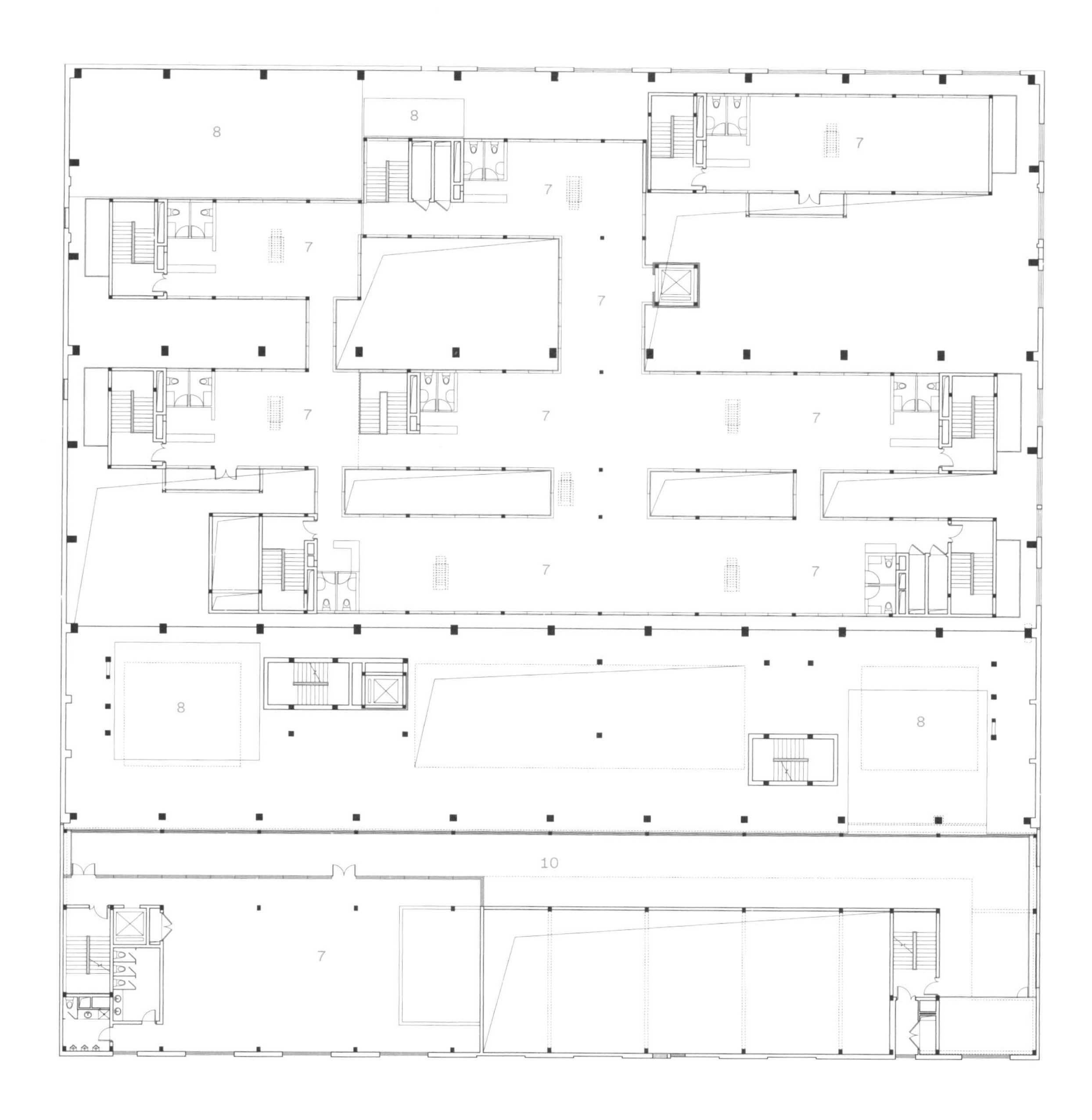

二层平面图
Second floor plan

8	非上人屋面	Unoccupied roof
9	上人屋面	Occupied roof
10	露台	Platform

Croissants de

悦阅书店
Yueyue Bookstore

17

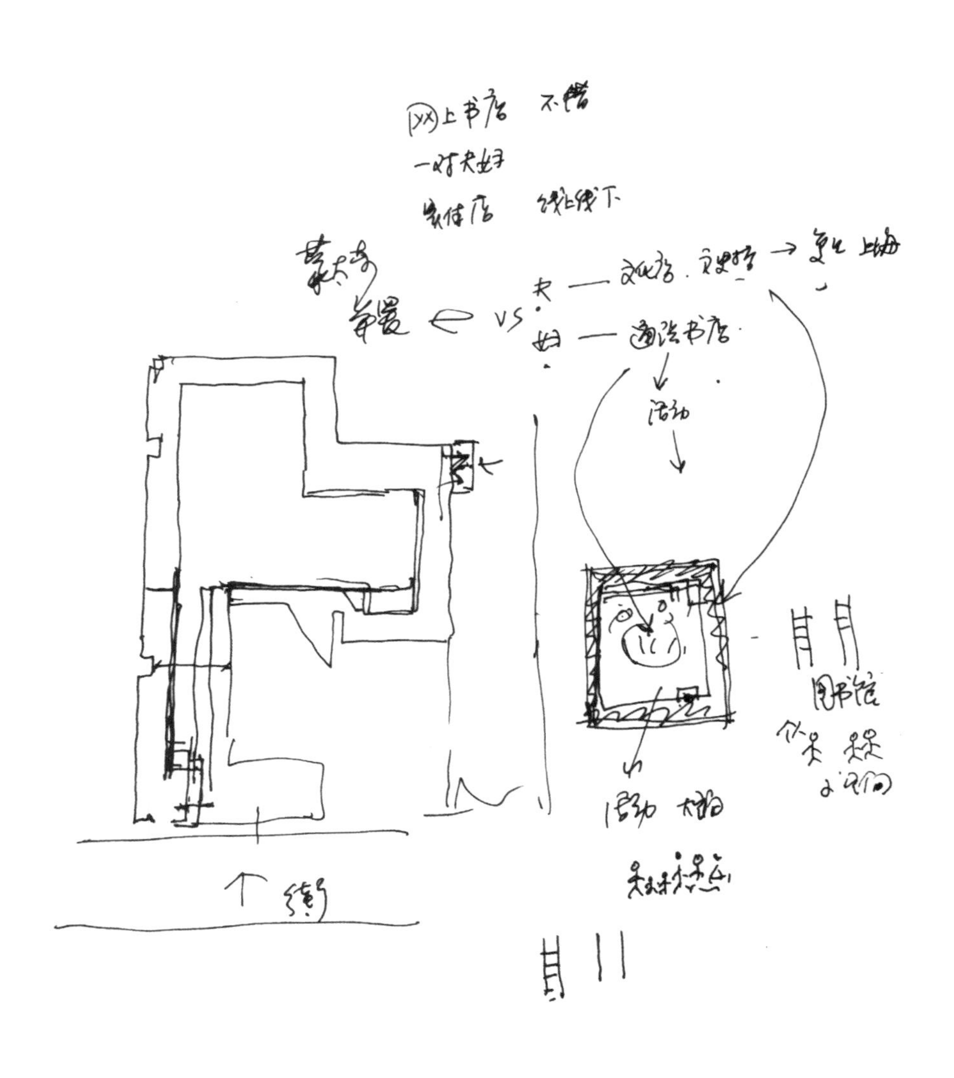

区域总平面图

Site plan

同一屋檐下，两扇像蝴蝶翅膀一样的连体门，通向 94 平方米空间里两个不同功能与调性的书店。文史哲书店如同曲折幽深的图书馆长廊，通识书店宛若轻松包容的共读大厅。不同需求的阅读者在此栖息、偶遇。

国权路

悦阅 / 志达书店位于国权路 525 号。树荫浓密的国权路西段属于复旦大学的城市区域，周边有复旦大学的体育馆、出版社、经管学院和生活区，还有复旦附属的幼儿园、小学和中学。

国权路曾以聚集众多文史类实体书店而知名，随着网上书店的兴起，跟其他实体书店一样，国权路上的书店也大多凋零。然而近年来，城市中出现了新一轮的实体书店建设热，并且大多以书店 + 的模式出现。在国权路这样一个普通的城市街区里，是否还可能做一家单纯的书店，氛围简单日常，却又能在功能和细节上创造自己的个性？

两个书店

业主夫妇同时经营着一家成功的网上书店。因为市场原因，这个经营了十多年的实体空间——原志达书店虽得以保留，却不断被压缩。这里在改造前主要经营教辅书，空间高度和面积都有限，平面轮廓曲折，更糟糕的是，楼上搭建的厨房还不断漏水。

业主夫妇一致同意更新和升级这个门店，但对于书店的定位却持不同的意见。丈夫放不下人文情结，希望做一个文史哲书店。妻子更加务实，考虑到市场趋向，希望做一个通识书店，可以做一些线下交流、亲子活动等。两种意见相持不下却都有很好的理由，设计师于是提出：何不做成两个书店？

通识书店需要大一点的内部空间，以容纳交流活动。文史哲书店采用像图书馆书库一样的窄长空间也没问题。但两者如何组织混合是关键。

金属工艺与机械设计

为了实现灵活的布置，并且考虑以后可以拆卸到其他空间继续利用，设计师设计了标准化的定制钢支架作为书架的结构，两边书店根据各自调性，采用了黑白不同的颜色。

两扇门，每扇重量超过 150 公斤，为了顺利打开，我们同金属加工的专业工程师与工人一起精心研究了门的构造细节。两扇门被安装在一根竖向的钢轴上，形成了一个像书页一样可以同时打开的门，而门轴采用多层轴承嵌套的方式，以避免钢轴变形造成开启不顺滑。

无人书店

天猫未来店无人技术的加盟使志达书店成功升级成智慧型的未来书店。因为新技术的加入和需要，顾客改为从通识书店入口“扫脸”开始阅读体验，经过小展厅(由之前的接待室改造而来)，然后进入文史哲书店。如果开通了线上免密支付还可以随时、自由地带书离开。除了线上支付，书店仍为老人、孩子保留现金购物的渠道。

改造前的志达书店，2016

Zhida Bookstore before renovation, 2016

Under the same roof, two entrances extend like symmetrical butterfly wings. Two bookstores, taking 94m^2, feature in distinctive functions and ambiances. One bookstore is home for books on liberal arts. The books lie on shelves that form a deep hallway with seemingly endless turns, like those in a library. The other for books on general knowledge is designed as a common area. Readers are welcome to read, relax, and share. Here, readers of different interest may rest and encounter.

Guoquan Road

Yueyue/Zhida Bookstore is located on No.525 Guoquan Road. The west part of this lushly-shaded road is within the vicinity of Fudan University, surrounded by Fudan gymnasium, publishing offices, School of Economic and Management, student residential area, kindergarten, primary school and middle school affiliated to Fudan University.

Guoquan Road used to be famous for being a hub for humanities bookstores. As online bookstores got popular, they gradually perished, like those in other parts of the city. However, in recent years, they have reappeared in “bookstore+” mode pumped by upgraded consumption habits and polished lifestyles. In such an ordinary urban block as Guoquan Road, is it still possible to build a bookstore that is simple but with personality in both function and details?

Two Bookstores

The store owners run an online bookstore successfully as well. The physical bookstore, used to be called Zhida Bookstore, has stood there for over a decade. Though kept in existence, it has shrunk continually in size due to market competition. Before renovation, it mainly sold supplementary books for students. The design situation was limited in height and capacity, the floor plan had a twisted outline, and the kitchen upstairs was always leaking.

The couple both agreed to refurbish the bookstore but disagreed on its revived identity. The husband is attached to humanities, hoping to turn it into one for liberal arts. The wife is more pragmatic about market trend, hoping to turn it into a general bookstore, and the space can be used for chatting, family gathering and other activities. Neither would give in and both are well-reasoned. The designer thus proposed: why don't we build two bookstores then?

A general bookstore requires a bigger internal space for communicative activities. A bookstore on liberal arts can do with twisted and narrow hallways like those in a library. The key is how to coordinate and combine these two.

Metal Craft and Mechanical Design

For flexible and sustained use in other spaces in the future, the structure of the bookcases is designed with customized standardized steel framework. They are painted black and white based on the intended ambiences of the bookstores.

The doors at the entrance both weigh more than 150kg. To facilitate their opening, local engineers and workers specialized in metal processing have carefully studied the door structure. Two doors are connected to a vertical steel shaft, thus shaping the entrance like flipping book-pages that can open at the same time. The door shaft is designed with multi-layered bearings nested within each other to protect the steel shaft from deformation and thus avoid jagged opening.

Self-Service Bookstore

Unmanned technology provided by T-mall Future Store has successfully upgraded Zhida Bookstore into a smart future bookstore. At the entrance of the bookstore on liberal arts, customers enter via facial scanning system. They pass a small exhibit hall adapted from the former reception room and come to the world of liberal arts. If online quick payment is available to them, they can freely take a book, pay, and leave. Besides, cash check-out is still available for the elder and children.

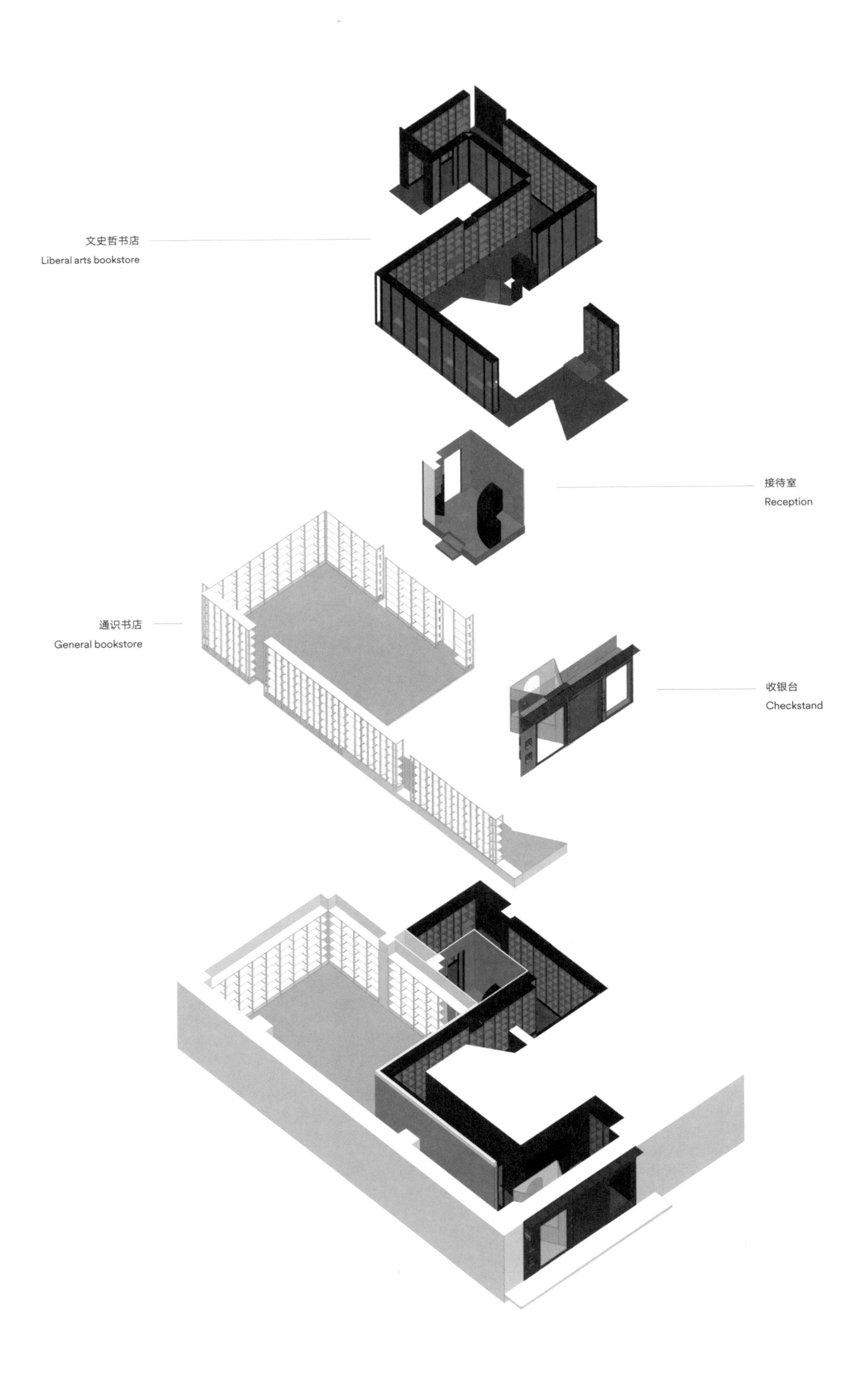
文史哲书店
Liberal arts bookstore
接待室
Reception
通识书店
General bookstore
收银台
Checkstand

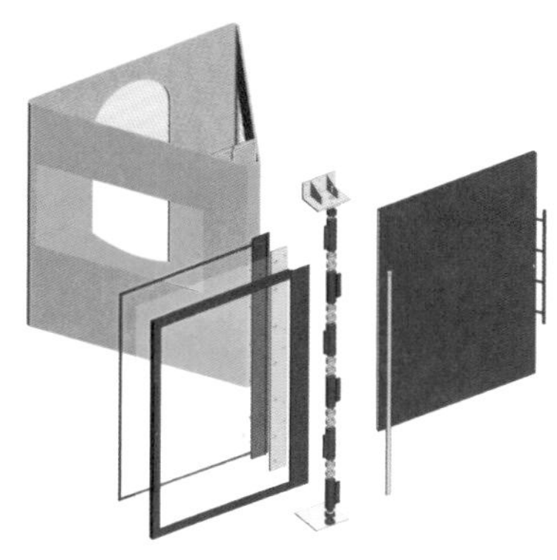

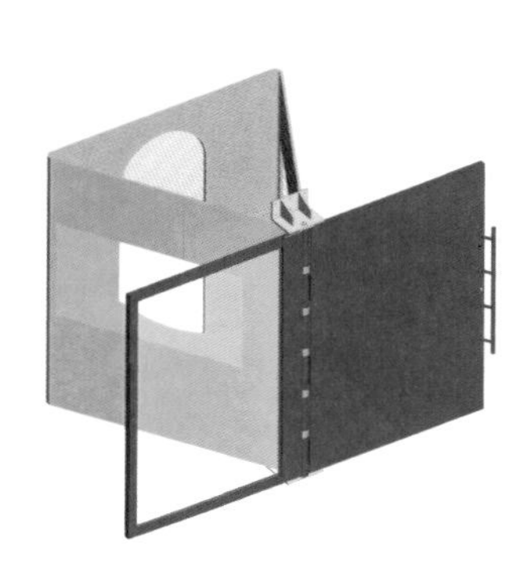

门与收银台组成的“分拣器”
"Sorting device" formed by the door and checkstand

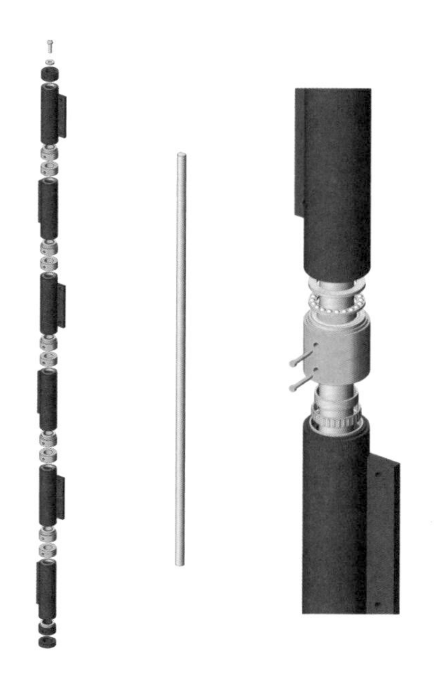

门轴分解示意
Gate shaft

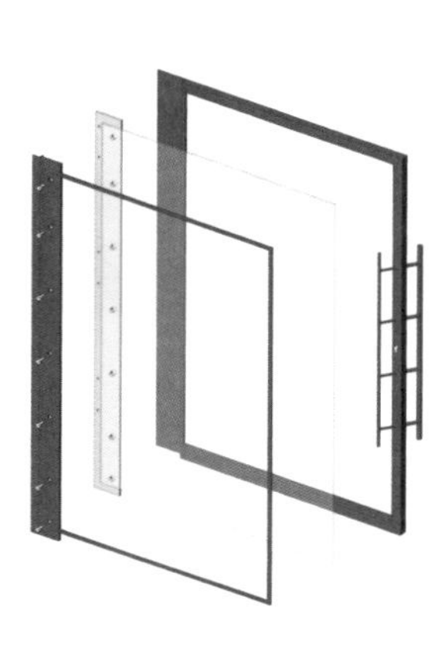

玻璃门分解示意
Glass door leaf

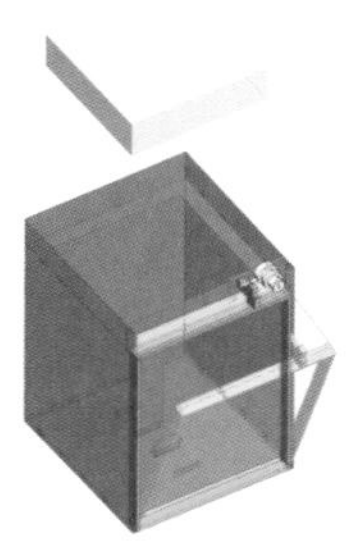

外立面展示窗口
Showcase in the facade

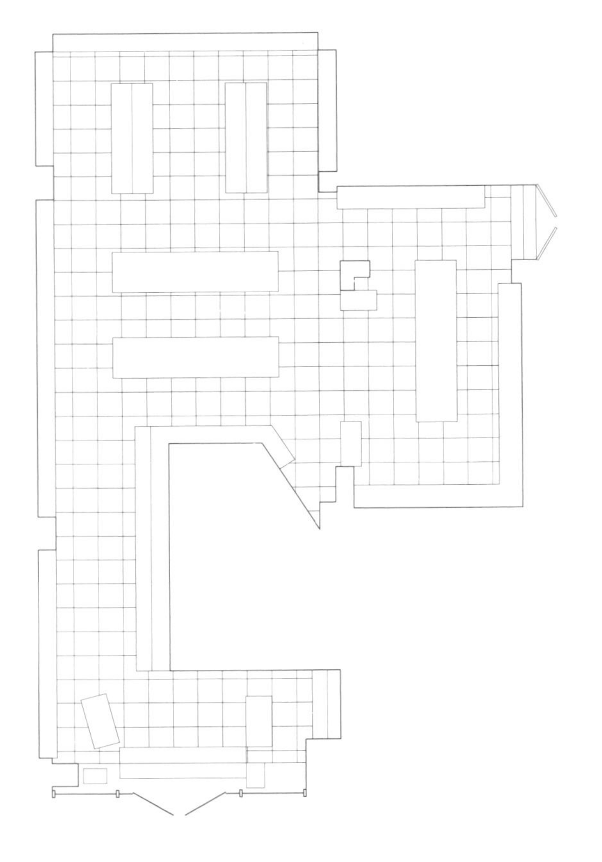

改造前的平面

Plan before renovation

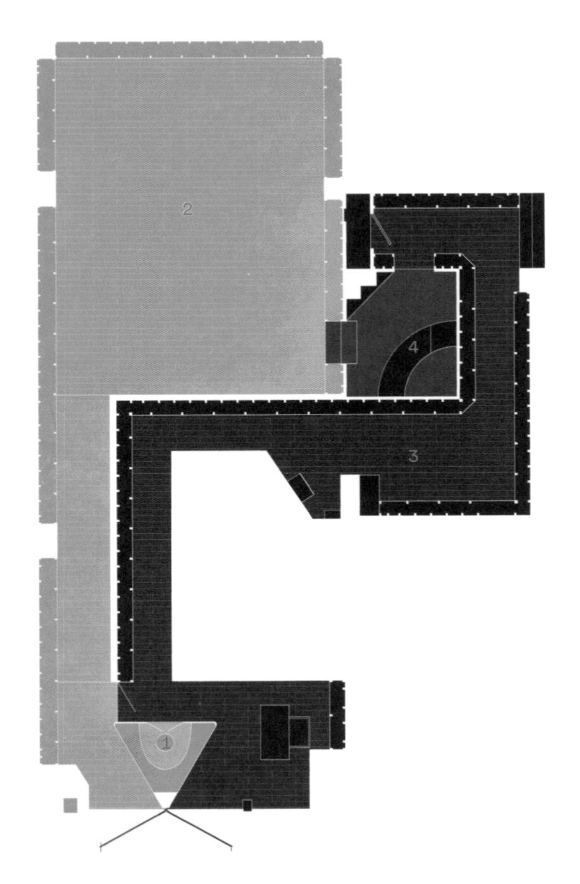

改造后的平面

Plan after renovation

1	收银台	Checkstand
2	通识书店	General bookstore
3	文史哲书店	Liberal arts bookstore
4	接待室	Reception

通识书店

General bookstore

文史哲书店
Liberal Arts bookstore

门轴现场放样

Door shaft on site lofting

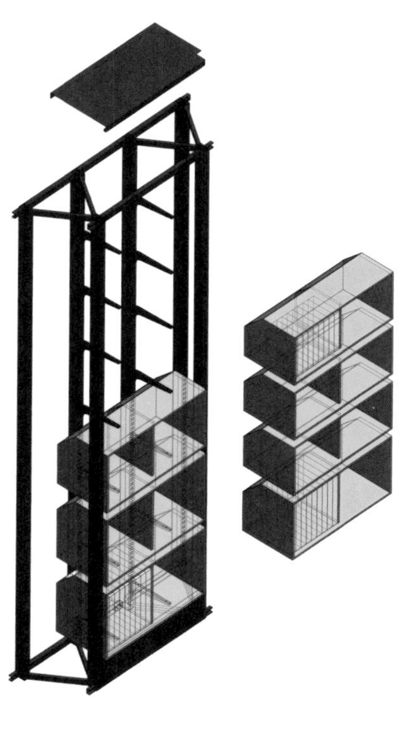

文史哲书店书架细节

Bookcases in the liberal arts bookstore

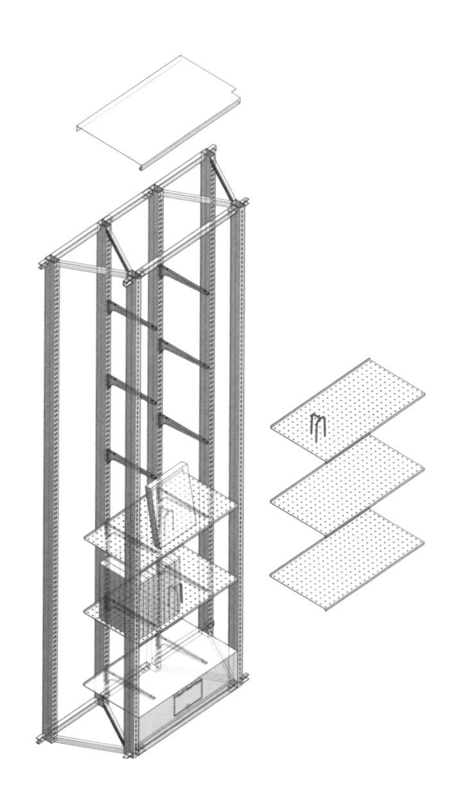

通识书店书架细节

Bookcases in the general bookstore

宝山贝贝佳欧莱幼儿园
Baoshan Beibeijia Olion Kindergarten

18

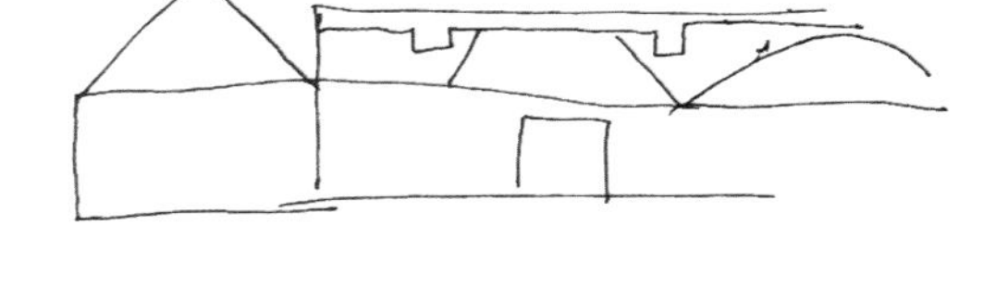

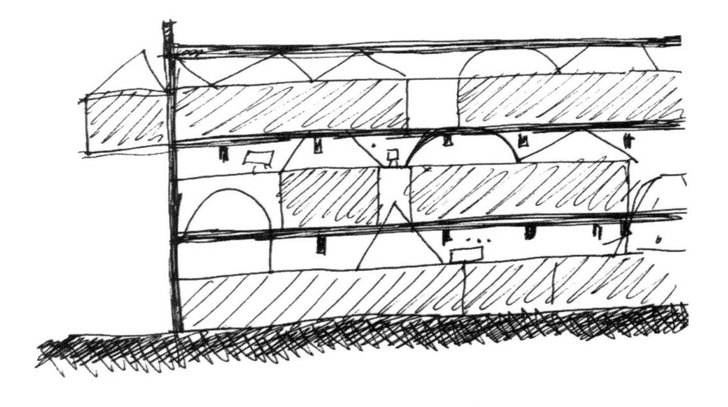

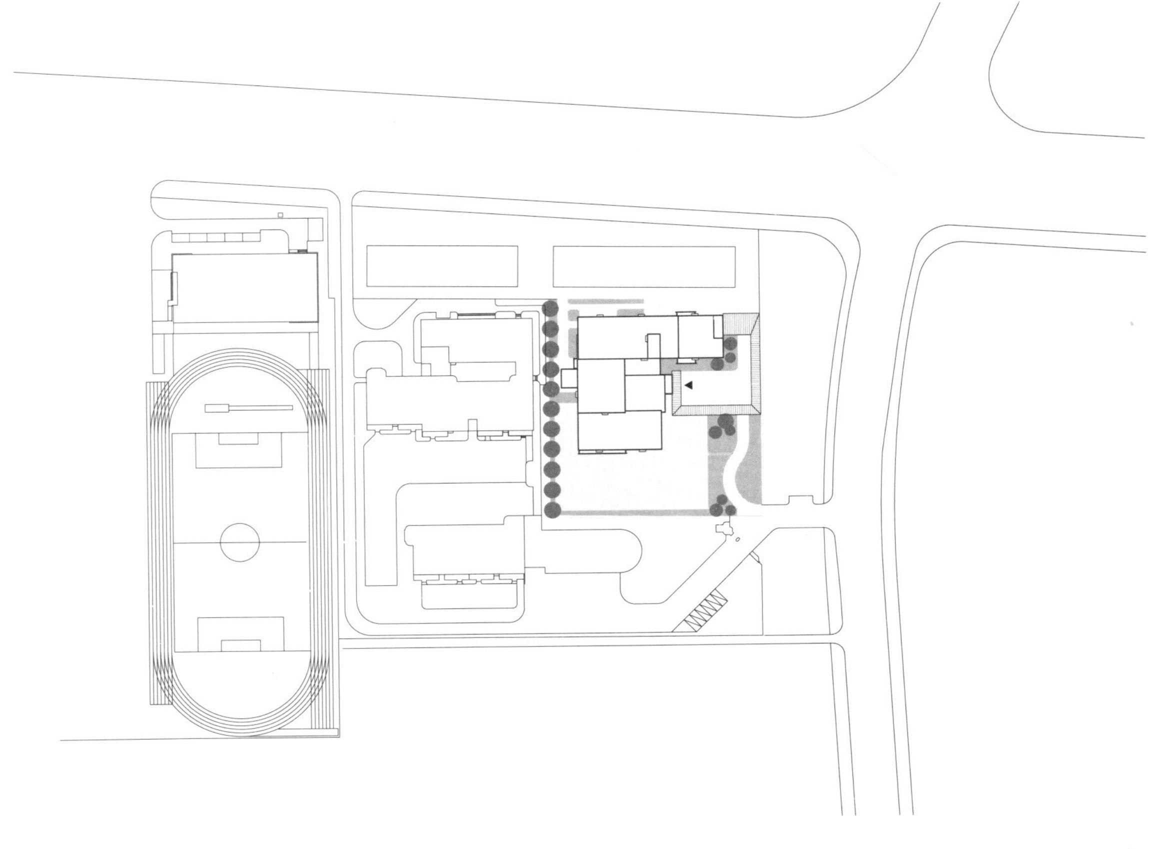

总平面图

General layout

欧莱幼儿园的改造对象是上海市宝山区宝菊路上一栋废弃的社区服务中心，房子非常普通，湮没在面无表情的城市里。本次改造需要采用一系列"调整"措施改变其空间结构，甚至可以通过激发既有空间未知的潜力从而创造一栋新的建筑。

剖面的潜力

在一些经典的改造项目中，建筑师往往采用塑造"间质空间"的策略，即通过将新建筑与旧建筑脱离，从而形成新、旧、新与旧之间的多样的空间体验的层次。对于非常局促的欧莱幼儿园来说，通过平面"建立"实际的间质空间在操作层面上存在困难，只能通过"调整"的方式"挤"出间质空间。这个问题迫使我们将注意力从平面转向剖面。从第二稿草图可以看出这个项目的决定性转变——我们试图设计一个包含室内"天窗"的吊顶体系：将吊顶的局部挖空并抬升至上层楼板，用光源或透光材料将其围合，形成室内局部净高接近层高的区域。即便并不明显，但"天窗"确实在局促的空间尺度下与原有结构拉开距离并强化了这种感觉，尤其在走道内部这种距离感可以清晰地被感知，这也赋予新分割出的功能空间以独立的特性。

天窗体系

在"天窗"体系中，如何平衡它的尺度是设计的关键。"天窗"的尺寸受限于三个因素：设备管线的布置、现状窗洞的位置、内部使用的逻辑。首先，"天窗"要尽可能地大，以争取更多净高舒适的区域；其次它要贴合使用的习惯，比如进门处和临近窗户的位置需要布置天窗，活动的区域天窗尺寸需要更大一些，班级之间需要天窗来打破隔断等，这就意味着要尽可能地压缩设备空间，并且设备管线的走向需要绕开"天窗"的布置。实际上，设计的主要工作都围绕着梳理管线展开，整个过程是让可使用的空间体积最大化。就像在装有石子的杯子里吹气球，气球会不断膨胀然后挤掉原本石子之间的空气，直到达到平衡。这一系列的操作，使得原本单纯的平层产生了不匀质的高低宽窄的变化，创造了新的空间体验。

尺度与颜色

幼儿园项目做到后期资金有限，我们甚至开始考虑将内隔墙留白，几经挣扎，最终还是坚持把局部的走道和洞口用木饰面来装饰。现场的效果很好，集中局促的资金去处理洞口这类面积不大的构造，着实是有效的。

在学校中学习的尺度往往是数字化、图解式的，幼儿园的施工让设计师产生了新的认知。吊顶 2.25 米的净高是在规范底线上游走，尤其是在施工过程中昏暗的光线、凌乱的骨架都会加重压抑的感觉，但最终隔墙吊顶安装涂刷齐整了，明朗的环境就让尺度适宜起来。此外，起初我们在临近窗口的位置试刷并确定涂料颜色，然而第一遍底涂结束，一些西向的教室，即便进深很浅也会在紧靠门的区域形成浓郁的色彩，让人觉得昏暗而逼仄。最终我们不得不整体调整颜色以改善入口区的尺度感。光线、质感这些具体环境最终组成了我们对于尺度的完整认知，这是通过观察建筑照片无法获得的。

改造前的建筑立面

Original building facade

改造前的室内

Original interior

Olion Kindergarten is renovated from a disused community service center on Baoju Road, Baoshan District, Shanghai. The design aims to change this banal structure with a series of adjustments and create a brand new building by exploring its unknown potential.

Exploring Potential from Section

It was very difficult to develop spaces on a plan due to the floor area restriction in the original building. Therefore, we have decided to try another way - "adjustment", shifting the focus from plan to section. A decisive change can be found in the second design draft - we have designed a suspended ceiling with interior "skylight" through hollowing part of the suspended ceiling and extending it to the upper floor. The whole ceiling is enclosed with light-wells or transparent materials, resulting in a space with a full story height under the ceiling. The "skylight" does achieve the effect that it is independent from the original structure and the interior space has been extended to a full story height. The expansion effect is even more explicit in the aisle. Moreover, this "skylight" has successfully made the new enclosed functional space more independent.

Skylight System

It is very important to determine the scale of this "skylight", though the work is limited to arranging pipelines, determining position where the "skylight" should be opened and finding the logic of inner usage. Firstly, the "skylight" should be opened as large as possible, because the bigger it is, the more spacious this place seems. Secondly, it had to fit in with the user's habits. For example, "skylights" are designed to be open at the entrance and near the window; "skylights" over the activity space should be larger; with "skylights" we can break the separation between classrooms. For the reasons mentioned above, it is necessary to narrow the equipment space and lay the pipelines away from the "skylight". In fact, this design is almost all about efficiently distributing pipelines to obtain maximum functional space. A series of operations bring in heterogeneous space with varied width and height for this originally flat floor, leading to new spatial experience through the changes in scale.

Scale and Color

With a limited budget, we hesitate whether we should leave the interior partitions undecorated. At last, we decide to decorate corridors and windows with wood veneer. Then, we realize that it is an absolutely excellent decision. Decorating structures like doors or windows is an effective way to make a better value from a limited budget.

We have received the knowledge of scale from numbers or graphics in college. In the process of construction, the architect has studied architectural scale from a fresh perspective. Considering construction standards, the clear height of the suspended ceiling is designed 2.25m. People might feel depressed in such a construction site with dim light and messy keel. However, when the suspended ceilings are completed and painted, the whole space looks so bright and clear. It indicates that in this project the architectural scale has been well controlled. In the beginning, we do a test on painting in the area close to windows and try to craft a proper color scheme. When the kindergarten is painted for the first time, we find that though the throat of some west-facing classrooms is short, the painting close to the door appears too strong, which makes the place dusk and narrow. So we can't do better than adopting a new color scheme to solve the problem. From what we have experienced in this project, we may finally draw the conclusion that as a spatial environment, light and texture are factors which may have direct impact on the architectural scale. This is what architectural photography won't tell us.

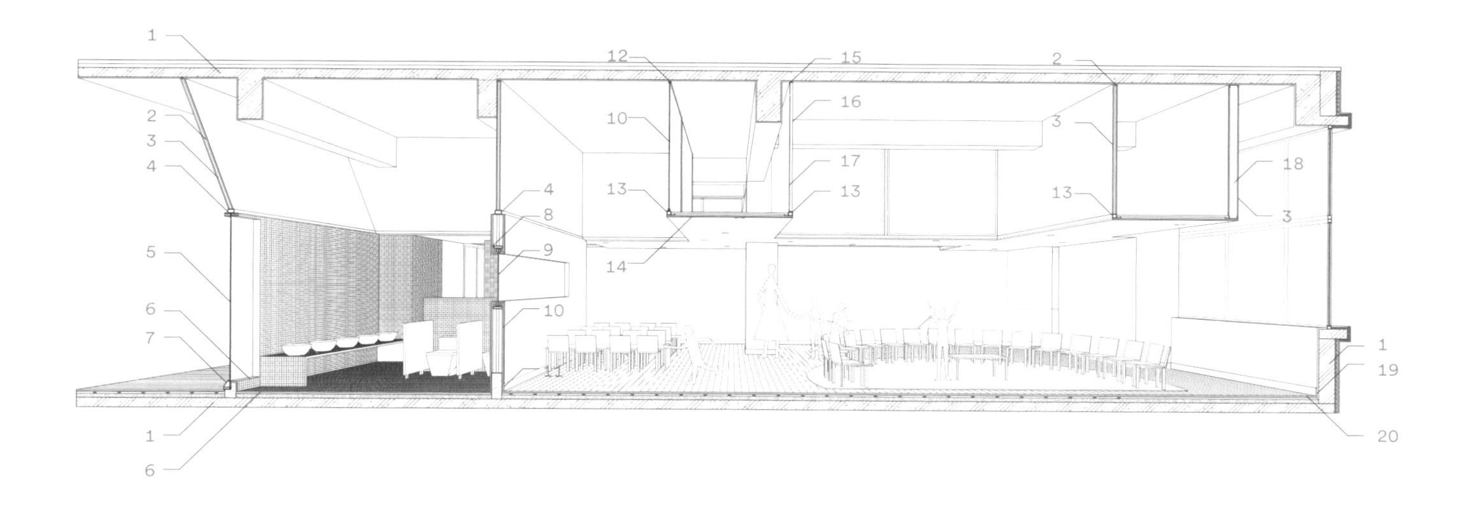

建筑剖透视
Perspective section

1	原有结构	Original structure
2	20/20 方钢管	20/20 square steel profile
3	12 厚耐火纸面石膏板	12mm refractory gypsum board
4	80/50 方钢管	80/50 square steel profile
5	F1.0 防火玻璃	F1.0 fireproof glass
6	马赛克	Mosaic
7	不锈钢扣板，木纹贴面	Stainless steel gusset plate, wood grain veneer
8	9 厚水泥纤维板	9mm cement fiberboard
9	6+6 钢化夹胶玻璃	6 + 6 toughened laminated glass
10	12+12 纸面石膏板	12 + 12 gypsum board
11	暗装踢脚线	Hidden skirting line
12	35/30 不锈钢角铁	35/30 stainless steel angle iron
13	50/50 方钢管	50/50 steel square profile
14	9.5 + 12 纸面石膏板吊顶	9.5 + 12 gypsum board ceiling
15	40 厚软膜天花角码	40mm soft film ceiling corner code
16	A 级软膜	A soft film
17	1.5 厚铝背板	1.5mm thick aluminum back plate
18	轻钢龙骨	Light steel keel
19	明装踢脚线	Exposed skirting line
20	实木复合地板	Solid wood composite floor

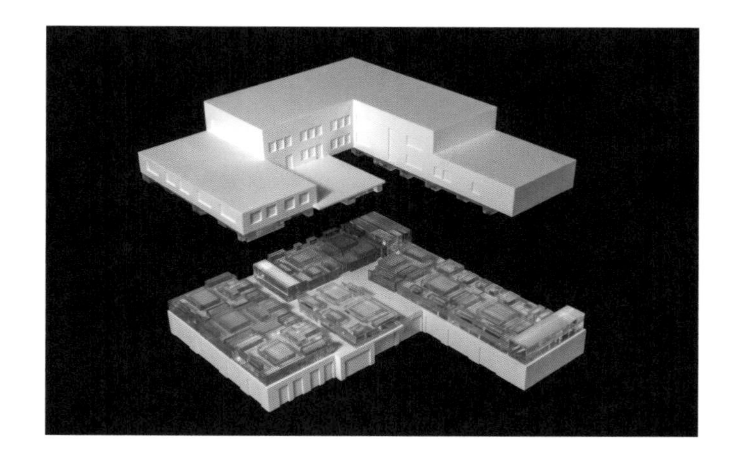

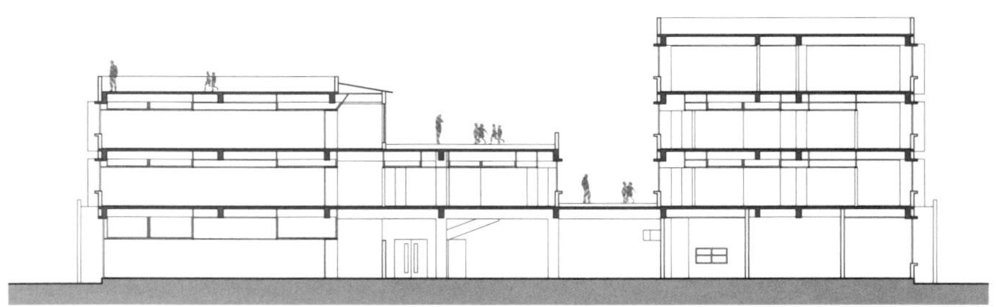

1–1剖面图

1–1 section

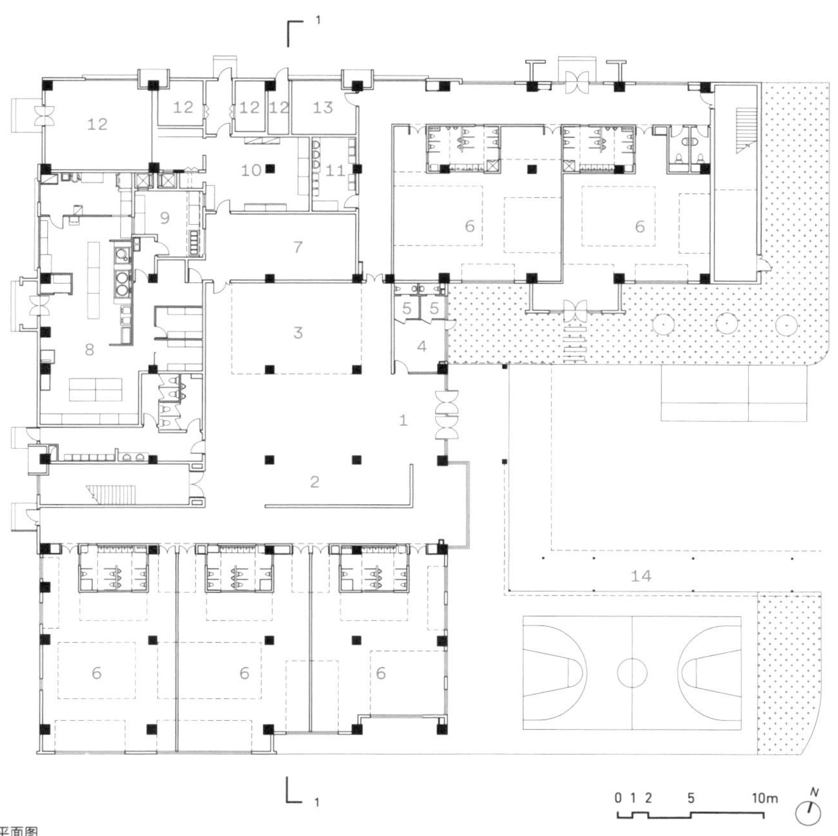

首层平面图

First floor plan

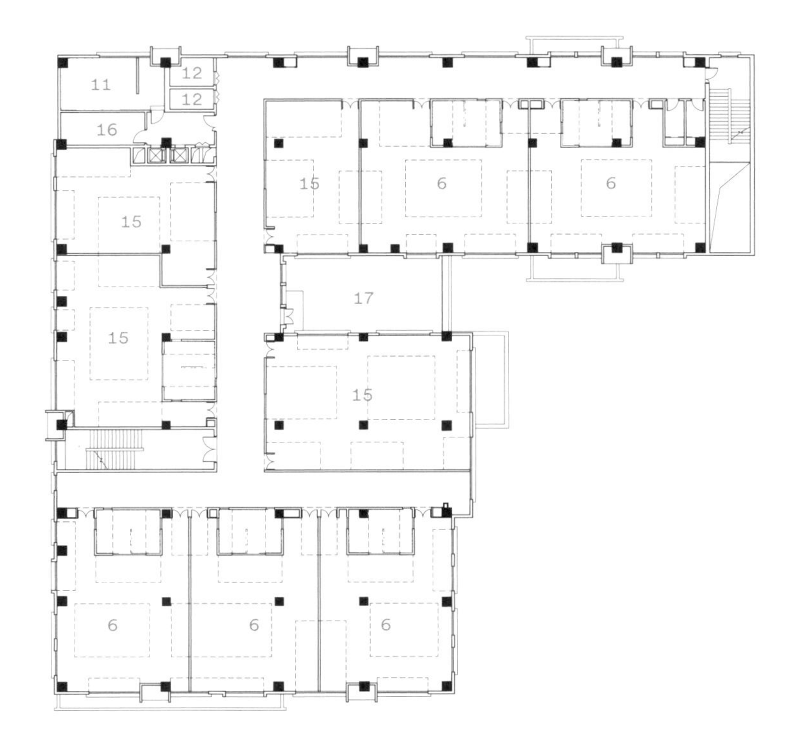

二层平面图

Second floor plan

1	门厅	Lobby
2	展览	Exhibition
3	阅览室	Reading area
4	晨检	Physical examination
5	医务室	Medical room
6	幼儿班级	Classroom
7	教工餐厅	Staff restaurant
8	厨房	Kitchen
9	配餐间	Pantry
10	洗碗间	Scullery
11	消毒间	Disinfect room
12	设备间	Device room
13	接待	Reception
14	风雨走廊	Corridor
15	活动室	Activity room
16	工具间	Tool room
17	室外活动场地	Outdoor playground

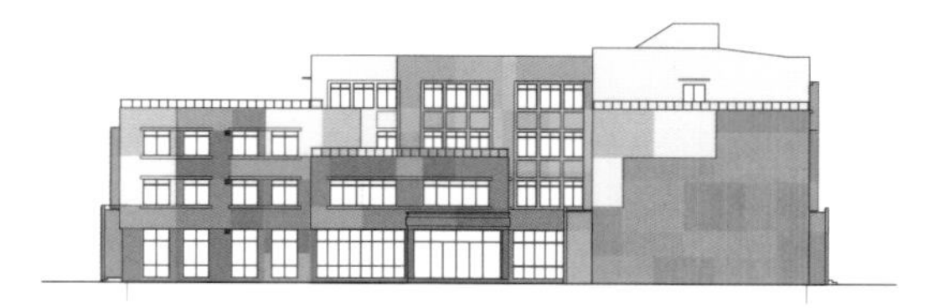

棉仓城市客厅
Cotton Lab Urban Lounge

19

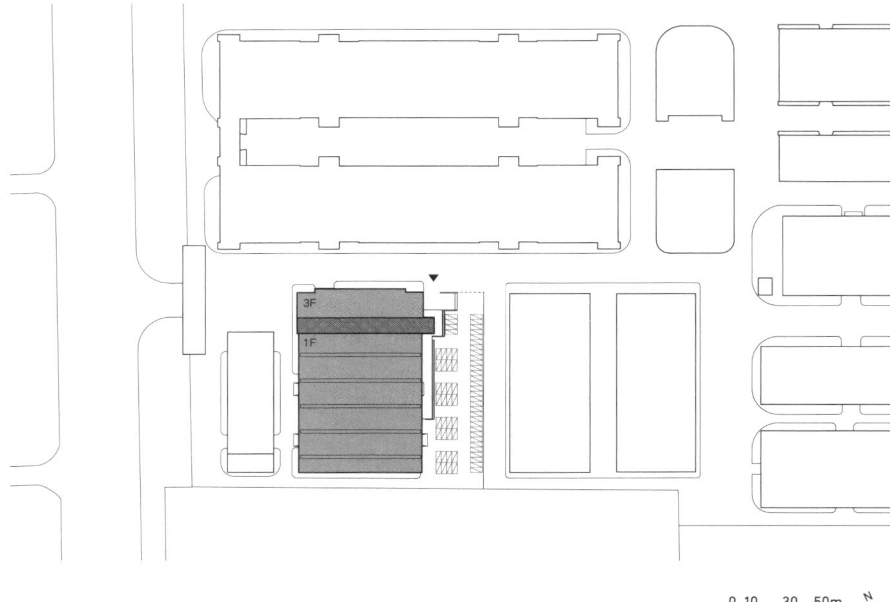

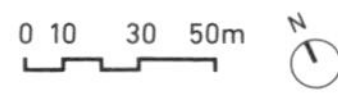

改造前的室内
Original interior

改造后的采光顶
Renovated lighting roof

新零售体验店

阿科米星设计的棉仓城市客厅位于江苏省常州市的一个科技园区里，原来是一处标准的工业厂房。业主在网上平台天猫商城经营着一家成功的服装店，2017 年他们计划创立一家融合服装零售、餐饮消费体验的生活美学店，并选了这个远离城市中心的地方打造新零售业态的线下品牌——棉仓。

在新的线上线下融合的商业模式中，虚拟空间中的网店与实体空间中的新零售体验店，相互成为展示性的“橱窗”和进行体验、消费的场所。这种新的“内外”关系正在成为当下中国建筑普遍存在的状况。

屋中屋

接受委托后，我们面临三大难题：第一，大尺度的厂房内部空间与零售、餐饮业的小尺度消费空间之间存在差异，如何协调与联系？第二，如何控制造价，创造吸引人的新空间？第三，原厂房单薄耗能，如何经济有效地实现新空间的环境舒适度？最后我们选择了“屋中屋”的设计策略。即在主体厂房内部建造完整的新形式的独立建筑物来容纳两个主要功能。

平行布置的两条南北走向、高大笔直的“屋中屋”透明舱体，它们的外轮廓一模一样，均采用尖顶双坡的标准断面，但构造方式完全不同。一条为钢结构的服装成衣舱体，通体白色，空灵富有韵律，其时尚感正好适合服装的展示。

另一条为钢木结构的餐饮空间和多功能活动空间，建筑师赋予这一驻留空间以温暖的调性，折线形的木杆重重叠叠地在头顶形成一个富有亲近感，且有一定分量感的覆盖。

主体厂房采用自然通风充当使用空间的第一道防护，形成初步的环境，两个新增构筑物采用新风系统形成各自定制的独立、全封闭的空调环境。设备系统的布置方式，不仅解决了功能问题，而且最终在空间中呈现出一套与结构完全结合的、完整的视觉造型体系。

建筑的入口大厅，既直接与外部空间相连通，也连接着整个“棉仓”的其他场所：餐厅、服装零售和室内活动空间。

内部中的内部

新增的两个舱体构筑物没有占据整个厂房空间，舱体之间有意留出一处带状空地。在这里放置了很多供儿童游戏，供大家运动和休憩的设施。大家在放松活动的时候往往会觉得自己置身于城市街道、广场。由于创造了“屋中屋”这种“内部中的内部”空间，使大家在厂房“内部”体验到了一种“温和的外部”。

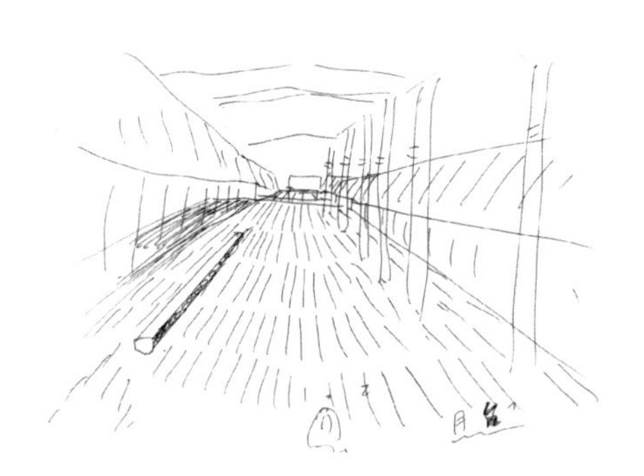

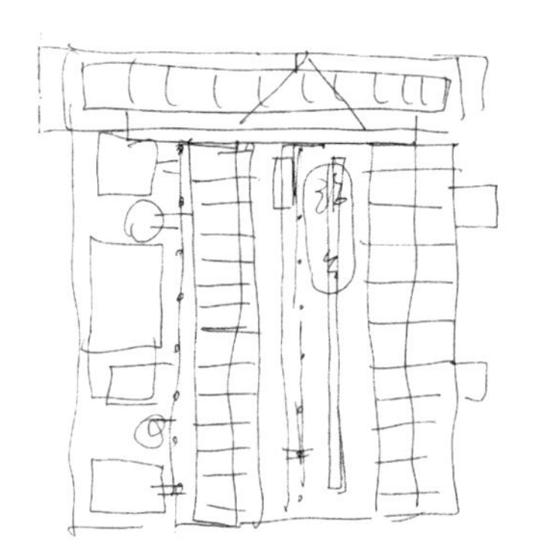

New Retail Experience Store

The Cotton Lab Urban Lounge is located in a technology park in Changzhou City, Jiangsu Province. It was a standard industrial plant. The owners have been operating a successful clothes store on Tmall. In 2017 they designed and created a store focusing on life aesthetics by integrating fashion retailing and catering consumption. They selected a place far away from the downtown to build a new retailing offline brand—“Cotton Lab”.

In this online/offline combined new commercial model, online stores and “new retail experience stores” are serving as display “windows” as well as experience and consumption places. This new “internal and external” relationship is prevailing in Chinese contemporary built environment.

House in House

There are three crucial challenges. The first is how to coordinate and bridge the gap between large inner space in a factory and small consumption space in catering and retail. The second is how to create a new and attractive space at a low cost. The third is how to economically make the new space comfortable when the original plant is simple and energy intensive. Finally, we pick a design called “House in House”, aiming to realize two major functions by establishing a complete independent new buildings inside the major plant.

We erect two “Houses in House”, two high, straight and transparent chambers in north-south direction. Although their appearance looks the same as the designer adopts a standard cross-section of arch shape with spire, they have a quite different structure. The chamber for clothing is of steel structure , entirely white, and appears so charming and fashionable for clothing display.

The other chamber of timber-steel structure consists of catering space and multifunctional activity space. The architect leaves many comfortable details in this space. With the folding line-shaped wooden poles hanging over head, people might find this chamber an intimate place with strong power.

The major plant applies natural ventilation as its first defense in order to create an initial environment. Meanwhile, in these two newly-built chambers, the designer has created a customized, independent and fully closed air-conditioning environment by the fresh air system. Other than this function, this system is visually and fully integrated with the structure.

The entrance hall connects directly to outer space as well as other sites of the “Cotton Lab”, including restaurant, fashion retailing and indoor activity space.

Interior in Interior

These newly-added chambers don't occupy the whole plant, because the designer intentionally leaves an open space between them. There are a lot of game facilities, sports facilities and lounges in this open space. People may feel wandering in urban streets or squares when they take leisure time there. Thanks to this design of “House in House”, space of “Interior in Interior”, people are able to experience “mild exterior” when they go into the plant.

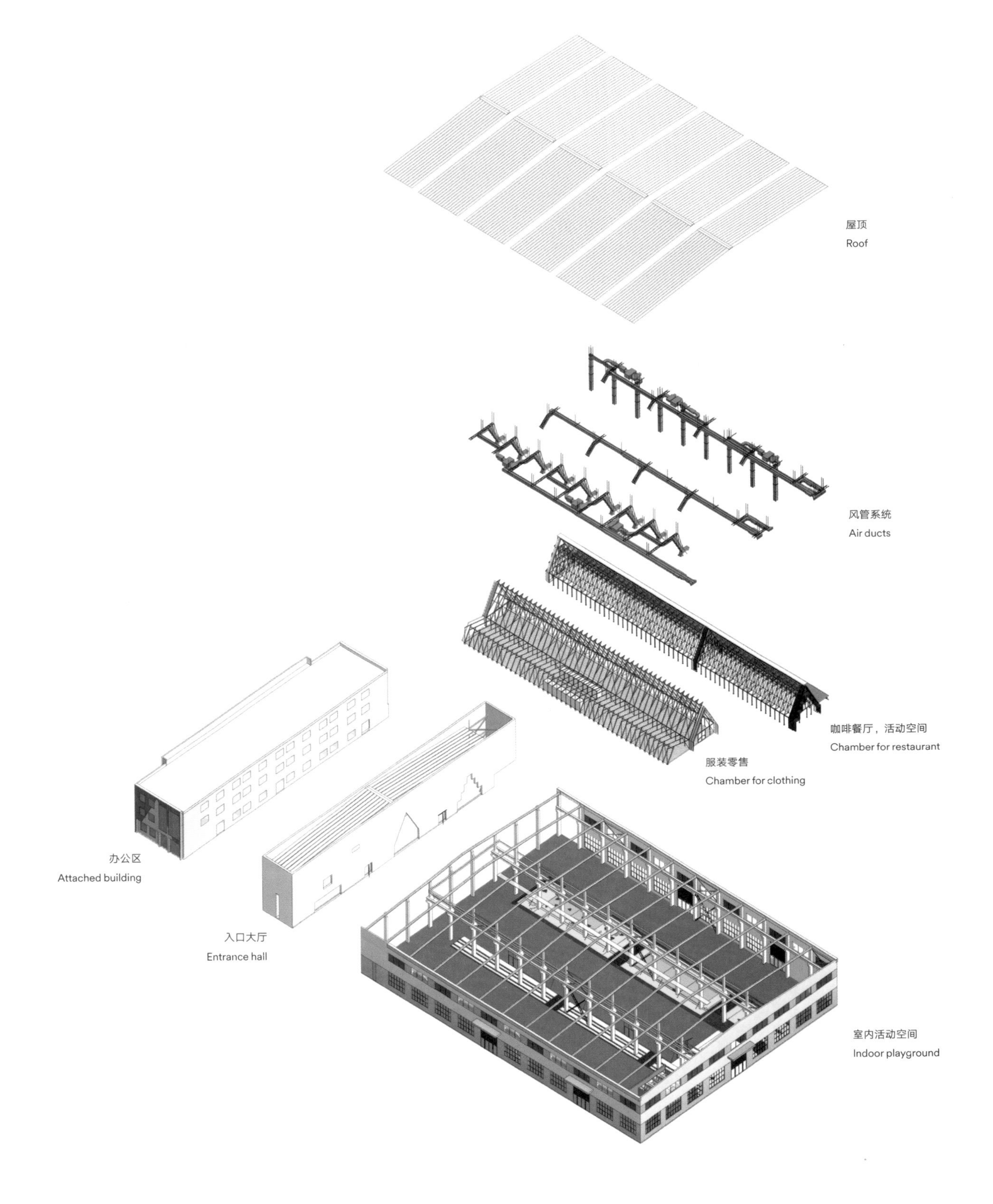
屋顶
Roof
风管系统
Air ducts
咖啡餐厅，活动空间
Chamber for restaurant
服装零售
Chamber for clothing
办公区
Attached building
入口大厅
Entrance hall
室内活动空间
Indoor playground

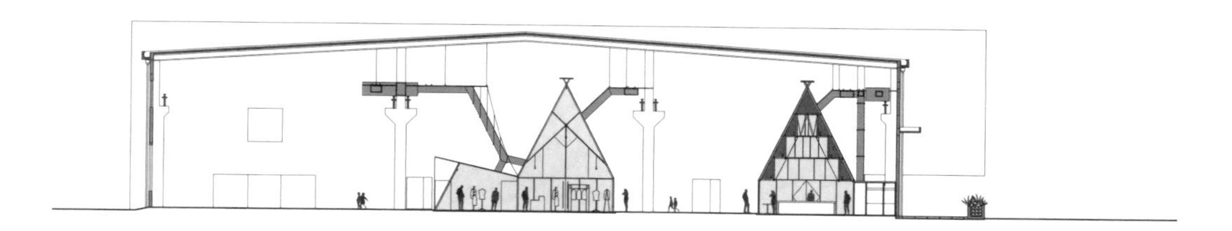

1-1 剖面图
1-1 section

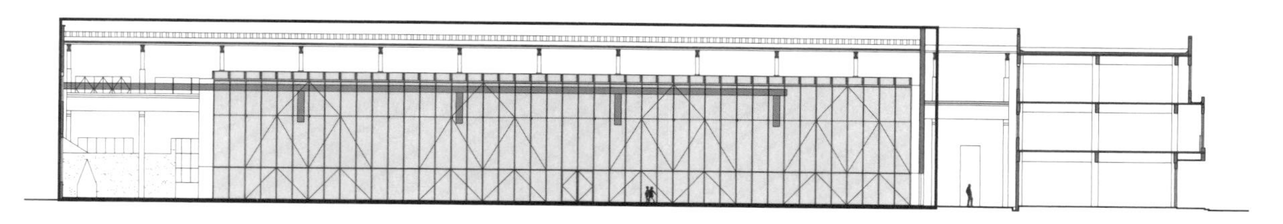

2-2 剖面图
2-2 section

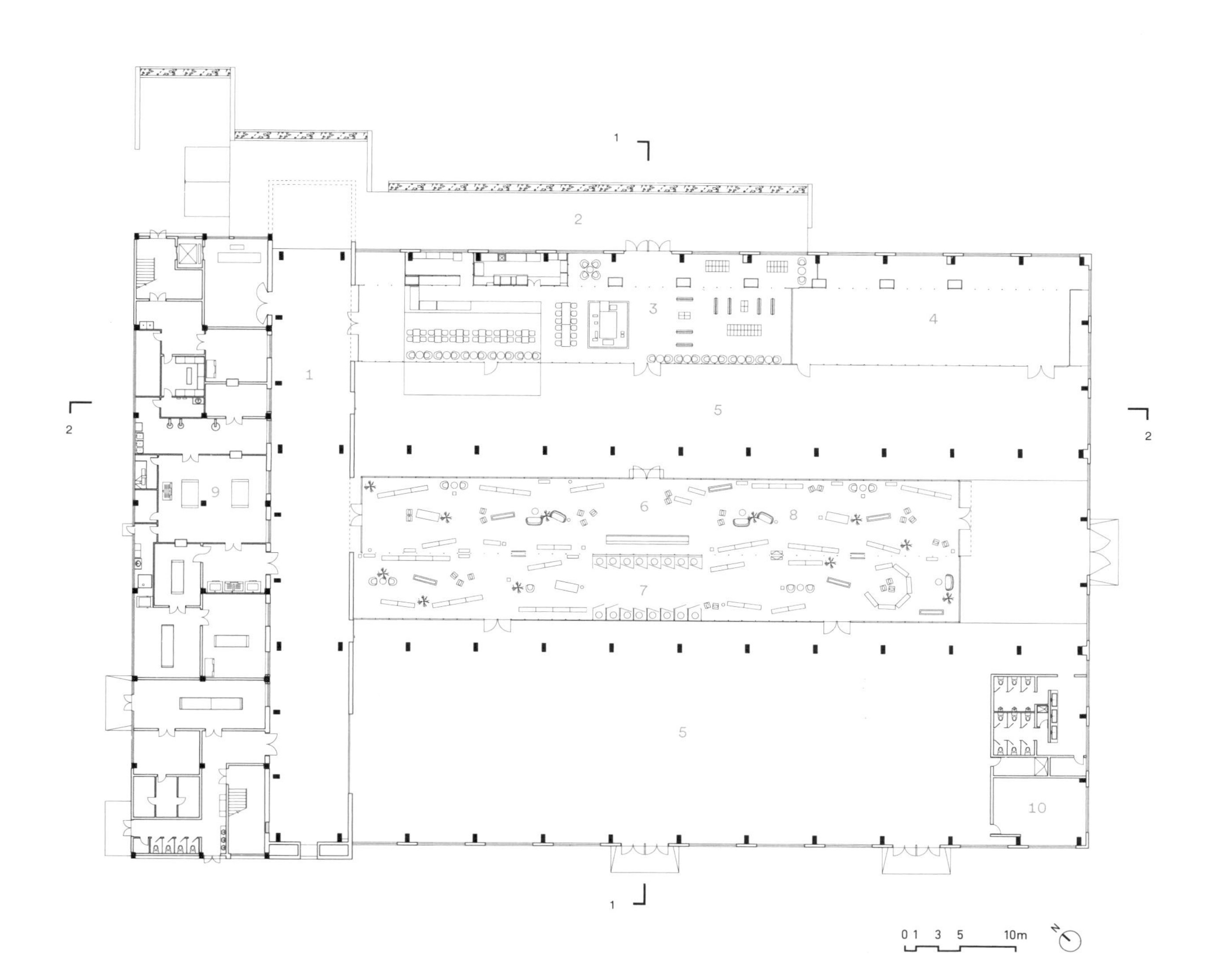

首层平面图

First floor plan

1	大天井	Entrance hall
2	户外场地	Outdoor venue
3	“初饭”餐厅	Chufan restaurant
4	活动室	Activity room
5	室内活动场地	Indoor playground
6	收银	Checkstand
7	试衣间	Fitting room
8	成衣店	Ready-to-wear clothes
9	烘培工作间	Baking workshop
10	设备	Device room

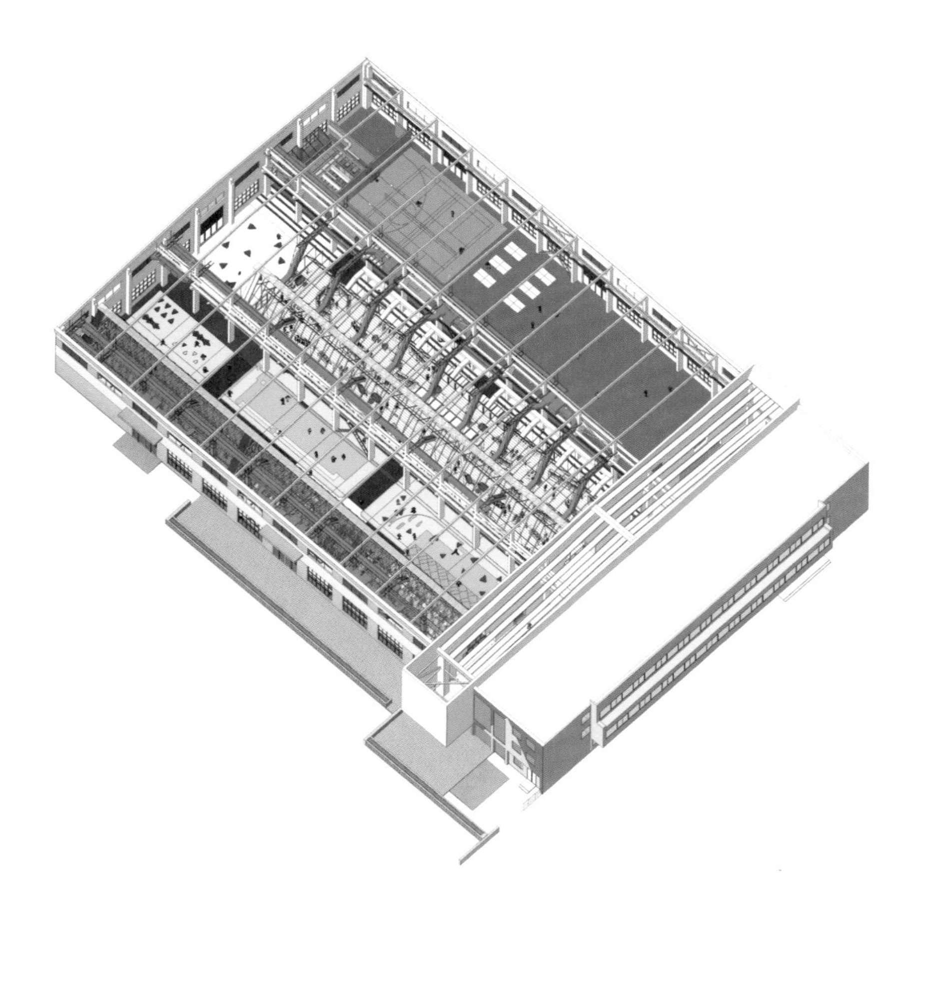

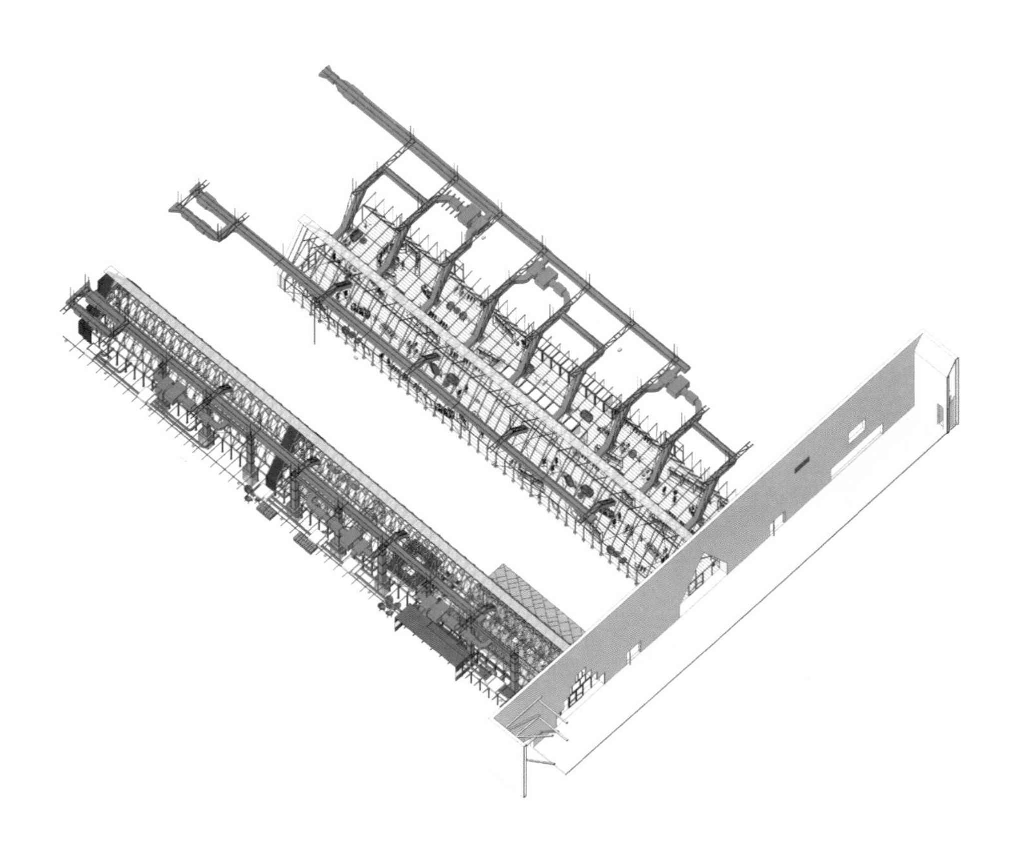

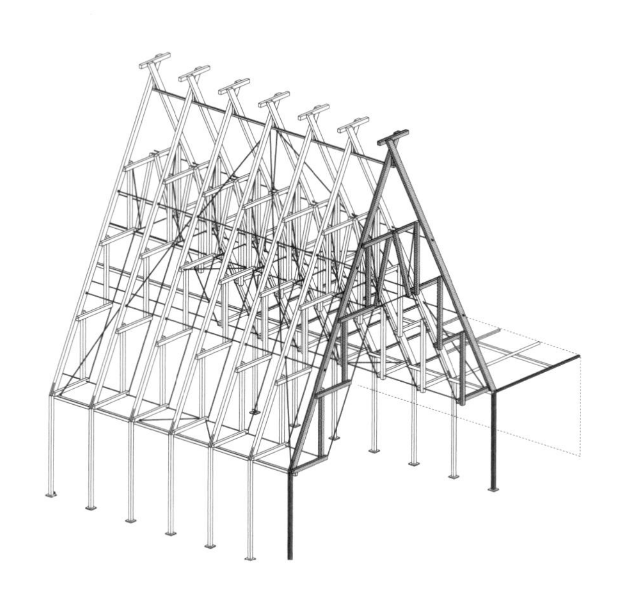

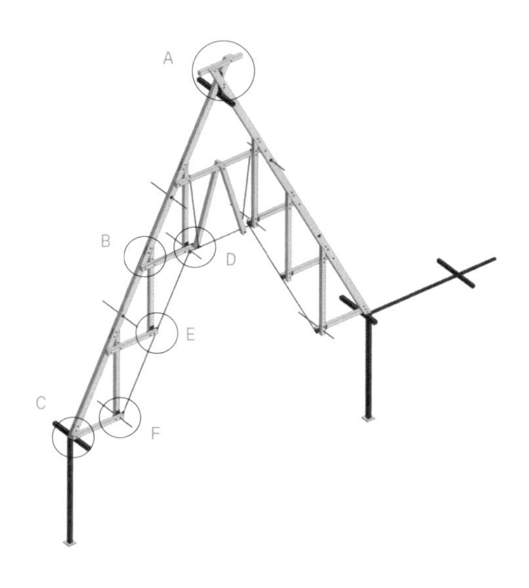

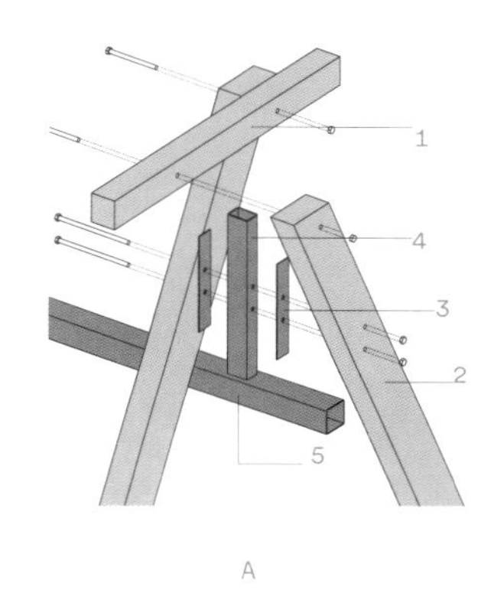

A

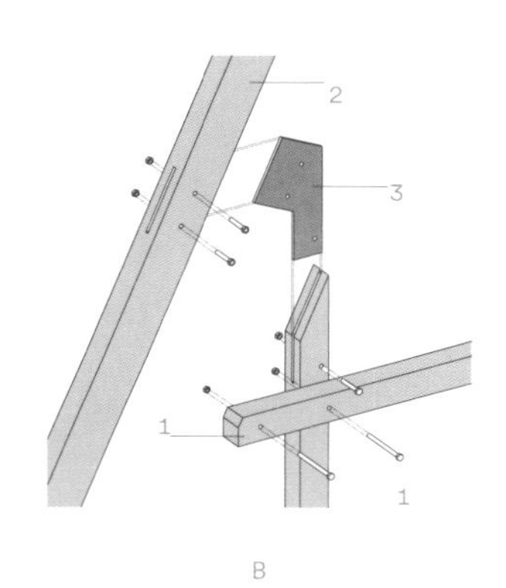

B

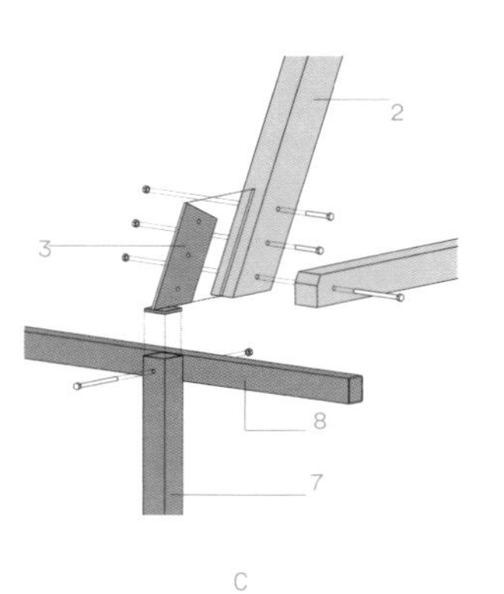

C

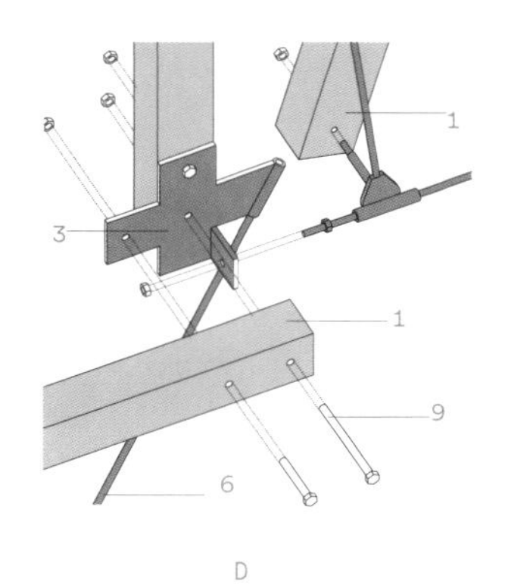

D

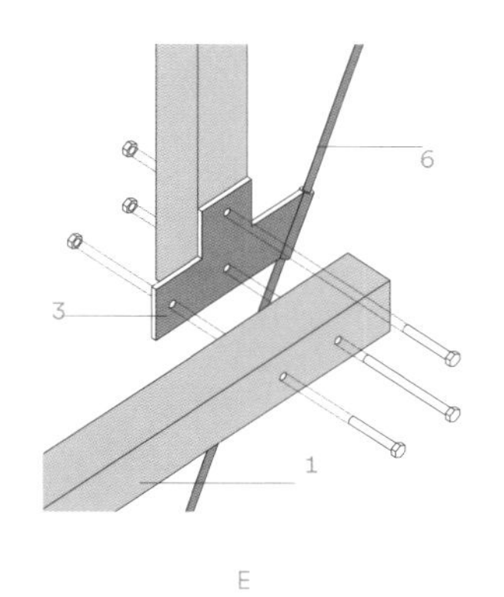

E

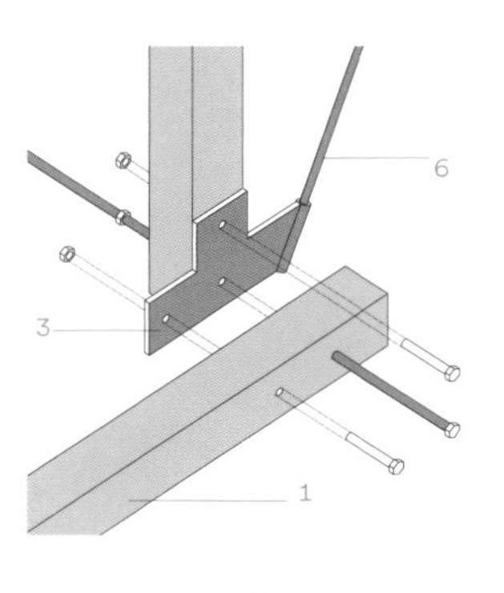

F

"初饭"舱体结构构造

Chamber structure for Chufan restaurant

1	90/76 木枋	90/76 timber tie
2	140/76 斜撑	140/76 oblique tie
3	金属连接件	Steel joint
4	50/76/4 钢立梃	50/76/4 steel pole
5	50/50/4 方钢	50/50/4 steel beam
6	φ16 钢索	φ16 cable
7	80/80/4 钢柱	80/80/4 steel column
8	80/80/4 方钢	80/80/4 steel beam
9	螺栓	Bolt

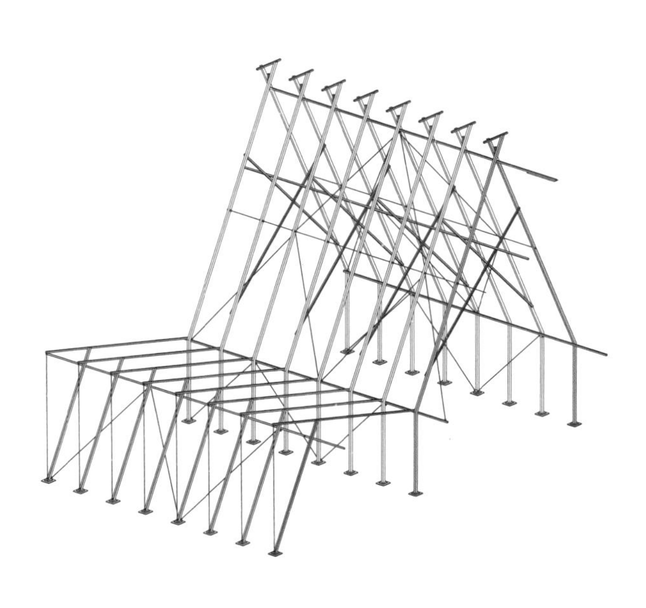

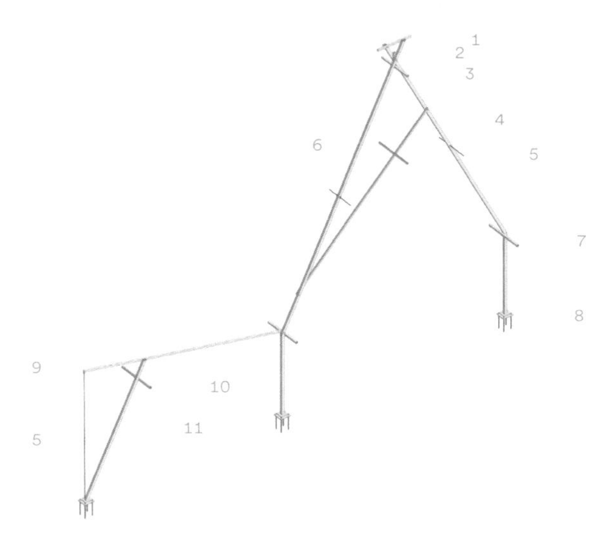

服装舱体结构构造
Chamber structure for clothing

1	50/50/4 方钢，长 1000	50/50/4, length 1000mm steel beam
2	50/50/4 方钢，长 400	50/50/4, length 400mm steel beam
3	50/50/4 方钢	50/50/4 steel beam
4	60/80/4 方钢	60/80/4 steel beam
5	φ10 钢索	φ10 cable
6	50/50/4 钢梁	50/50/4 steel beam
7	50/50/4 方钢	50/50/4 steel beam
8	20/320/340 基础（预埋钢板）	20/320/340 foundation (embedded steel plate)
9	50/50/4 方钢	50/50/4 steel beam
10	50/50/4 方钢	50/50/4 steel beam
11	60/80/4 方钢	60/80/4 steel beam

智慧欧莱托育园
Wisdom Olion Nursery

20

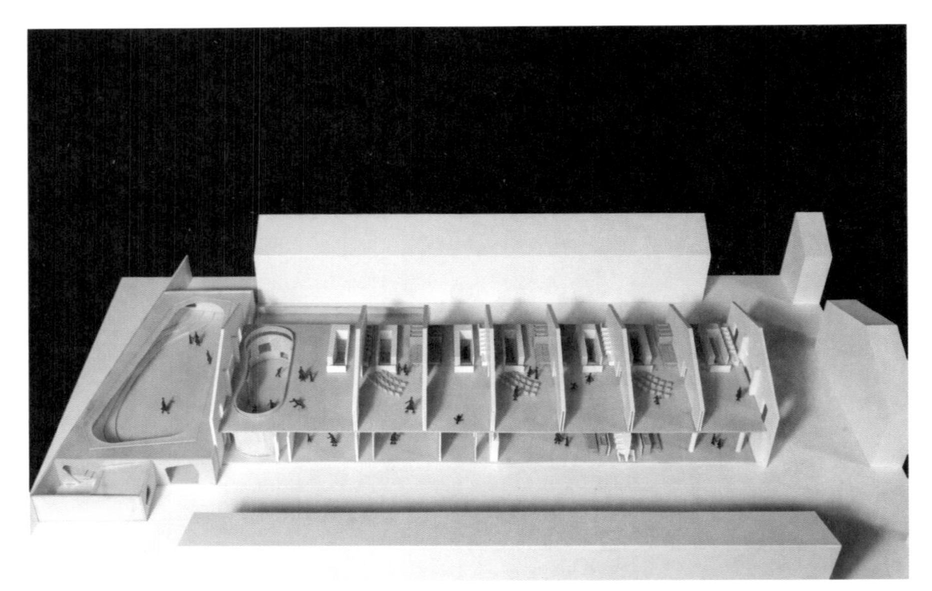

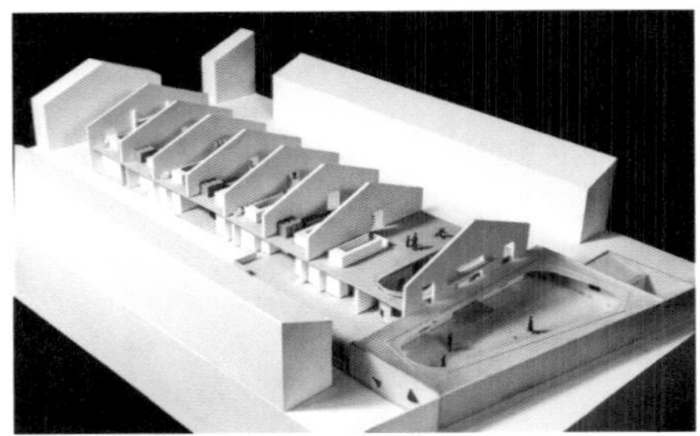

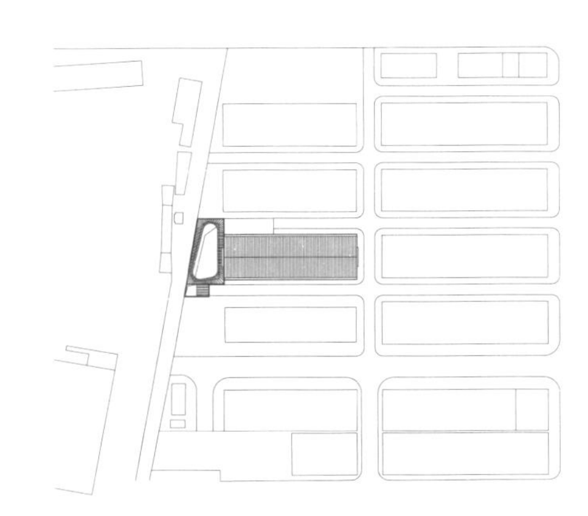

改造前的室内
Original interior

智慧湾幼儿园是由一座位于园区内的长长的普通厂房改建而成的。幼儿园在使用上希望实用的房间多多益善，同时对于造价的控制十分严格。设计这个幼儿园，面临的最大问题是用地范围内基本没有户外的活动空间。如何能够获得更多教室空间，并保证使用空间具有特色？是否能够解决活动场地的问题？这两个看似无关的问题或矛盾被看作设计的契入点。

我们采用的方式是把厂房原来的高空间隔成两层，把单个教室设计为南北通长的条状，密排在上层，每个教室从底层楼梯进入，以取消二层公共走廊，从而使上层的空间变成简洁的并置模式。相互并置的各教室之间的墙体开大洞口，设置大移门，用打开与关闭来调整空间，形成不同的使用空间模式。当所有的移门同时打开时，小朋友们就能够获得一个接近外部操场尺度大小的室内大活动空间，他们能够在这里进行游戏、跑步、追逐、骑车等各种活动；当两两或者数个教室相互连通起来时，就能扩大教室，或者形成剧场和舞台模式，来举办小戏剧演出等。之前的两个主要问题，以这样的方式获得了解决，并得到了有趣的空间。

Wisdom Olion Nursery is transformed from a long factory building. The client asks for many classrooms, but the budget is tight. In addition, there is barely any field around the building fit for the children to do outdoor activities. How to provide classrooms as many as possible without losing their quality? How to define an outside space for physical exercises? Our design results from these two seemingly irrelevant questions.

The solution is to divide each story height into two levels. Each classroom occupies one big bay from south to north, and all the classrooms are sitting side by side on the upper floor, each entering through an independent staircase from the ground floor. The corridor on the upper floor is saved and the room number has been doubled. Between two classrooms, there is a large sliding door that can change the whole structure of space through its opening and closing. When one door is open, two adjacent rooms then merge into a larger one to provide space for play and stage performance. When all doors are open, all rooms will be connected to form a big space, literally an interior playground for the kids to play, run or even ride. The flexibility of its interior structure enables the project to go beyond its limits and achieve interesting spaces.

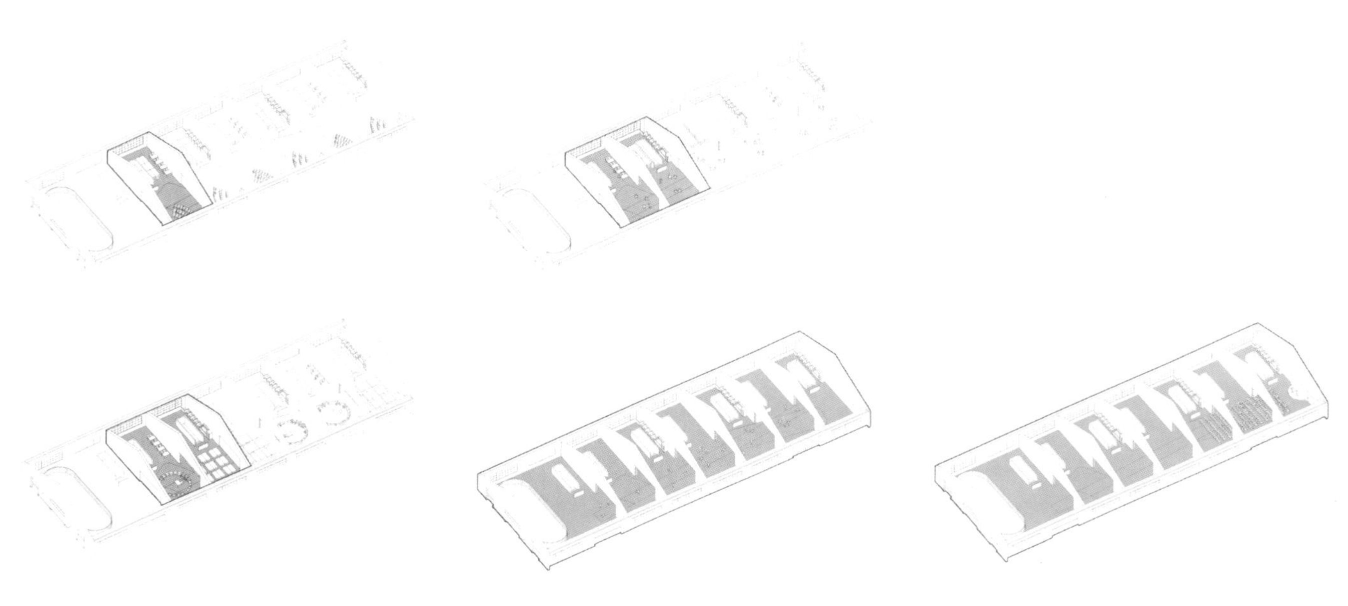

二层活动单元的组合使用分析

Analysis of the combined use of units on the second floor

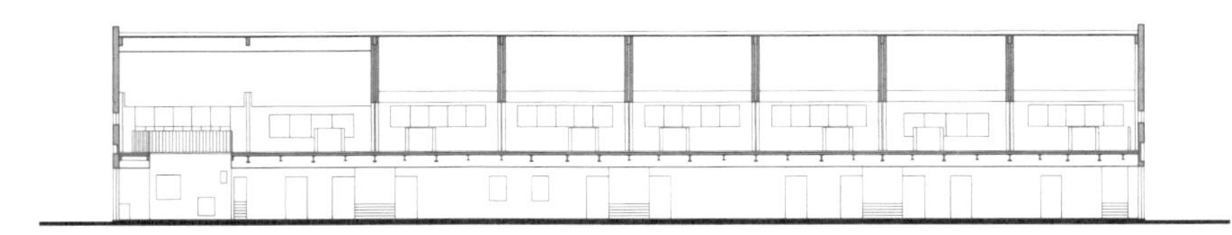

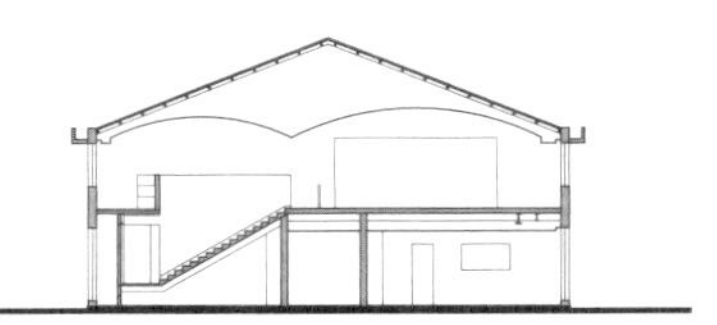

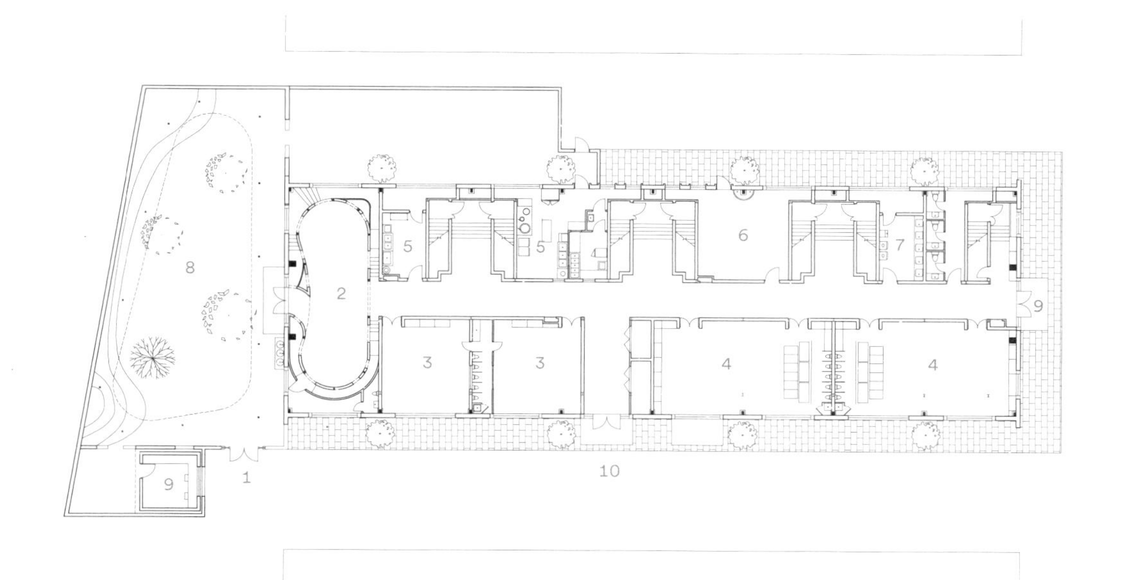

首层平面图
First floor plan

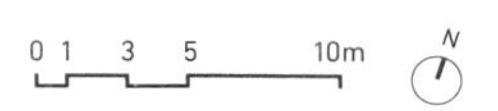

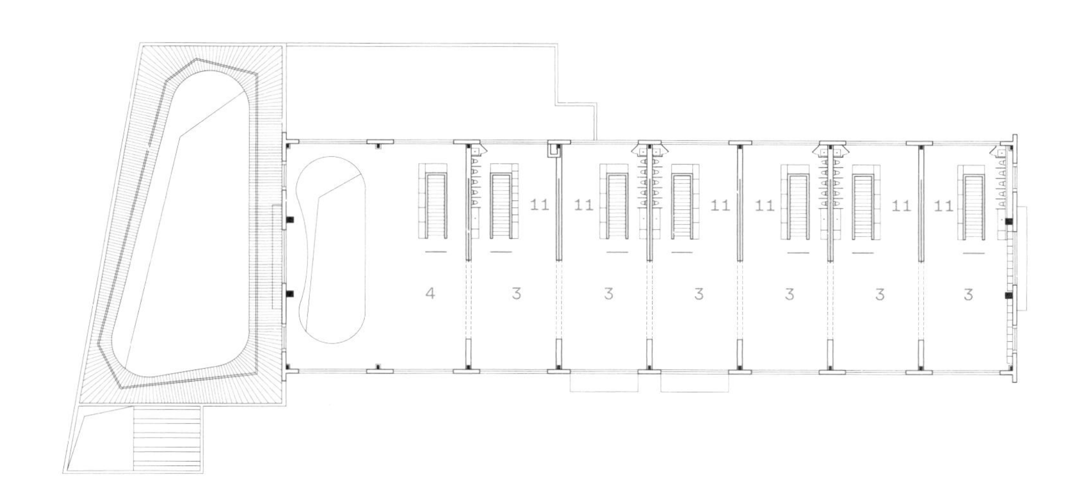

二层平面图
Second floor plan

1	入口	Entrance
2	门厅	Lobby
3	教室	Classroom
4	活动室	Activity room
5	厨房	Kitchen
6	办公室	Office
7	消毒室	Disinfect room
8	室外活动	Outdoor playground
9	门卫	Gate room
10	疏散出口	Exit
11	辅助空间	Accessory space

21

宝山陈化成纪念馆移建改造
Removal Renovation of Chen Huacheng Memorial

区域总平面图
General layout

0 10 25 50m N

1	陈化成纪念馆	Chen Huacheng Memorial
2	大成殿	Confucian Temple
3	淞沪抗战纪念馆	Songhu Battle Memorial

在大成殿前晨练的人们
Morning exercises in front of the Confucian Temple

控制设计的力量和欲望

上海宝山陈化成纪念馆原址是上海宝山临江公园内一座巍峨的孔庙，因为孔庙恢复，纪念馆需要移建，业主要求利用公园内一个折尺形的小型附属用房做立面改造。对于设计师来说，纪念馆本身前后空间秩序和体量规格落差太大，仅作立面改造，无论形式多么新颖，也不能避免凑合之感。于是说服业主，在相同的造价下，改用空间整理的方式来重新组织流线及氛围，以实现历史纪念馆应有的纪念性，同时跟公园周边环境互动，形成积极而融洽的日常公共活动场所。

不一般的普通围廊：公共性与纪念性

熟悉与安静，是建筑师为这个纪念馆建筑营造的日常基调。设计主要引进了公园中最常见，也是比较节省材料的一种建筑类型——开敞围廊。四条长短不一、宽窄各异的单坡顶敞廊环绕在既有建筑的周围，形成连续的柱廊空间，并与原有建筑曲折的边界围合成大小、形状不一的庭院。一方面，它有效地扩大和规整了纪念馆的空间和体量，最大限度地拉长了出入口流线，为这个不起眼的小建筑赋予了端庄体面的外观形象和富有韵律的空间序列，营造出必要的严肃氛围；另一方面，在保证原有建筑封闭外墙（这对于馆内陈设是必需的）的前提下，实现了与公园环境融合的开放边界。这样一来，无论是否进馆参观，公园游人都可以在纪念馆所辖的立面空间中穿行、逗留，纪念馆因此成了一个自由的日常公共空间而不是限定的爱国教育场所。

通过控制实现日常感和严肃感

除了传统的坡顶围廊形式外，水泥砂浆抹面的主体和外侧廊柱，深色的内侧廊柱、木梁、木檩条、木椽子，砖望板和小青瓦的构造，被刷成黑色的精心设计的现代钢木节点，这些习见普通的形式，低沉调性的处理是为了使新建筑形象让公园的日常使用者感觉不到视觉冲击。在阿科米星看来，针对日常环境中普通建筑的改造，控制设计的力量和欲望，强调建筑学的运用而非创造，不仅是一种值得重视的态度和方法，也蕴含着揭示平常生活本身张力的新契机。

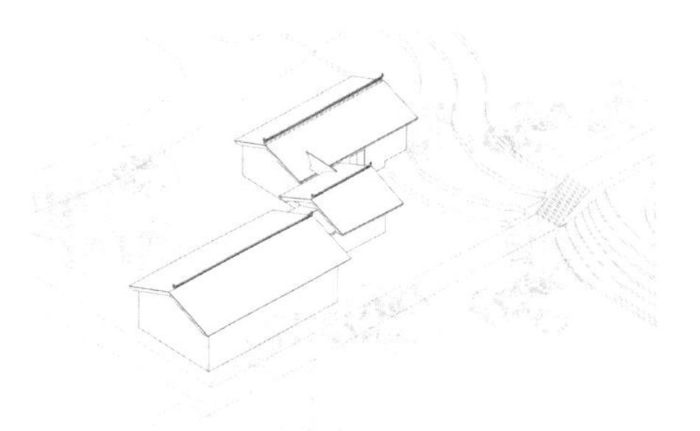

保留建筑
Exisiting building

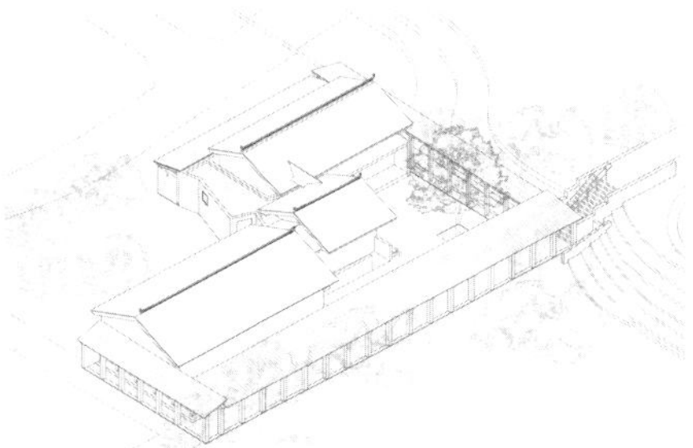

改造增设围廊
Adding galleries

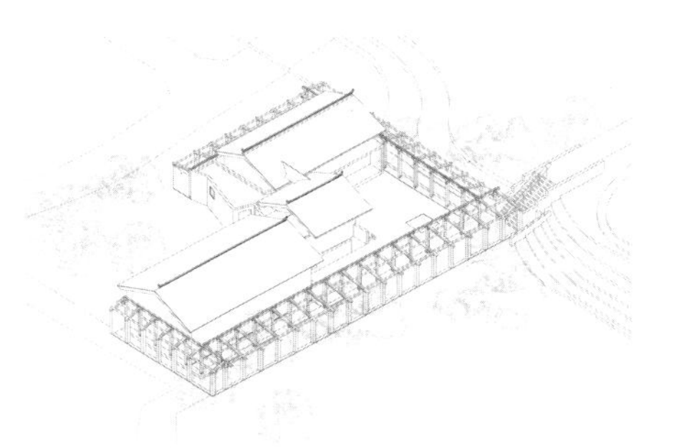

围廊结构
Structure of the galleries

Controlling the Design Power and Desire

Chen Huacheng Memorial was originally located in a magnificent Confucian Temple in Linjiang Park, Baoshan District, Shanghai. When the Confucian Temple was planned to reopen, this memorial had to be moved to a small subsidiary building with a zigzag form. We are asked to make facade renovation of this building. However no matter how creative the renovation will be, it is impossible to fill the giant gap between these two structures according to spatial sequence and building volume. Therefore, we persuade the owner to change his plan without extra charges. This plan includes reorganizing the visiting route and atmosphere, making the memorial a real place for commemorating, and creating a positive everyday public space by integrating with the surrounding environment.

Unusual Ordinary Galleries: Publicity and Monumentality

This memorial is designed to be a familiar and tranquil place. Open veranda, a popular and material saving building typology is used in this project. Four sloping roof linear galleries with different length and width enclose the major building, forming a continuous colonnade, and constituting courtyards of various sizes and shapes with the zigzag boundaries. On the one hand, it effectively enlarges and regulates the space and volume of the old building. Besides, it maximally extends the visiting route, providing this humble building with a formal and decent appearance, rhythmic spatial order and a proper serious atmosphere. On the other hand, by retaining the original blank walls (necessary for interior exhibition), the memorial is integrated with the surrounding environment without boundaries. People are able to walk through or stay in the veranda or courtyards without entering the exhibition hall. This memorial hall has become an open public space rather than an enclosed place for patriotism education.

Create an Everyday and Solemn Memorial Through Accurate Control

In order to avoid everyday users' visual shock in this park, the design adopts not only traditional building forms and tectonic but also common materials and modest treatment. The exterior walls and concrete columns are covered with plain cement plaster, the timber columns, beams, purlin and rafters are oiled dark, brick roof boarding and grey tiles are following the traditional tectonics, in harmony with those black modern and accurately connected timber-steel joints. Atelier Archmixing believes that for renovations of ordinary buildings in everyday environment, controlling the design power and desire with an emphasis on architectural application instead of creation is a valuable attitude and approach, where we can find new potential to explore intrinsic strength of everyday life.

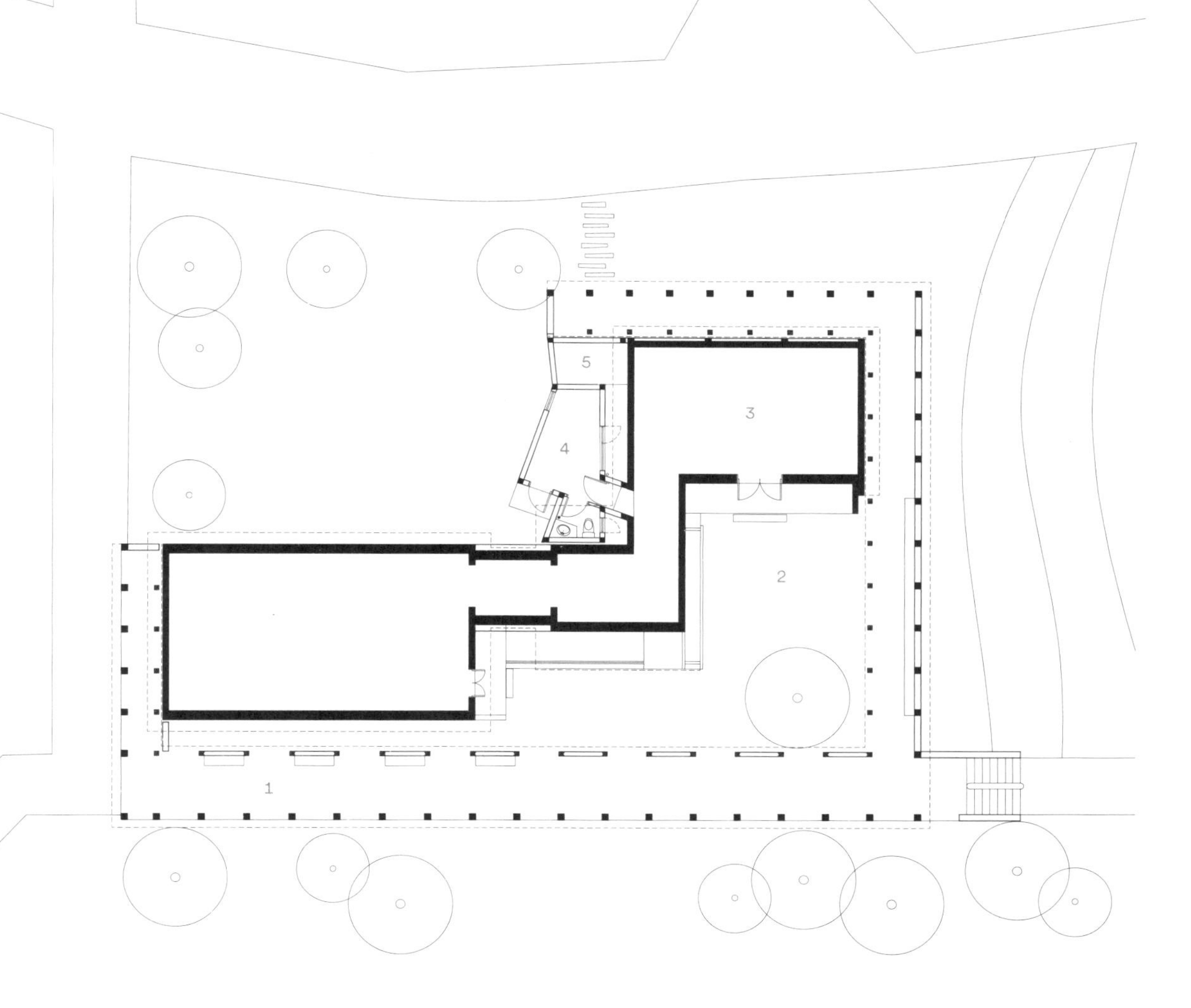

1 新建游廊 New gallery
2 入口庭院 Entrance courtyard
3 保留建筑（改建为展厅） Preserved building (renovated into exhibition hall)
4 新建管理办公室 New office
5 设备庭院 Equipment yard

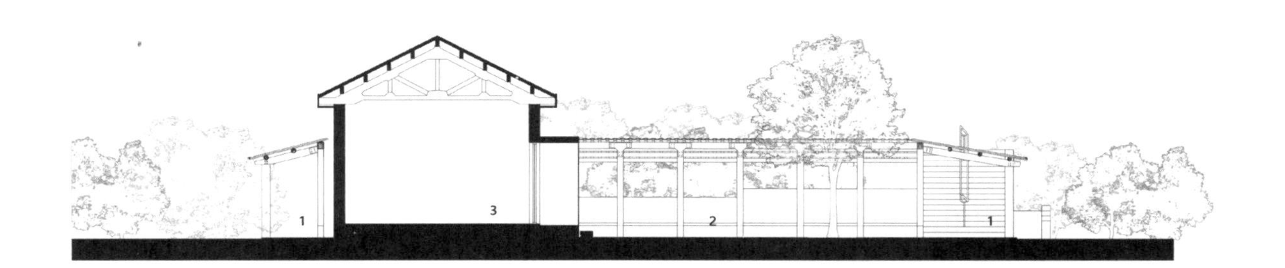

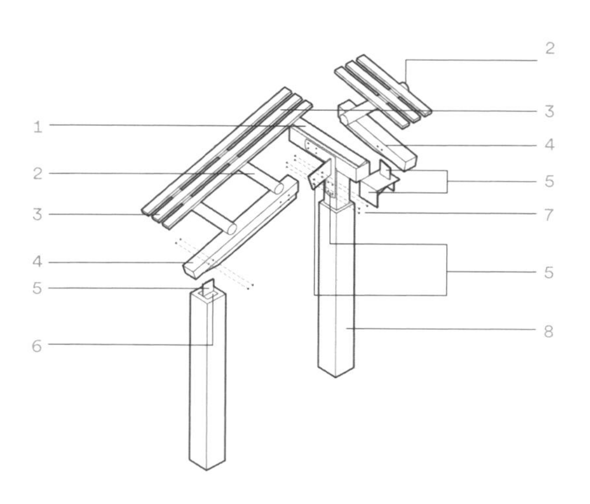

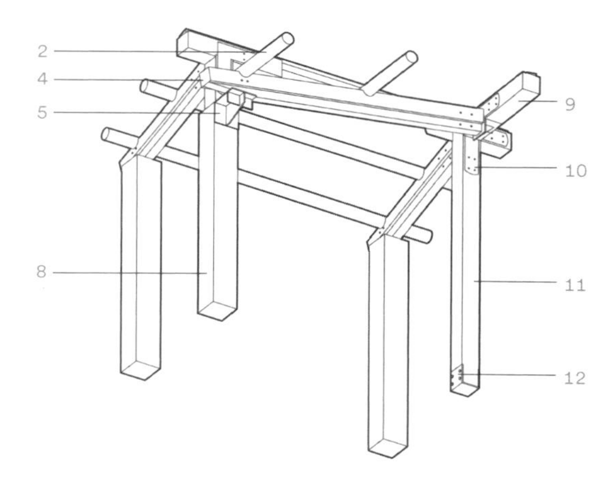

1	支撑椽子的木垫片	Rafter support
2	圆木檩条	Purlin
3	木椽	Rafter
4	斜木梁	Oblique tie
5	金属连接件	Metal joint
6	金属预埋件	Metal embedded part
7	螺栓	Bolt
8	混凝土立柱	Concrete column
9	水平木梁	Beam
10	T 型金属件	T type metal joint
11	木立柱	Wooden column
12	U 型金属件	U type metal joint

舟山青龙山公园入口改造
Qinglong Mountain Park Entrance Renovation

22

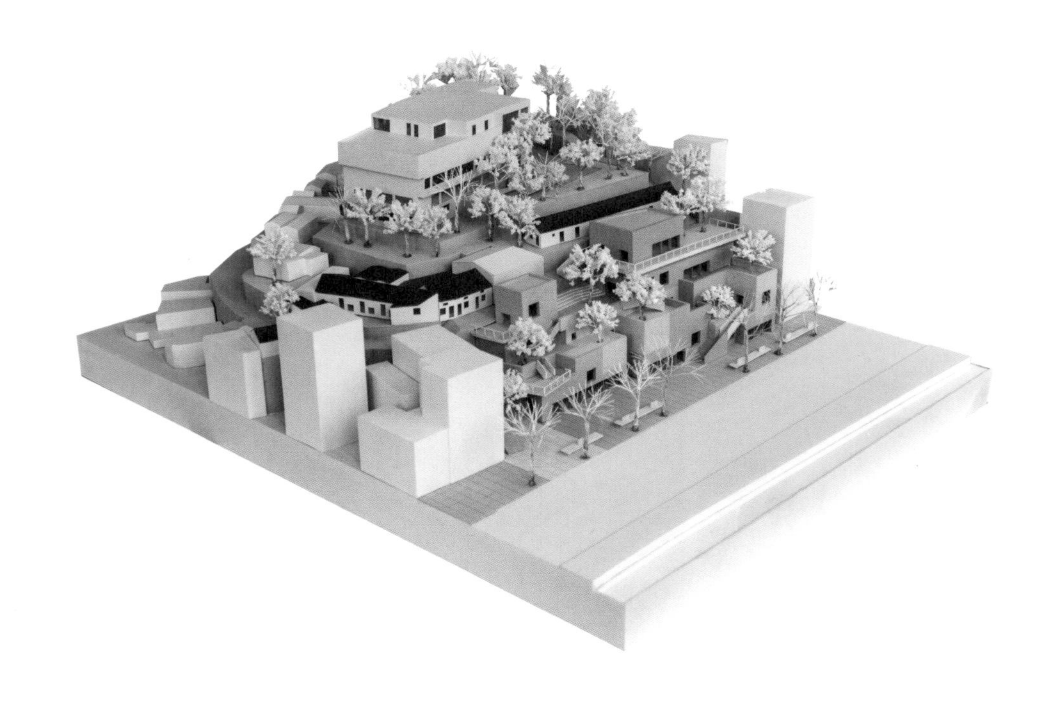

连接山体与海岸

舟山普陀区沈家门的滨港路，是一条富有当地气息、热闹的滨海街道。道路的北面一侧是紧挨着的立面丰富欢快的连排多层建筑，底层有各色饭店餐厅与其他商业；道路南面是滨水人行道与外面的渔港，对面是鲁家峙。这里天气好的时候一派明亮的渔港风情，平时很多渔船或聚集于此，或忙碌于往来进出，海风拂面，空气清新。在连排的多层建筑后面是山地，只是一直都被沿街密不透风的多层连排遮挡，因此无法被走在街道的人感受到。这样的问题在其他地方同样存在，在舟山政府的总体规划中，未来要创造条件实现更多的山体与海岸的自然连接。青龙山原来沿着滨港路有一个入口，一条连续的上山石阶夹在两栋多层大楼之间。为了能在视线上将山体入口打开，地方政府拟拆除原入口一侧一栋老的七层楼房，在那里重新设计一个包含部分房屋使用空间的青龙山入口。

地景式的迷宫

设计设想拆除原建筑物后将被遮蔽的青龙山显露出来，在原基地处建设一座地景式的综合体建筑——由一系列跌落体块构成，像山坡一样从山体延伸到路边。整个入口建筑体是一座集公共活动空间与公共功能空间为一体的市民立体景观公园，内外分为两种性质的空间，内部空间是独立的使用空间，外部由表皮立面与平台形成另外一个世界。建筑体一方面提供了实用的空间，可以容纳公共展示空间、市政配套的公共厕所、一系列的大小商业点，另一方面提供了从下而上的一系列景观平台，为上下青龙山的人们所用，同时平台上种植绿化树木，供休息与商业服务所用，将青龙山与滨港道路、水岸在空间景观动线与体验上连接起来。

动线的组织上，将原来的石头台阶通道保留组织在整体的空间形态里面；同时，利用跌落的形体与平台，设计了另外几条上下的通道，用一段段台阶相连，形成一个系列性的交通网络，希望上下的人们能够漫步在不同的路径与平台间，获得新鲜的上下体验，这样也可增强建筑的场所个性。小小的迷宫式的构思是想创造一种开始的陌生感，而在渐渐熟悉的过程中整个地方给人烙下印象。

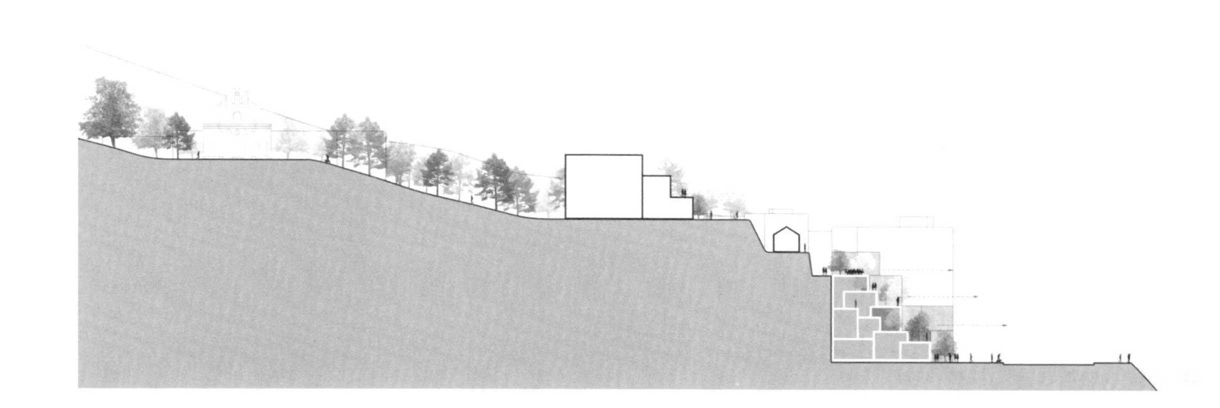

Connecting Mountains with Coast

Bingang Road is a busy coastal street with a smell of local prosperity in Shenjiamen, Putuo District of Zhoushan. The north side of the road is lined with multi-story buildings with bright and vivid facades one after another. On the ground floor of these buildings serve various restaurants and other businesses. The south side connects the waterfront walkway and fishing port, facing towards Lujiazhi Island. On sunny days, with gentle wind and fresh air, the fishing port looks very beautiful. Numerous fishing boats dock together or come and go. By the hills and mountains behind are completely blocked by these multi-story buildings from the passers-by. Although this is a common problem troubled many cities, based on the overall planning, Zhoushan Government wants to establish more natural connection between mountains and the coast.

On Bingang Road, there is originally an entrance to Qinglong Mountain formed by continuous stone steps between two multi-story buildings. In order to highlight the mountain entrance, the local government plans to remove an old 7-story building beside the original entrance, and create a new entrance to Qinglong Mountain, which also includes functional spaces.

Landscape Maze

The designer intends to uncover the invisible Qinglong Mountain by replacing the original building with a landscape-like complex. With a series of stacked blocks combining public activity spaces and functional spaces, the complex will stretch to the road from the mountain like a slope, creating a three dimensional landscape park for local people. The inner and outer spaces are different in nature. The interior is of independent functions, while the facade and platforms are designed like another world. On one hand, the complex provides practical space for public exhibitions, public toilets and commercial outlets. On the other hand, it provides sightseeing platforms for visitors to Qinglong Mountain. Meanwhile, the platforms support greenery, relaxation and small business, connecting the Qinglong Mountain, Bingang Road and the waterfront through spatial circulation and experience.

As for circulation, the original stone stepped pathway is kept and merged into the new spatial sequence. In order to establish a network of circulation, several new pathways are designed by making use of those old stacked blocks and platforms. Visitors can then enjoy a fresh experience when walking on different pathways and between platforms. The landscape complex thus accomplishes a distinctive character. This maze-like place might offer a rather strange atmosphere at first, then become more and more impressive when people were familiar with it.

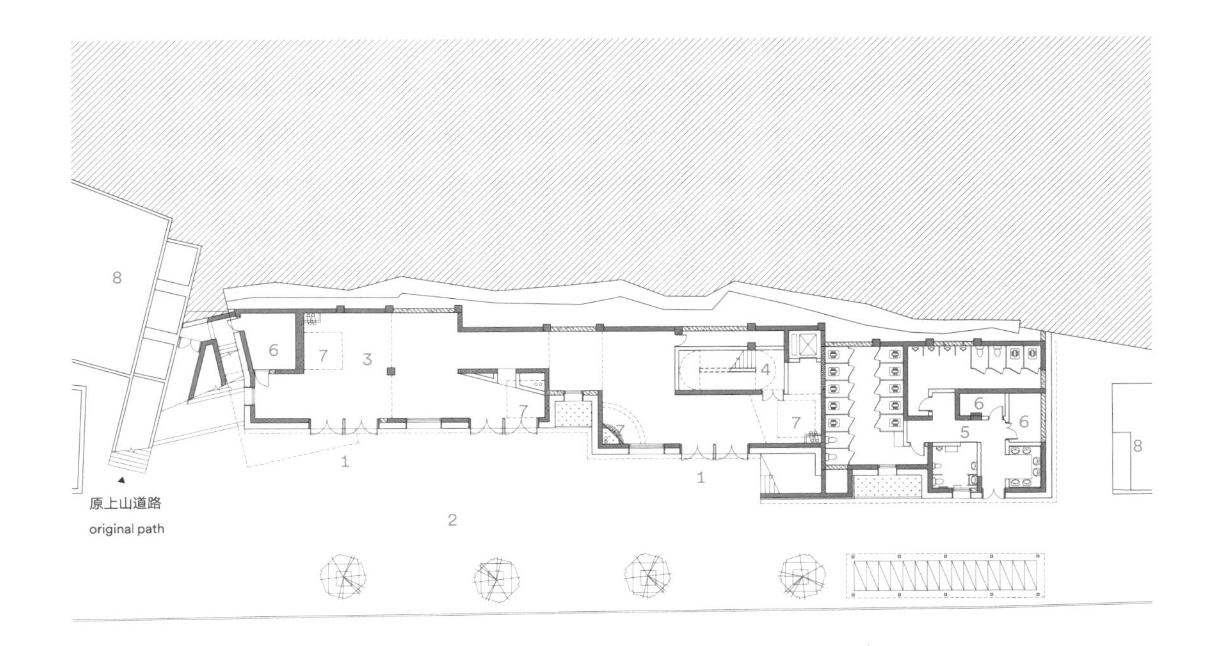

±0.000 平面图

±0.000 level plan

1	入口	Entrance	8	原有建筑	Existing building
2	城市广场	Plaza	9	露台	Terrace
3	展厅	Exhibition	10	树池	Planting
4	楼梯间	Stairs	11	咖啡	Cafe
5	公共卫生间	Public restroom	12	特色商业	Retail
6	管理 / 设备用房	Management/Equipment room	13	后勤用房	Service room
7	树池洞口投影	Planting projection			

建筑东侧露台

East terrace

城市广场

City plaza

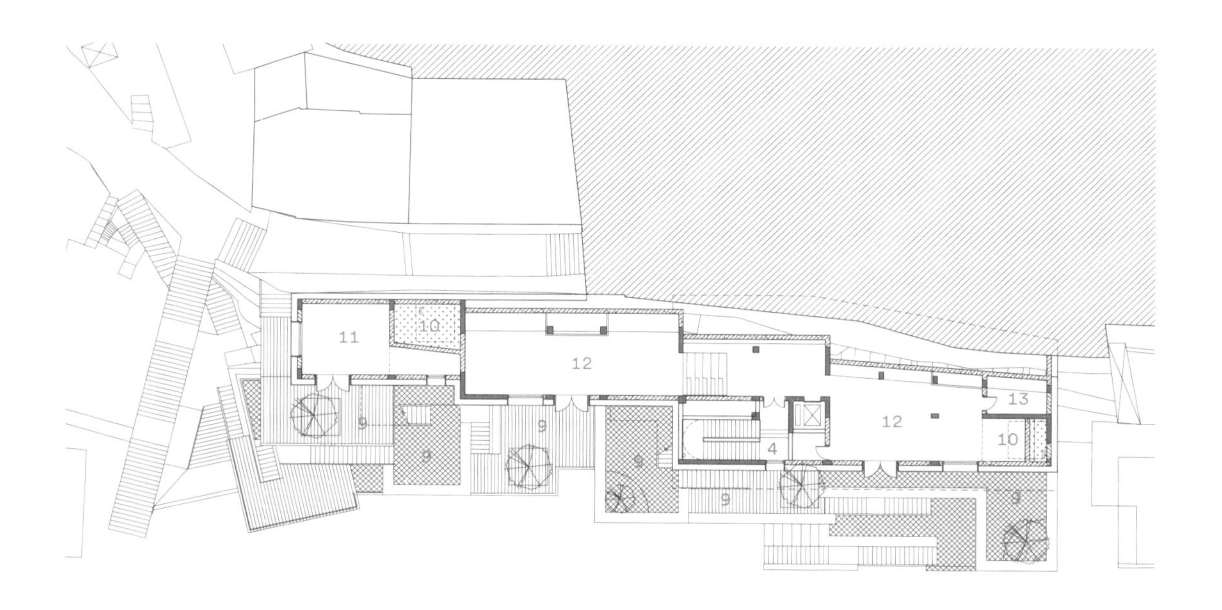

10.500 平面图

10.500 level plan

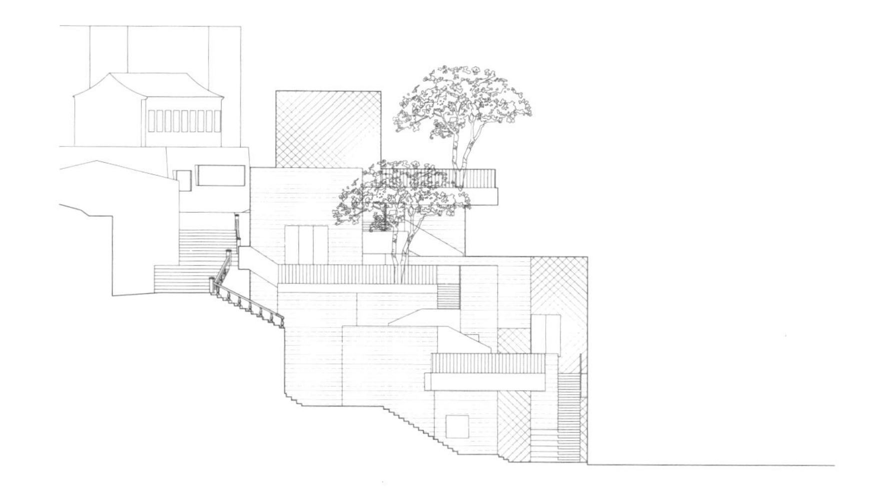

永嘉路口袋广场
Pocket Plaza, Yongjia Road

23

基地原貌
Original site

口袋广场鸟瞰图
Aerial view of Pocket Plaza

用敞廊围合的小广场

永嘉路位于上海老城核心区，周边分布着众多老住宅区，街道尺度宜人，绿树浓密，沿街多为生活配套小商业，生活氛围浓郁。309 弄口袋公园的基地位于永嘉路中段，借老城区风貌整治的契机，区政府部门决定拆除原用地内存在消防隐患的两排残旧住宅，将其改造为服务于周边居民的城市公共空间。基地周边除了街道，缺乏属于居民使用的公共场所，因此设计一个尺度宜人、可供日常使用的开放广场对于这里是一件好事。

拆除老建筑后的用地大致呈长方形，沿街宽约 18 米，纵深约 40 米，与街道几乎垂直，东南西三面均被住宅围合，空间呈口袋状。公共空间被设计为由敞廊围合的小广场，通过将地坪抬高 0.5 米增强场所感，同时与街道之间形成有趣的抬高视线关系。广场区域设置旱喷，根据不同时段选择开启或关闭，软性控制广场的使用方式。广场南侧尽端布置两个隐蔽的辅助用房，与广场之间用一道连续曲折的耐候钢板墙隔开，钢板墙成为广场尽端的边界。

广场开放面通过缓坡与街道连接，紧贴人行道为标志墙与低矮的灌木花池，东北角设置便于管理的铁移门，移门与灌木高度控制在 1.2 米高，保证视线贯通。广场东西面紧贴的均为住宅山墙，东面山墙按原样重新粉刷，西面山墙在重新粉刷之余还特地保留了一段老墙的原有肌理，隔着敞廊后面的竹丛若隐若现。

钢木结构与红面砖

敞廊是主要的构筑物，四段长廊呈风车状，围合出规整的长方形内院，廊下高度刻意压低至 2.1~2.7 米，形成亲切的尺度，同时整体与周边建筑在高度上形成合适的比例。敞廊形式上采用精致的钢木混合结构，考虑到居民公共空间的属性，在理性的形式基础上特意增加了具有戏剧性的、富有张力的细节。每个典型的钢木梁架单元均由钢柱、木梁、钢拉杆组成。钢柱由两片扁钢组成，主木梁由成束的叠退木方组成，向中心广场出挑承托轻盈的金属屋面。木梁加载于钢柱上，屋面的重量由梁传给钢柱，梁后端的钢拉杆用于平衡前端出挑的重力。木梁与钢柱的连接节点被设计为可拆卸装配节点，这个节点专门设计为单向转动模式，反向限制梁的转动，用以平衡风对于屋面的上掀力。

周边住宅区以上海城区常见的暖色调涂料或清水砖为主，因此广场地面铺砖选择了最常见的红色透水砖，木梁、屋面底部及坐凳采用浅色原木，耐候钢板墙则是锈红色。整体暖调的氛围里，钢柱则采用了明亮的鲜艳绿色氟碳喷涂，反差色也带来了轻松的日常气质。

被住宅包围的基地

The site surrounded by houses

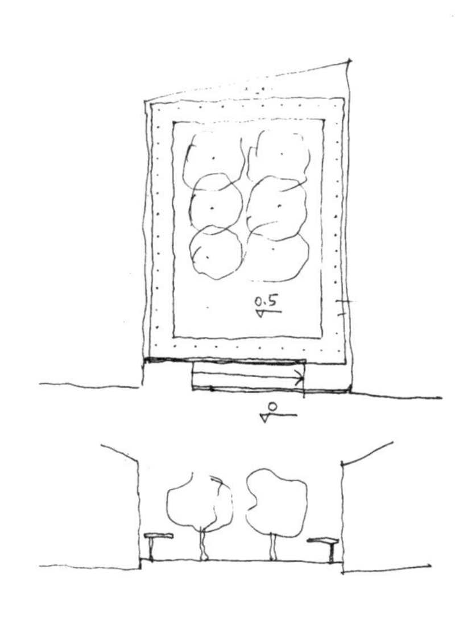

A Small Plaza Enclosed with Galleries

Yongjia Road is located in a historical downtown district in Shanghai, with many old residential blocks scattered in its neighborhood. This road is of pleasant scale and luxurious greenery, crowded with small retails. One can sense a strong living atmosphere here. The Pocket Plaza in lane 309 sits in the middle of Yongjia Road. Taking the opportunity of old quarter renovation, Xuhui District Government decided to dismantle two rows of shabby residences with fire hazards and transform the site into a public urban space. Based on observation, we believe that an open plaza with appropriate scale would be most necessary and suitable for this neighborhood.

After demolishing those old buildings, the site turned out to be a rectangular area, 18m in width and 40m in depth, almost at a right angle with the street. It is enclosed on three sides (the east, south and west sides), shaped like a pocket. We created here a small square fully enclosed with open galleries. The ground of the square is 0.5m higher than Yongjia road, providing a sense of territory and an interestingly elevated view of the cityscape. Inside the square, a dry fountain can be switched on or off on required conditions, which serves as a soft management strategy to control the usage. The south end stands two secluded auxiliary rooms of continuous folded surface, separating from the square as a notable boundary, covered with weathering steel wall.

The entrance connects to the street with a gentle slope, defined with a logo wall and a framed flower bed with low shrub beside the sidewalk. An iron sliding gate was installed at the northeastern corner to achieve better management. The iron-gate and the shrubs keep the height at about 1.2m as a boundary for the plaza. The residential pediments of both the east and west sides were repainted in its original appearance, for the west side, the old surface texture was also preserved. They are partially hidden behind the bamboo grove planting along the open galleries.

Timber-Steel Structure and Red Tiles

Four galleries with a basic beam frame formed a windmill shape, enclosing the rectangular plaza. The ceiling of the open galleries is intentionally lowered to 2.1m–2.7m, a friendly height harmonious with neighboring buildings. The open gallery adopted a timber-steel structure. Based on its rational architectural form, we added dramatic details to build public attention. Each typical unit of the timber-steel beam frame is composed of steel pillar, wooden beam and steel cable. The steel pillars consist of two pieces of flat steel. The wooden beams are multi-layered. Their cantilevered parts are used to bear the light metal roofing. The weight of the building is transferred from the wooden beams to the steel pillars. The end of the beams is fixed with steel cables in order to keep its balance. All joints are assemblage ones. These joints are rotatable only in one direction, so that they can restrain the rotation of beams and counteract the wind force pushing roof upwards.

The residence in the neighborhood is mainly painted with warm-toned color or built with ganged bricks, which are very common in Shanghai's urban area. Therefore, the plaza ground was paved with common red water permeable brick; its wooden beams, soffit and benches are made of light color logs and the weathering steel wall is painted rust red. In this warm-toned context, the steel pillars were painted into astonishing bright green by fluorocarbon spraying. Such a sharp color contrast has brought a kind of everyday relaxing atmosphere.

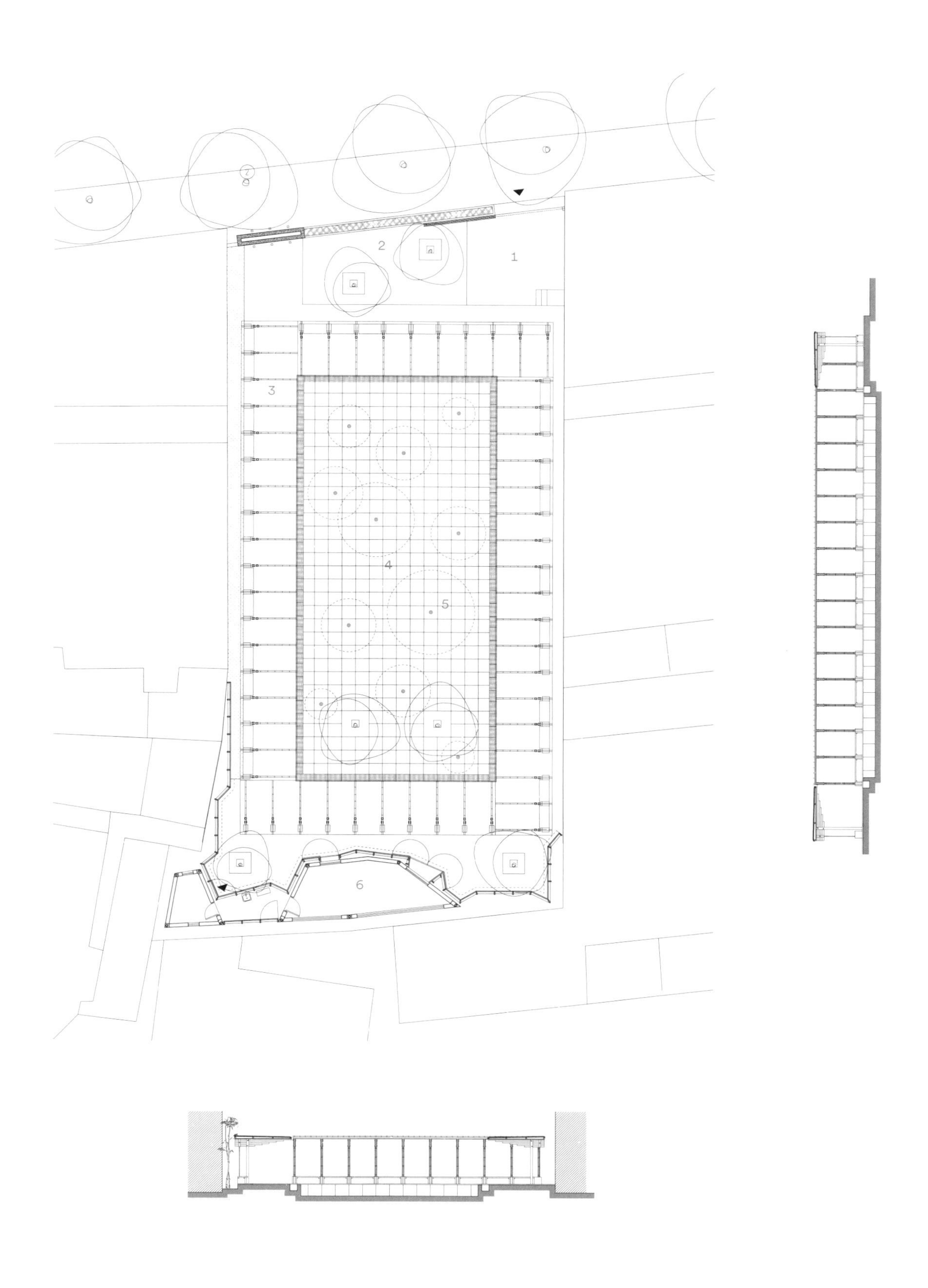

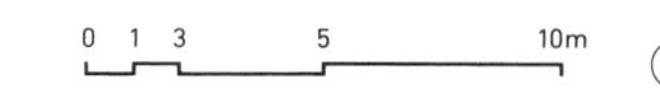

1	入口	Entrance
2	坡道	Ramp
3	廊	Gallery
4	庭院	Courtyard
5	喷泉	Fountain
6	辅助用房	Service room

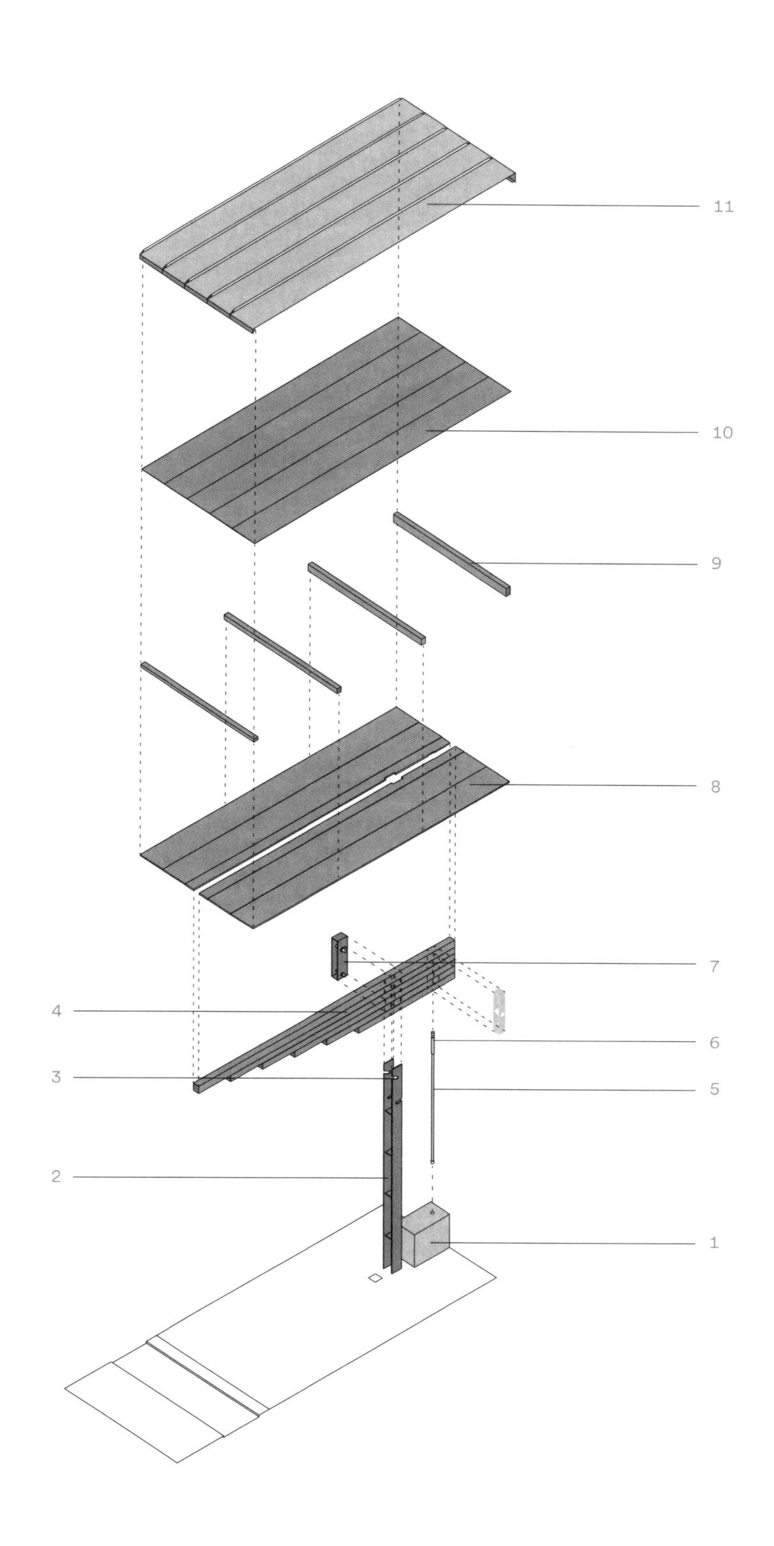

1	混凝土块	Concrete block
2	扁钢柱	Flat steel columns
3	活动锁扣	Latch
4	70/76 木方	70/76 timber tie
5	钢拉杆	Steel tie
6	松紧调节杆	Adjusting rod
7	钢箍	Steel hoop
8	屋面底板	Roof backboard
9	木檩条	Timber purlin
10	屋面垫板	Roof backing panel
11	直立锁边金属屋面	Seaming metal roof

城市内的工作室
Work Within the City
24

经常有人问我们，阿科米星是什么意思？后来找出 2009 年 6 月的原始笔记，又想起我们曾经为即将成立的工作室起过很多名字。

其中有无厘头的 R2D2 工作室、飞行器、低空飞行、漂浮工作室，大概是星战电影看多了；又有受到消费文化影响，或怀疑物质化状态的三明治工作室、生产线工作室、狂欢工作室、短暂工作室、移动影像工作室、临时物件工作室；还有考虑退休时间倒计时模式的 2039 工作室、2029 工作室，等等。慢慢地，“混合”作为关键词出现了，我们设想了混合 + xx（建筑、组织、体、空间、事情、时间、系列……）工作室的各种可能性，最后决定就叫混合建筑工作室（Atelier Archmixing）。这个英文拼凑的生造词很贴合我们主张的设计理念和策略，但是中文名字我们又不想显得那么严肃专业，希望更大众文化一些，于是想到了直接采用音译，在“艾克美星”“埃科美星”“阿克米星”“阿科米星”等写法中，最终选定“阿科米星”作为公司的大名并注册成功。

在设想第一个工作室时，2009 年 6 月 14 日的笔记上是这么写的：“选址：接近社区，与社会连接最佳。混合工作室是一处多重功能的模糊场所。室内空间能够灵活地改变成日常工作室、展示厅、沙龙、社会活动场所、教室、青少年课余兴趣活动场所。工作室多用途空间要够大，有庭院或露台更佳，底层方便为佳，整个面积应该在 400 平方米以上。工作室平时工作，周末或假期可以改变为社会活动的开放场所，免费；可以吸引一些有品质的设计界、艺术界朋友作为小型展示场所；可以发动有兴趣的人组织网友艺术沙龙；可以与周边小区联系普及建筑知识，做义务的建筑学讲座推广。共享的工作室！临时的工作室！”

今天，当无意间发现当时随手写下、早已遗忘的这段话时，我们很惊讶：当时我们在第一个长居五年的工作室里没有意识到，之后应该也没有意识到，阿科米星工作室后来的成长经历，一开始就已在我们的潜意识里了。

阿科米星的故事 2009.6.17

THE STORY OF ARCHMIXING

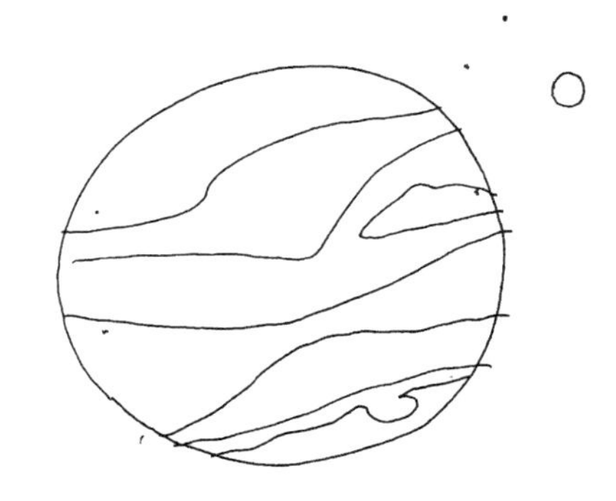

阿科米星是一颗变化的星球……

我们是阿科米星人.

ARCHMIXING

阿科米星

We are often asked what exactly “阿科米星(ā kē mǐ xīng)” means. We looked over the original notes and found that we had given the studio to be established lots of names for discussion in June 2009.

Some of the names resulted from our obsession with Star Wars: R2D2 Studio, Air Vehicle, Low-Altitude Flying, Floating Studio; some of them were derived from the influence of consumer culture or the skeptical attitude toward materialization: Sandwich Studio, Production Line Studio, Carnival Studio, Momentary Studio, Moving Image Office or Temporary Object Studio; some of them had taken retirement date into consideration: Studio 2039, Studio 2029, etc.. Gradually, it occurred to us that Mixing could be an appropriate keyword. We tried compound words in the form of “Mixing + noun.”, exploring all sorts of possibilities (architecture, organization, body, space, matter, time, series...). Then we picked Archmixing as a final decision for this English compound phrase perfectly fitted our design philosophy and strategy. Instead of being serious and professional, we would rather the Chinese counterpart be kept close to mass culture, so this English name was directly transliterated into Chinese. Finally “阿科米星(ā kē mǐ xīng)” fended off other homophonic Chinese characters and successfully got registered.

Imaging the first office, according to the documented files on Jun. 14th, 2009: “As far as location is concerned, we are seeking a place near the community. A place that adjoins the community would be our first priority. This Mixing Studio shall be a multi-purpose space and can be flexibly changed into places for everyday work, exhibition, public meeting, social activities, class and extracurricular activities of teenagers. Therefore, this space needs to be large enough. A place of more than 400m², on the ground floor with courtyard or terrace may give us the utmost satisfaction. It serves as a studio on a workday and free open space for social activities on weekends or holidays where quality designers or artists can make small displays; people interested can organize art salon; neighbors can acquire architectural knowledge in voluntary lectures on architecture. It will be a shared studio, a temporary studio.”

Today, when this long-forgotten passage is inadvertently brought to light, we finally realized that Atelier Archmixing’s growing process had already been in our subconscious from the very start. To our surprise, actually we have never been aware of this in our first studio where we worked for five years, neither have we later in other offices.

“一年一个工作室”，搬家地图

“New Year, New Office” map, Shanghai

从2014–2019年，阿科米星采用了“一年一个工作室”的工作方式，这主要是源于阿科米星的设计研究。我们越来越关注建筑的改变与使用。之所以这么关注改变，并非是从理论上的推理而来，而是基于我们多年来的实践。我们工作的背景是近40年中国的快速城市化，这期间，一方面空间数量惊人增长，另一方面，也留下了大量需要被重新认定价值与利用的空间。在日常工作里，我们越来越多地接触到建筑物的使用改变。不仅在设计实践里，而且在设计研究与空间使用实验里面。因此，我们不仅希望从一个项目类型的角度去思考这样的情况，也希望从建筑学的认知上去思考这样的问题。不断在城市里把工作室搬来搬去，不仅是设计研究的需要，也是一项空间使用实验，更像是一次次冒险。

In 2014, Archmixing launched a “New Year, New Office” plan and has moved five offices in Shanghai till now. This strategy is rooted in Archmixing's design research program, which highlights the change and usage of existing buildings. Decades of architectural practice has led Archmixing to focusing on changes, rather than theory. With four decades of rapid urbanization in China, architectural practice and urban expansion grow in a neck-breaking speed. Meanwhile, numerous built areas need reevaluation and reuse. Architects are increasingly exposed to architectural changes, knowing how people used buildings will affect not only design practice, but also design research and experiments on the spatial usage. Therefore, we want to study it not only in the perspective of spatial types, but also aiming at new architectural cognition and design strategies. The frequent relocations of the workplace within the city transcend a simple need for design research, which represents a spatial experiment, rather like series of adventure.

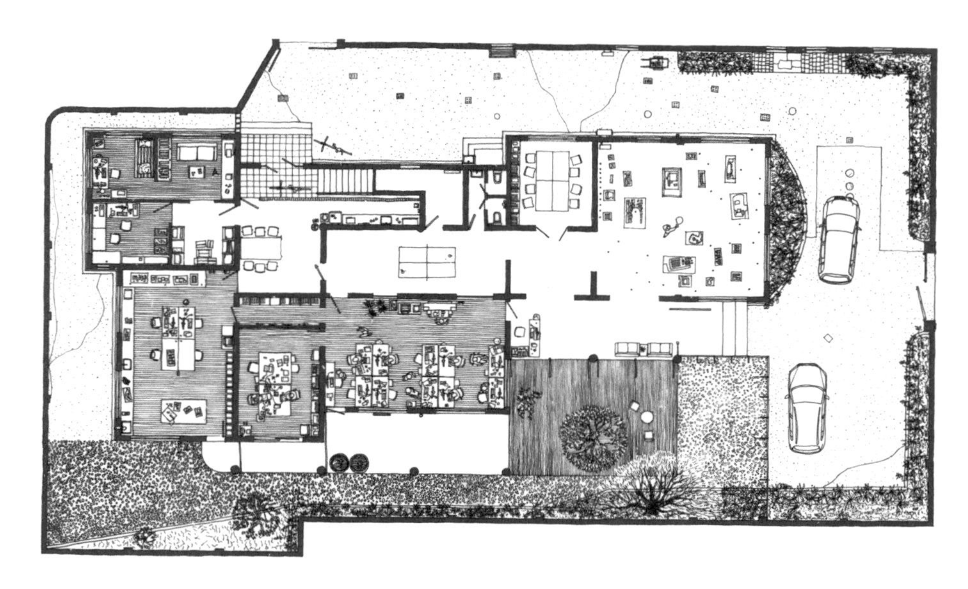

第一个工作室（2009–2014）：社区中的幼儿园

阿科米星的第一个工作室开始于 2009 年盛夏，我们租下了老式居民区内一间闲置的小型托儿所的一层空间并加以改造。这个原来作为小区配套的托儿所，房屋所处的居民区很有上海本地特点，是标准化的多层住宅，经过 20 多年，住宅楼随处可见居民自行搭建和改造后生机勃勃的成果。这个工作室是宜人、有特色的，经年成长后四时变化的景象替代了视觉化的固定形式。近 100 平方米的院落被重新整理：靠围墙栽种各种树木爬藤；在新门厅南侧，利用施工垃圾堆出了一个木平台，中植香樟，与原来活动室外抬高的凉廊连成一体；凉廊外搭起了葡萄架。春天，爬山虎悄悄铺满整个墙面与窗户，影子水墨画般地印在白色的窗帘上，蔷薇、杜鹃盛开，樟树换叶；夏秋，凌霄、桂花渐次开放，葡萄藤蔓茂密，果实累累。与此相伴的，还有一季一回的讲座沙龙。这个普通小区的庭院在很多员工和访客心里都是一个美好的场所。在这个混杂的居住小区，我们安定工作了 5 年，不知不觉成为了社区的一员和旁观者。

工作室的搬家始于 2014 年，每年能够搬家的具体的方法如下：①租一处永久便宜的仓库，归类储藏我们的模型、不用的图纸、档案、书籍等；②将原来的活动讲堂另找个固定的地方（事实上，它变成了另外一处空间试验场——那行空间）；③完成办公的轻质化组织，包括设计便于拆搬但牢靠的家具，将电脑换成一体机，改有线网络为无线，培养好团队简化的工作习惯；④三周找到场地，两周“格式化”，租期一年。

First Office (2009–2014): Community Nursery

Archmixing set up its first studio in the summer of 2009 by renting and renovating the first floor of a small vacant nursery inside an old residential area. This residential area is a typical Shanghai community with standard, multistory buildings. Having been used for more than two decades, informal buildings and renovations pepper the area.

The first studio has a pleasant and distinctive feeling, boasting scenery that has developed over years and changed with the seasons. The courtyard, which measures nearly 100m² in size, is reorganized by planting trees and vines along the wall and building a wooden platform on the south side of the entrance hall to connect to the elevated loggia. Grape trellises grow outside the loggia, while a camphor tree is planted in the center of the platform. Vine plants quietly cover the walls and windows in the spring, casting their shadows onto the white curtains like an ink-and-wash painting. Roses and rhododendron bloom and the camphor leaves change colors. In summer and fall, Chinese trumpet vine and sweet osmanthus gradually blossom and bunches of grapes hang from the thick vines. A series of lectures are held to complement this seasonal change, and staff members and visitors flock to the courtyard. The studio remains a part of the community for five years and becomes observers of the daily life of the community.

Archmixing began the annual moving in 2014, having developed a set of strategy. First, rent a permanent, cheap warehouse to store models, unused drawings, files, and books. Next, find a fixed venue for the seasonal lectures. (This multi-functional space actually transforms into another spatial experiment, NEXTMIXING.) Then, minimalize the office. This includes designing movable furniture, installing all-in-one computer systems, converting to a wireless network, and cultivating a simplified team working habit. Finally, find a new space within three weeks, reformat it within two weeks, and rent it for one year.

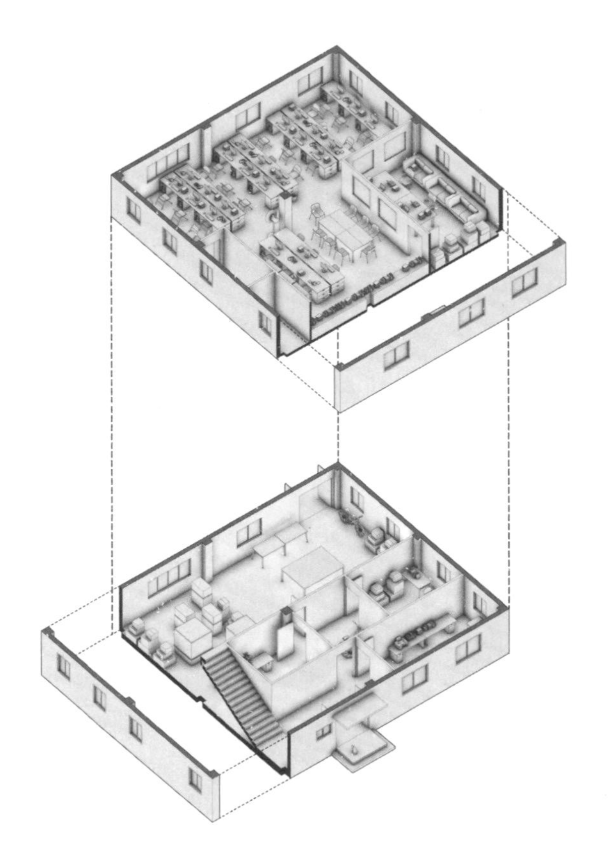

第二个工作室（2014–2015）：无人厂区的辅助用房

这是一个乏味的厂区的辅助用房，唯一的好处是无人打扰。在这个几个月的过渡空间里，工作室被压缩在一个小小的房间里，工作空间少了固定的氛围，简易的环境更像是一种匆匆的撤离。搬家把以往掩盖在光鲜外衣下的现实情绪暴露出来，办公室充满了临时与变化带来的某种隐约的不安感，也催生出越来越冷静的现实感知。

Second Office (2014–2015): Auxiliary Room in an Abandoned Factory

Archmixing's second office is an auxiliary room in an abandoned factory area. The only advantage of the site is the quietness. It has served as a transitional office for only ten months, so everything is compressed into a small space, and the lifeless environment leaves a sense of hasty evacuation.

The architects face the same feeling of the place as many urban citizens. The past can enhance the significance and atmosphere of a place, but likewise, designs can be void of that significance, just like a building can be stripped of all social and cultural context. That makes it hard to adapt and accept the building. Frequent office relocations often expose the sentiment that lies beneath the more colorful layer of the past. For instance, transience and change can lead to unease, but it can also cultivate a calmer approach to reality.

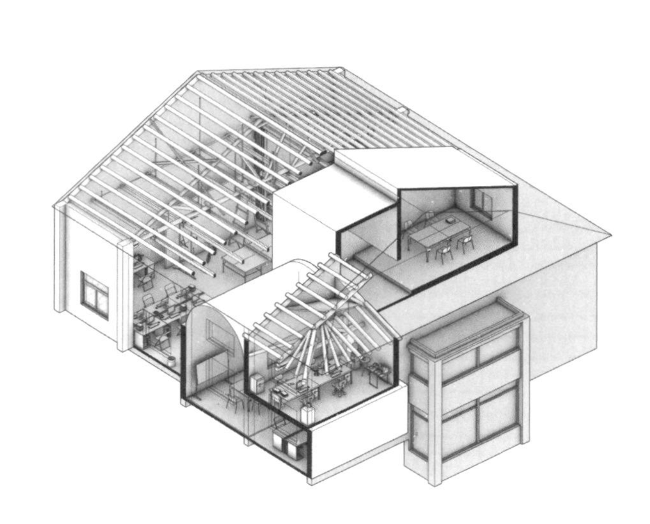

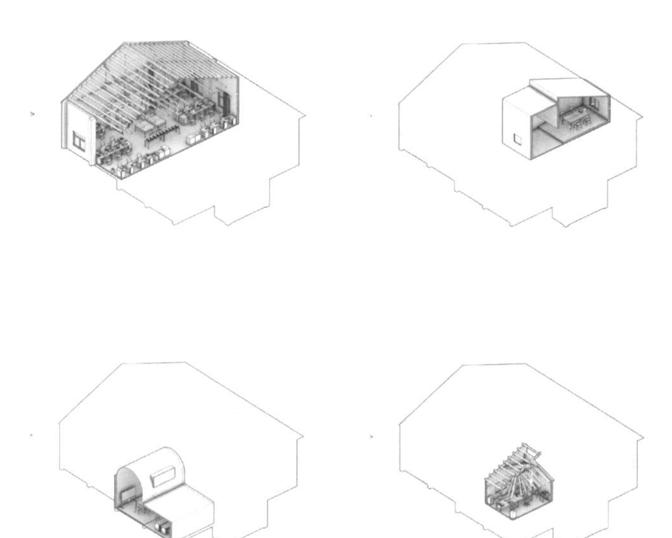

第三个工作室（2015–2016）：紧挨轻轨的摄影棚

第三个工作室是一个由厂房改造的摄影棚。这是一个残留着过往使用痕迹的城市内部空间：近6米层高的厂房大空间被从事婚纱摄影的前租客分隔得高高低低、上上下下。室内很多窗口有趣地贯通着，坐在办公室里，时不时就能看见、听见紧挨的城市轻轨呼啸而过。调整设计基本保留了原来的空间格局，只在局部设计了新的功能区。这些局部的改变不求风格统一，有点各自为政，以期形成不同体验的空间效果。新的布局再次影响了我们的工作方式：主持建筑师们变得高高在上，可以透过楼上的控制窗瞭望楼下的大工作间和会议区，若要直接交流却需要经过数次上上下下的空间转换。

Third Office (2015–2016): A Photography Studio Neighboring Light Rail

Archmixing's third office was a factory-cum-photo studio—an urban space with visible traces of a wedding photography company. The previous tenant has separated the large building—nearly 6m high—into upper and lower areas. Its many windows look out to stunning views. The sounds of the light rail trains filter through the office.

Archmixing largely retains the spatial layout, adding only a few functional areas. In an effort to create a different spatial experience, these small changes are intended to be independent of the style, rather than constrained to it. Once again, the new layout influences the way the architects work. The main architects sit upstairs and can look down through the control windows into the large working and conference area. At the same time, they still need to move through the space several times a day for personal communication with the employees below.

"一天 · 世界"展览

A Day A World exhibition

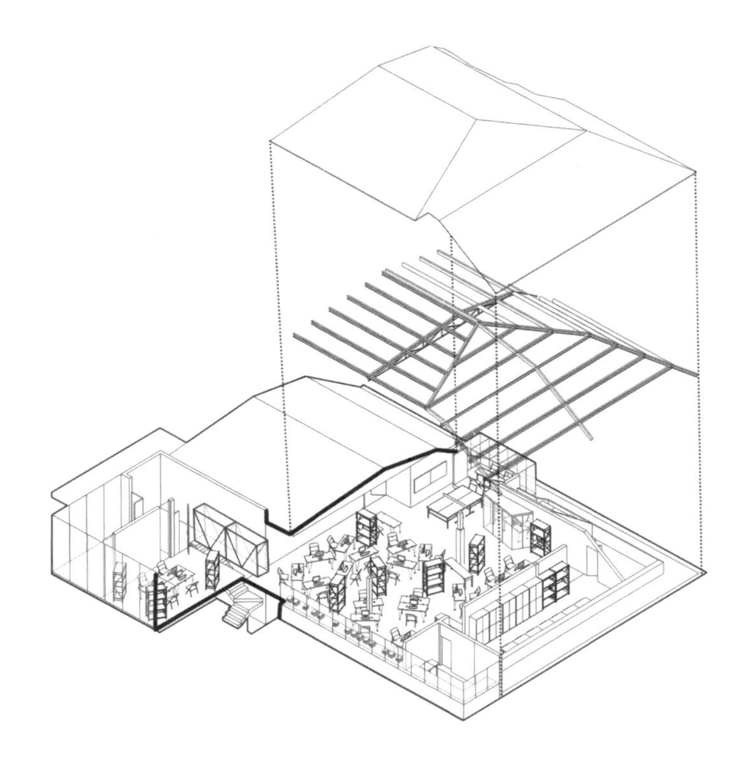

第四个工作室（2016–2017）：瞭望虹口的阁楼

这个工作室位于虹口西江湾路，周围紧邻着虹口体育场、虹口公园、龙之梦商业综合体、虹口的历史文化街区，以及大大小小各种日常商业、餐饮丰富的街道与社区，这是一个工作环境相当舒适的工作室。工作室的主体是一个阁楼，可以眺望周围的环境，与前一个工作室相似，地铁三号线还在不远的视野内，每天充满了活力。

工作室的另外一部分与主体原来并不连通，我们做了一条小小的架空甬道将二者联系了起来。这个工作室采用了随意散布、随时改变与组合的桌面布置，这个布置深受大家欢迎，让大家感到交流更方便，环境更轻松。

在这个工作室，我们完成了历次中最大规模的城市调研——“虹口 1617 逆向还原”，研究了周边各种类型的城市建筑的局部与片段的改变。在结束这个周期的工作室时，我们在再次搬空的工作室里举办了一个展览。

Fourth office (2016–2017): A Loft Overlooking Hongkou District

Located at West Jiangwan Road in Hongkou District, Archmixing's fourth studio sits adjacent to several major points: Hongkou Stadium, Hongkou Park, Longemont Commercial Complex, and the historical and cultural blocks of Hongkou, along with the streets and communities lined with businesses and restaurants. The studio provides a comfortable working environment. Its main body features a loft overlooking the surroundings and within sight of the metro, similar to the previous studio. The studio is a vibrant environment every day.

One part of the studio is originally separated from the main body, but it has been later connected with a small overhead gangway. A flexible desk arrangement can be easily changed or combined, which many staff members prefer as it makes the area more comfortable and easier for communicating with one another.

During the time working in this studio, Archmixing has completed the largest urban survey in the firm's history, "Hongkou 1617 Analytic Restoration," along with studying various partial and fragmented changes of surrounding urban buildings. Before moving to the next venue, we held an exhibition *Hongkou 1617* in the vacated studio.

连通合伙人办公室与办公区的桥

The "bridge" connecting main work space to the principals' office

“虹口 1617”展览
Hongkou 1617 exhibition

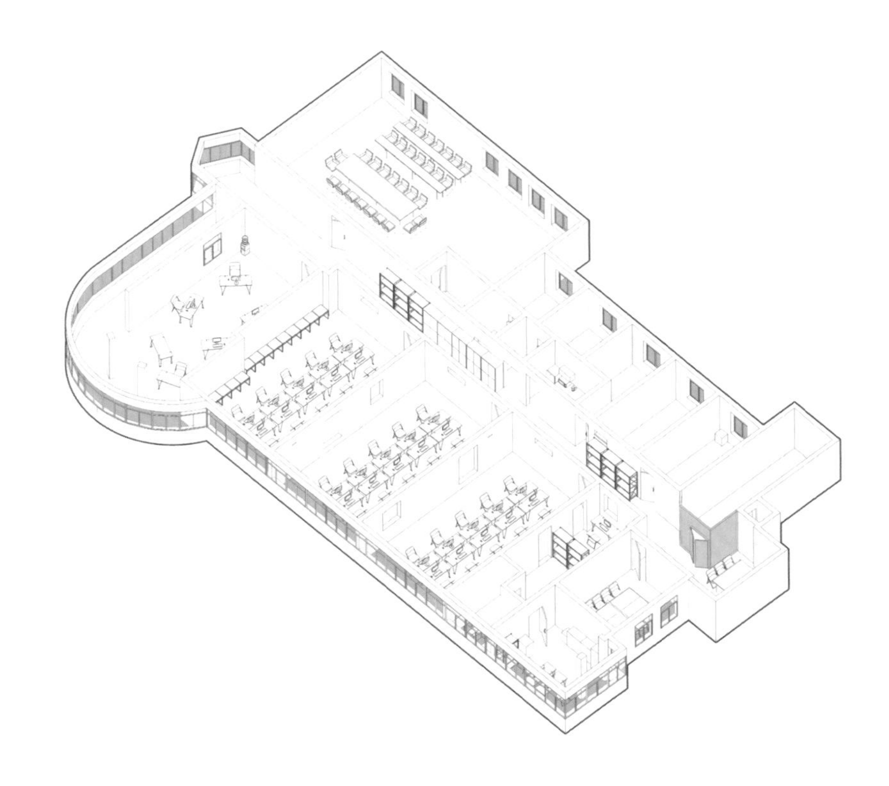

第五个工作室（2017–2018）：粉红色墙面的幼儿园

在结束了对于建筑局部的改变研究之后，我们计划进行建筑的内部改变研究，于是这个工作室的选址放在了徐家汇的商圈，研究这里庞大的城市地下空间和地面商业空间的内部。工作室租了一个待拆迁的幼儿园二层，紧挨着港汇广场，在地铁 18 号口的正上方。幼儿园室内是一连串有着粉红色墙面的教室，我们拆除了护墙板，任由斑驳的墙面暴露出来，在三个相邻教室的隔墙上打开一些洞口，使彼此能跨越连通，长长的连续日光灯斜着直直地穿过相邻的房间，一直延伸到走道上。设计师把这里的房间戏称为“包间”，但大部分人还是思念虹口聚在一起的散乱的桌面，认为那样更好些。

Fifth Office (2017–2018): Kindergarten with Pink Walls

After studying the partial building changes, Archmixing launched a research plan focused on internal changes and consequently set up a studio in the business district of Xujiahui in order to better study interior commercial spaces and vast urban underground spaces. We rented the second floor of a kindergarten, marked for impending demolition and located next to the Grand Gateway Shanghai, directly above the metro exit. The kindergarten comprises a cluster of classrooms with pink walls. The wainscots are removed to expose the mottled walls underneath, and holes are cut in the partition walls to connect three adjacent classrooms, through which long fluorescent lamps extend at an angle into the corridor. The designers dub these areas as “private rooms”, but most of the staff still miss the scattered desk arrangement of the Hongkou studio which they felt has made for a better working environment.

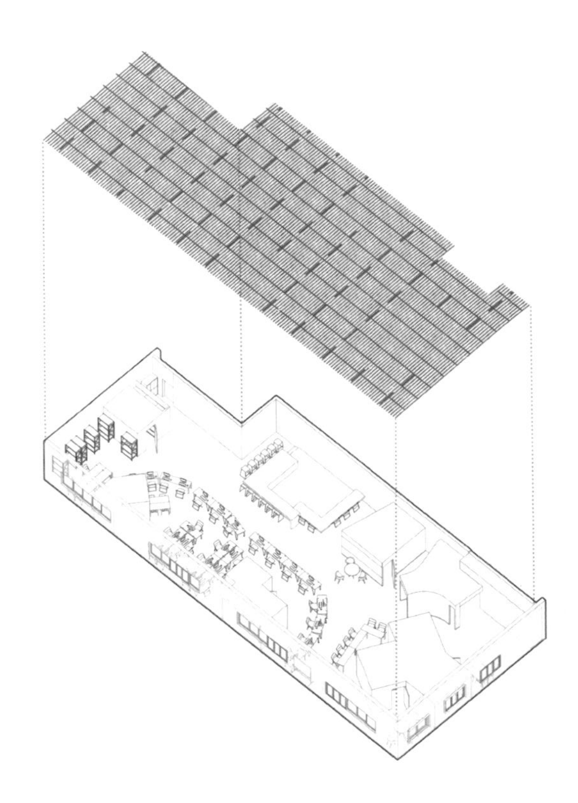

第六个工作室（2018 至今）：小白楼

小白楼是一栋厂房，整个工作室是一个无柱的大空间，这是一个新的开始。由于我们开始试验个人的移动办公，团队的远程云上办公，所以我们尝试把这里做得更像一个做工作坊的总部。总的想法是能创造更多的可以讨论或者独处的角落。因此，合伙人办公室、财务、储藏、司机休息室等被布置为既满足自己最小使用，又能创造角落的形状。我们还布置了一个十分大的咖啡吧台，能够在这里聚会与交流。同时，这次，我们又再次把桌面布置回了自由的散乱的模式。

Sixth Office (2018–): Little White House

Archmixing is currently located in a pillarless, spacious studio within a factory building. It tries to make the space akin to studio headquarters because of its recent experimentation with personal, mobile offices and remote cloud offices, all with an eye on creating more spaces to foster discussions or work solo. As a result, each part of the office—executive office, finance office, storage, break room, and more—is designed with two requirements in mind: suitability for the barest amount of usage and suitability for encouraging collaboration. Archmixing also adds a large coffee bar for staff to meet and chat, as well as reverting to a freestyle desk arrangement once again.

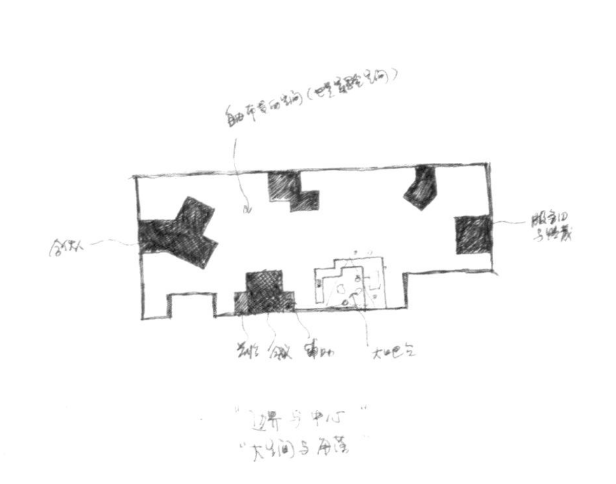

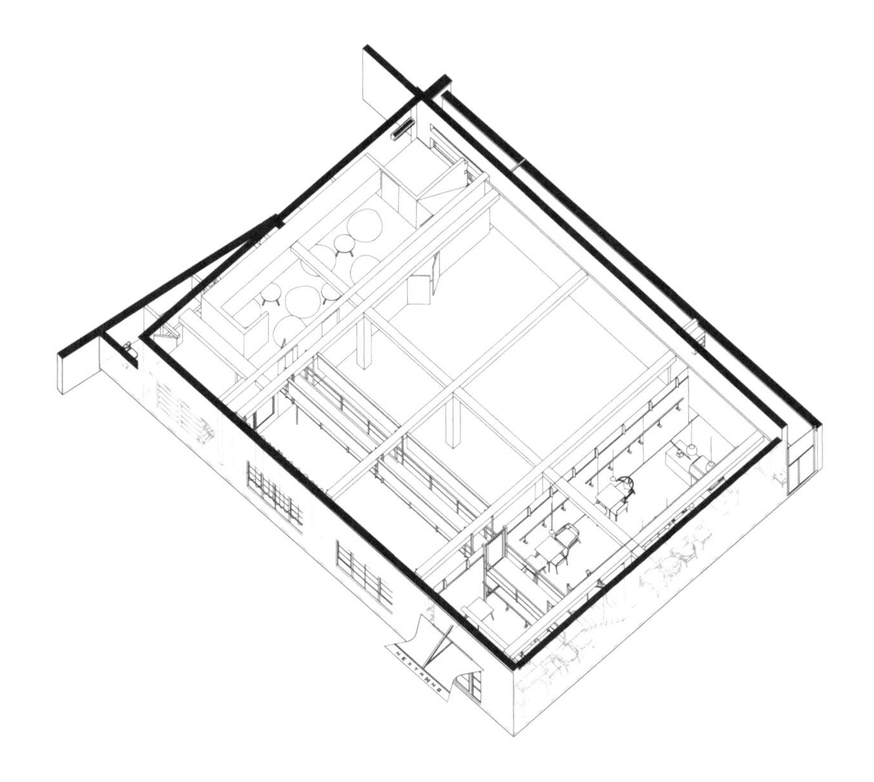

那行文化（2016–2019）：愚园路上的文化生活驿站

“那行文化”是“一年一个工作室”计划衍生的文化交流空间，由阿科米星与朋友们共同创建和运营。在工作室频繁辗转于城市不同街区时，原来的多功能厅功能板块被独立出来，放在交通便利的市中心，不仅举行自己的四季沙龙和展览、活动，更在三年里逐渐成长为一个设计和艺术领域的跨界实践平台。

“那行文化”选址在愚园路历史街区一个弄堂工厂改造的创意园区，需要穿过狭窄曲折的里弄小巷才能辗转抵达。4.4 米层高的厂房空间被分割为黑白灰三厅：黑厅为咖啡厅，满足办公、活动及等候时段的休闲所需；面积最大的白厅简洁明了又暗藏玄机，是各种活动的主要场所；黑白两厅之间以大面积透明玻璃相隔，通过影音设备，两个空间可分可合；位于夹层的灰厅，由一座轻巧的渡桥与黑厅连接，提供较为私密的休息与谈话空间，也可用作展示。在这里，每年以近百场的频率举办着设计、插画、声音、表演、文学、美食等各种活动。

“那行文化”的运营实践使阿科米星多了一重就近观察空间使用方式的维度。

NEXTMIXING (2016–2019): A Cultural Hub on Yuyuan Road

As a positive result of “New Year, New Office”, NEXTMIXING, established and run by Atelier Archmixing and its friends, was born to be a space for cultural exchange. When the Studio started its urban nomadism, the multi-functional hall was set apart and relocated to the city center where public transportation is more convenient and we can hold seasonal lectures, exhibitions and other cultural activities. With three years, NEXTMIXING has grown into a stage for cross-border practice in design and art.

NEXTMIXING is built in a Creative Park on Yuyuan Road which used to be a lane plant in this historic district. To get access to this place people need to go through the winding alleys. The plant of 4.4m high is divided into three halls: black, white and gray. The black hall is NEXTMIXING cafe where people spend time for working, waiting and other leisure activities; the white hall is the largest which is simply and skillfully designed as a major space for various activities; the black and white are separated by large clear glass, but these two can also be combined into a larger hall through multi-media facilities; a light bridge connects the black hall with the gray hall on the interlayer which is a private space for lounge, talk or exhibition. Every year, approximately 100 events are held in NEXTMIXING including design, illustration, sound art, performance, culture and delicacy, etc.

NEXTMIXING has been another pair of eyes that enable Atelier Archmixing to make close observation on space usage.

附录
Appendix

项目信息 Project Data

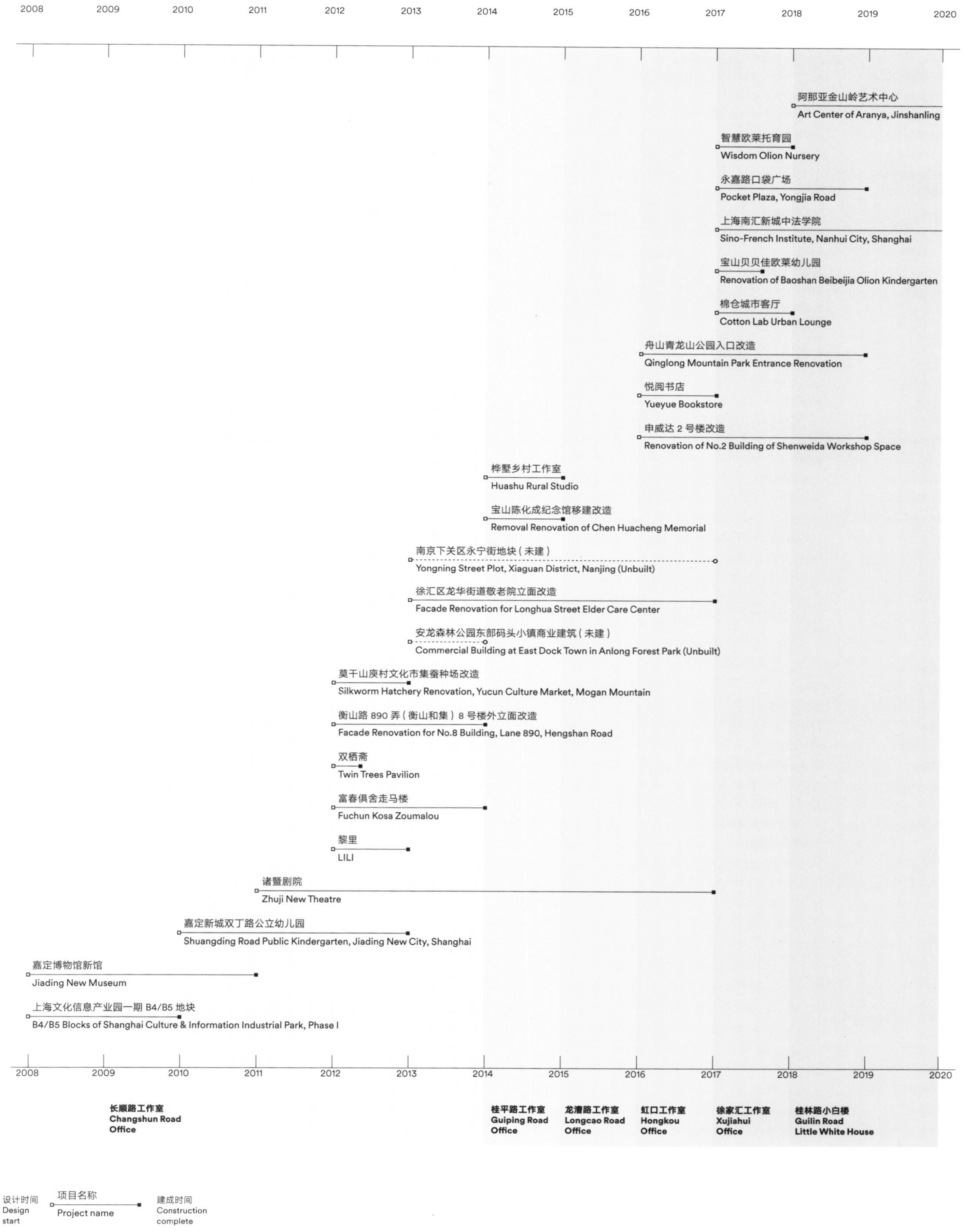

2008 2009 2010 2011 2012 2013 2014 2015 2016 2017 2018 2019 2020

阿那亚金山岭艺术中心
Art Center of Aranya, Jinshanling

智慧欧莱托育园
Wisdom Olion Nursery

永嘉路口袋广场
Pocket Plaza, Yongjia Road

上海南汇新城中法学院
Sino-French Institute, Nanhui City, Shanghai

宝山贝贝佳欧莱幼儿园
Renovation of Baoshan Beibeijia Olion Kindergarten

棉仓城市客厅
Cotton Lab Urban Lounge

舟山青龙山公园入口改造
Qinglong Mountain Park Entrance Renovation

悦阅书店
Yueyue Bookstore

申威达 2 号楼改造
Renovation of No.2 Building of Shenweida Workshop Space

桦墅乡村工作室
Huashu Rural Studio

宝山陈化成纪念馆移建改造
Removal Renovation of Chen Huacheng Memorial

南京下关区永宁街地块（未建）
Yongning Street Plot, Xiaguan District, Nanjing (Unbuilt)

徐汇区龙华街道敬老院立面改造
Facade Renovation for Longhua Street Elder Care Center

安龙森林公园东部码头小镇商业建筑（未建）
Commercial Building at East Dock Town in Anlong Forest Park (Unbuilt)

莫干山庾村文化市集蚕种场改造
Silkworm Hatchery Renovation, Yucun Culture Market, Mogan Mountain

衡山路 890 弄（衡山和集）8 号楼外立面改造
Facade Renovation for No.8 Building, Lane 890, Hengshan Road

双栖斋
Twin Trees Pavilion

富春俱舍走马楼
Fuchun Kosa Zoumalou

黎里
LILI

诸暨剧院
Zhuji New Theatre

嘉定新城双丁路公立幼儿园
Shuangding Road Public Kindergarten, Jiading New City, Shanghai

嘉定博物馆新馆
Jiading New Museum

上海文化信息产业园一期 B4/B5 地块
B4/B5 Blocks of Shanghai Culture & Information Industrial Park, Phase I

2008 2009 2010 2011 2012 2013 2014 2015 2016 2017 2018 2019 2020

长顺路工作室
Changshun Road Office

桂平路工作室
Guiping Road Office

龙漕路工作室
Longcao Road Office

虹口工作室
Hongkou Office

徐家汇工作室
Xujiahui Office

桂林路小白楼
Guilin Road Little White House

设计时间 Design start — 项目名称 Project name — 建成时间 Construction complete

上海文化信息产业园一期 B4/B5 地块

类型：办公
地点：上海市嘉定区
时间：2008/2010
建筑面积：24 900 m²
业主：上海东方文信科技有限公司
建筑团队：庄慎、任皓、彭旭、李珺、邱梅、杨舒婷
合作设计：大舍建筑设计事务所
结构机电：天功建筑设计有限公司
摄影：张嗣烨

嘉定博物馆新馆

类型：文化
地点：上海市嘉定区
时间：2008/2011
建筑面积：9 600 m²
业主：上海市嘉定区文化广播影视管理局
建筑团队：庄慎、任皓、朱峰、邱梅、黄东、杨舒婷
合作设计：大舍建筑设计事务所
苏州园林设计院
结构机电：上海市建筑材料设计研究院
施工单位：上海第五建筑工程有限公司
监理单位：中国建筑技术集团有限公司
摄影：唐煜

嘉定新城双丁路公立幼儿园

类型：教育
地点：上海市嘉定区
时间：2010/2013
建筑面积：6 100 m²
业主：上海嘉定新城发展有限公司
建筑团队：庄慎、任皓、华霞虹、彭旭、邱梅、蒋卓希、崔惠子、龚佳佩、梁博
结构机电：江苏省第一工业设计院有限责任公司上海分公司
施工单位：上海嘉定娄塘建筑安装有限公司
摄影：唐煜

诸暨剧院

类型：文化
地点：浙江省诸暨市
时间：2011/2017
建筑面积：32 000 m²
业主：诸暨市城市建设投资发展有限公司
建筑团队：庄慎、任皓、唐煜、彭旭、邱梅、周渐佳、蒋卓希、梁博
合作设计：浙江舞台设计研究有限公司
诸暨建筑设计院
结构机电：同济大学建筑设计研究院（集团）有限公司
施工单位：诸暨市城市建设投资发展有限公司
景观设计：浙江省建筑设计研究院
幕墙顾问：浙江中南建设集团有限公司
幕墙施工：浙江省武林建筑装饰集团有限公司
室内设计：同济大学建筑设计研究院（集团）有限公司
摄影：苏圣亮

南京下关区永宁街地块

类型：商业、办公、住宅
地点：江苏省南京市
时间：2013
建筑面积：370 000 m²
状态：方案设计
业主：江苏证大商业文化发展有限公司
建筑团队：庄慎、朱捷、唐煜、姚文轩、王侃、孙闻良、陈龙、方昱
合作设计：江苏省建筑设计研究院有限公司

B4/B5 Blocks of Shanghai Culture & Information Industrial Park, Phase I

Type: Office
Location: Jiading District, Shanghai
Date: 2008/2010
Size: 24,900 m²
Client: Shanghai Dongfang Wenxin Technology Co., Ltd.
Design Team: Zhuang Shen, Ren Hao, Peng Xu, Li Jun, Qiu Mei, Yang Shuting
Collaborator: Atelier Deshaus
Structural & Engineer: Tiangong Architectural Design Co., Ltd.
Photographer: Zhang Siye

Jiading New Museum

Type: Cultural
Location: Jiading District, Shanghai
Date: 2008/2011
Size: 9,600 m²
Client: Shanghai Jiading New City Development Co., Ltd.
Design Team: Zhuang Shen, Ren Hao, Zhu Feng, Qiu Mei, Huang Dong, Yang Shuting
Collaborator: Atelier Deshaus
Suzhou Institute of Landscape Architectural Design Co., Ltd.
Structural & Engineer: Shanghai Building Material Industrial Design & Research Institute
Constructor: Shanghai No.5 Construction Engineering Co., Ltd.
Supervision: China Building Technique Group Co., Ltd.
Photographer: Tang Yu

Shuangding Road Public Kindergarten, Jiading New City, Shanghai

Type: Educational
Location: Jiading District, Shanghai
Date: 2010/2013
Size: 6,100 m²
Client: Shanghai Jiading New City Development Co., Ltd.
Design Team: Zhuang Shen, Ren Hao, Hua Xiahong, Peng Xu, Qiu Mei, Jiang Zhuoxi, Cui Huizi, Gong Jiapei, Liang Bo
Structural & Engineer: First Industrial Design Institute of Jiangsu Province Shanghai Branch Co., Ltd.
Constructor: Shanghai Jiading Loutang Construction Installation Co., Ltd.
Photographer: Tang Yu

Zhuji New Theatre

Type: Cultural
Location: Zhuji, Zhejiang Province
Date: 2011/2017
Size: 32,000 m²
Client: Zhuji City Construction Investment Co., Ltd.
Design Team: Zhuang Shen, Ren Hao, Tang Yu, Peng Xu, Qiu Mei, Zhou Jianjia, Jiang Zhuoxi, Liang Bo
Collaborator: Zhejiang Research Institute of Stage Design Co., Ltd.
Zhuji Architectural Design Co., Ltd.
Structural & Engineer: Tongji Architectural Design (Group) Co., Ltd.
Constructor: Zhuji City Construction Investment Development Co., Ltd.
Landscape: Zhejiang Province Institute of Architectural Design and Research
Facade Consultant: Zhejiang Central South Construction Group Co., Ltd.
Facade Construction: Zhejiang Wulin Building Decoration Group Co., Ltd.
Interior: Tongji Architectural Design (Group) Co., Ltd.
Photographer: Su Shengliang

Yongning Street Plot, Xiaguan District, Nanjing

Type: Commercial & Office & Residence
Location: Jiangsu, Nanjing Province
Date: 2013
Size: 370,000 m²
Status: Schematic design
Client: Jiangsu Securities University Commercial Culture Development Co., Ltd.
Design Team: Zhuang Shen, Zhu Jie, Tang Yu, Yao Wenxuan, Wang Kan, Sun Wenliang, Chen Long, Fang Yu
Collaborator: Jiangsu Provincial Architectural D&R Institute Co., Ltd.

上海南汇新城中法学院

类型：教育
地点：上海市浦东新区
时间：2017/
建筑面积：29 000 m^2
状态：施工中
业主：上海港城开发（集团）有限公司
建筑团队：庄慎、朱捷、吴启晶、尹济东、孙荃、徐帆
结构机电：上海天功建筑设计有限公司
施工单位：上海建工七建集团有限公司

Sino-French Institute, Nanhui City, Shanghai

Type: Educational
Location: Pudong District, Shanghai
Date: 2017/
Size: 29,000 m^2
Status: Under construction
Client: Shanghai Harbour City Development (Group) Co., Ltd.
Design Team: Zhuang Shen, Zhu Jie, Wu Qijing, Yin Jidong, Sun Quan, Xu Fan
Structural & Engineer: Tiangong Architecture Design Co., Ltd.
Constructor: Shanghai Construction No. 7 (Group) Co., Ltd.

双栖斋

类型：建筑改造
地点：江苏省苏州市
时间：2012
建筑面积：56 m^2
业主：张弘
建筑团队：庄慎、田丹妮
施工：上海开乐建设工程有限公司
摄影：唐煜

Twin Trees Pavilion

Type: Building Renovation
Location: Suzhou, Jiangsu Province
Date: 2012
Size: 56 m^2
Client: Zhang Hong
Team: Zhuang Shen, Tian Danni
Constructor: Shanghai Kaile Construction Engineering Co., Ltd.
Photographer: Tang Yu

莫干山庾村文化市集蚕种场改造

类型：建筑改造
地点：浙江省湖州市
时间：2012/2013
建筑面积：3 000m^2
业主：清境旅游投资管理有限公司
建筑团队：庄慎、孔锐、王侃、姚文轩、贺珊
模型制作：王侃、陆津硕、王世亮、姚文轩、田丹妮、刘羽云、刘霞、蒋玲玲、黄莹（实习）、徐渐佳（实习）
合作设计：上海东联设计集团
施工单位：上海开乐建设工程有限公司
摄影：唐煜、庄慎

Silkworm Hatchery Renovation, Yucun Culture Market, Mogan Mountain

Type: Building Renovation
Location: Huzhou, Zhejiang Province
Date: 2012/2013
Size: 3,000 m^2
Client: Qingjing Tourism Investment Management Co., Ltd.
Design Team: Zhuang Shen, Kong Rui, Wang Kan, Yao Wenxuan, He Shan
Manual Model: Wang Kan, Lu Jinshuo, Wang Shiliang, Yao Wenxuan, Tian Danni, Liu Yuyun, Liu Xia, Jiang Lingling, Huang Ying (Intern), Xu Jianjia (Intern)
Collaborator: Donglian Design Group (Shanghai)
Constructor: Shanghai Kaile Construction Engineering Co., Ltd.
Photographers: Tang Yu, Zhuang Shen

富春俱舍走马楼

类型：建筑改造
地点：浙江省杭州市
时间：2012/2014
建筑面积：464m^2
业主：上海宏泰文化传播有限公司
建筑团队：唐煜、姚文轩、蒋卓希、黄莉敏、杨毓琼、陆津硕、梁博
合作单位：上海源规建筑结构设计事务所（普通合伙）
摄影：唐煜

Fuchu Kosa Zoumalou

Type: Building Renovation
Location: Hangzhou, Zhejiang Province
Date: 2012/2014
Size: 464 m^2
Client: Shanghai Hongtai Culture Communication Co., Ltd.
Design Team: Tang Yu, Yao Wenxuan, Jiang Zhuoxi, Huang Limin, Yang Yuqiong, Lu Jinshuo, Liang Bo
Constructor: Shanghai Wildness Structural Des. Firm Inc. (General partnership)
Photographer: Tang Yu

安龙森林公园东部码头小镇商业建筑

类型：商业
地点：浙江省杭州市千岛湖
时间：2013
建筑面积：2 900 m^2
状态：方案设计
业主：杭州华联置业有限公司
建筑团队：庄慎、唐煜、王侃
合作设计：同济大学建筑设计研究院（集团）有限公司

Commercial Building at East Dock Town in Anlong Forest Park

Type: Commercial
Location: Hangzhou, Zhejiang Province
Date: 2013
Size: 2,900 m^2
Status: Schematic design
Client: Hangzhou Hualian Real Estate Co., Ltd.
Design Team: Zhuang Shen, Tang Yu, Wang Kan
Collaborator: Tongji Architectural Design (Group) Co., Ltd.

阿那亚金山岭艺术中心

类型：商业
地点：河北省承德市
时间：2018/2020
建筑面积：875 m^2
状态：施工中
业主：承德阿那亚房地产开发有限公司
建筑团队：庄慎、朱捷、龚晨曦、邱梅、黎家泓、徐帆、张灏宸、李文伟（实习生）、梁俊（实习生）、孙曦梦（实习生）
结构设计：上海源规建筑结构设计事务所（普通合伙）
机电单位：同济大学建筑设计研究院（集团）有限公司
施工单位：佳木斯三江建筑工程有限责任公司
幕墙顾问：上海熙玛工程顾问有限公司

Art Center of Aranya, Jinshanling

Type: Commercial
Location: Chengde, Hebei Province
Date: 2018/2020
Size: 875 m^2
Status: Under construction
Client: Chengde Aranya
Design Team: Zhuang Shen, Zhu Jie, Gong Chenxi, Qiu Mei, Li Jiahong, Xu Fan, Zhang Haochen, Li Wenwei(Intern), Liang Jun(Intern), Sun Ximeng(Intern)
Structural: Shanghai Wildness Structural Des. Firm Inc. (General partnership)
Engineer: Tongji Architectural Design (Group) Co., Ltd.
Constructor: Jiamus Sanjiang Construction Engineering Co., Ltd.
Facade Consultant: Shanghai Ciam Facade Engineering Consultants Co., Ltd.

黎里

类型：建筑改造
地点：江苏省黎里镇
时间：2012
建筑面积：200 m^2
业主：张弘
建筑团队：庄慎、田丹妮、杨云樵
施工单位：上海开乐建设工程有限公司
摄影：唐煜

衡山路 890 弄（衡山和集）8 号楼外立面改造

类型：立面改造
地点：上海市徐汇区
时间：2012/2014
建筑面积：230 m^2
业主：上海衡复置业有限公司
建筑团队：庄慎、王侃、杨云樵、解文静（实习）
施工单位：上海徐房建筑实业公司
上海日朗门窗有限公司
上海盖珂实业有限公司
摄影：唐煜

徐汇区龙华街道敬老院立面改造

类型：立面改造
地点：上海市徐汇区
时间：2013/2017
建筑面积：10 500 m^2
业主：上海市徐汇区人民政府龙华街道办事处
建筑团队：庄慎、朱捷、王侃
结构机电：上海徐汇规划建筑设计有限公司
施工单位：上海汇成建设发展有限公司
幕墙施工：上海苡顺幕墙系统工程有限公司
摄影：唐煜

桦墅乡村工作室

类型：建筑改造
地点：江苏省南京市周冲桦墅村
时间：2014/2015
建筑面积：252 m^2
业主：上海东联设计集团
建筑团队：庄慎、唐煜、王迪、陈向鹏、吴奇韬
结构设计：上海源规建筑结构设计事务所（普通合伙）
施工单位：南京正玲古典园林建筑有限公司
摄影：唐煜、陈平楠

申威达 2 号楼改造

类型：建筑改造
地点：上海市徐汇区
时间：2016/2019
建筑面积：6 674 m^2
业主：上海净鑫置业有限公司
建筑团队：朱捷、庄慎、姚文轩、邱梅、李立德、邓健
结构设计：上海源规建筑结构设计事务所（普通合伙）
机电单位：上海华东建设发展设计有限公司
施工单位：上海华洲建设发展有限公司
摄影：吴清山

悦阅书店

类型：室内改造
地点：上海市杨浦区
时间：2016/2017
建筑面积：94 m^2
业主：上海悦悦图书有限公司
建筑团队：庄慎、华霞虹、游诗雨、赵鹏宇（实习生）
施工单位：上海杜雅建筑装饰有限公司
上海申信五金制品有限公司
摄影：吴清山

LILI

Type: Renovation
Location: Lili, Jiangsu Province
Date: 2012
Size: 200 m^2
Client: Zhang Hong
Design Team: Zhuang Shen, Tian Danni, Yang Yunqiao
Constructor: Shanghai Kaile Construction Engineering Co., Ltd.
Photographer: Tang Yu

Facade Renovation for No.8 Building, Lane 890, Hengshan Road

Type: Facade Renovation
Location: Xuhui District, Shanghai
Date: 2012/2014
Size: 230 m^2
Client: Shanghai Hengfu Properties Co., Ltd.
Design Team: Zhuang Shen, Wang Kan, Yang Yunqiao, Xie Wenjing (Intern)
Constructor: Shanghai Xufang Construction Industry Company
Shanghai Rilang Co., Ltd.
Gainker (China) Building Technology Company
Photographer: Tang Yu

Facade Renovation for Longhua Street Elder Care Center

Type: Facade Renovation
Location: Xuhui District, Shanghai
Date: 2013/2017
Size: 10,500 m^2
Client: Shanghai Xuhui District People's Government Longhua Street Office
Design Team: Zhuang Shen, Zhu Jie, Wang Kan
Structural & Engineer: Shanghai Xuhui Planning Architectural Design Co., Ltd.
Constructor: Shanghai Huicheng Construction and Development Co., Ltd.
Facade Construction: Shanghai Yishun Facade Engineering Co., Ltd.
Photographer: Tang Yu

Huashu Rural Studio

Type: Building Renovation
Location: Nanjing, Jiangsu Province
Date: 2014/2015
Size: 252 m^2
Client: Donglian Design Group (Shanghai)
Design Team: Zhuang Shen, Tang Yu, Wang Di, Chen Xiangpeng, Wu Qitao
Structural: Shanghai Wildness Structural Des. Firm Inc. (General partnership)
Constructor: Nanjing Zhengling Classical Garden Architecture Co., Ltd.
Photographers: Tang Yu, Chen Pingnan

Renovation of No.2 Building of Shenweida Workshop Space

Type: Building Renovation
Location: Xuhui District, Shanghai
Date: 2016/2019
Size: 6,674 m^2
Client: Shanghai Jingxin Real Estate Co., Ltd.
Design Team: Zhu Jie, Zhuang Shen, Yao Wenxuan, Qiu Mei, Li Lide, Deng Jian
Structural: Shanghai Wildness Structural Des. Firm Inc. (General partnership)
Engineer: Shanghai East China Construction Development Design Co., Ltd.
Constructor: Shanghai Huazhou Construction Development Co., Ltd.
Photographer: Wu Qingshan

Yueyue Bookstore

Type: Interior Renovation
Location: Yangpu District, Shanghai
Date: 2016/2017
Size: 94 m^2
Client: Shanghai Yueyue Book Co., Ltd.
Design Team: Zhuang Shen, Hua Xiahong, You Shiyu, Zhao Pengyu (Intern)
Constructor: Shanghai DUYA Construction & Decoration Shanghai Co., Ltd.
Shanghai Shenxin Hardware Products Co., Ltd.
Photographer: Wu Qingshan

宝山贝贝佳欧莱幼儿园

类型： 建筑改造
地点： 上海市宝山区宝菊路 377 号
时间： 2017
建筑面积： 5 400 m^2
业主： 上海贝贝佳欧莱幼儿园
建筑团队： 庄慎、朱捷、张灏宸、李立德、陈弘邦、丁心慧、王子潇（实习生）
结构设计： 上海源规建筑结构设计事务所（普通合伙）
机电单位： 上海亚新工程顾问有限公司
施工单位： 上海维方建筑装饰工程有限公司
摄影： 吴清山

Baoshan Beibeijia Olion Kindergarten

Type: Building Renovation
Location: Baoshan District, Shanghai
Date: 2017
Size: 5,400 m^2
Client: shanghai Beibeijia Olion Kindergarten
Design Team: Zhuang Shen, Zhu Jie, Zhang Haochen, Li Lide, Chen Hongbang, Ding Xinhui, Wang Zixiao (Intern)
Structural: Shanghai Wildness Structural Des. Firm Inc. (General Partnership)
Engineer: MAA Engineering Consultants (Shanghai) Co., Ltd.
Constructor: Shanghai Weifang Construction Decoration Engineering Co., Ltd.
Photographer: Wu Qingshan

棉仓城市客厅

类型： 建筑改造
地点： 江苏省常州市长江路 25 号三晶科技园 8 号
时间： 2017/2018
建筑面积： 6 300 m^2
业主： 江苏棉仓服饰有限公司
建筑团队： 庄慎、唐煜、朱捷、王迪、邓健、叶阳、游诗雨、张灏宸、黎家泓、许晔（实习生）、周炜楠（实习）
结构设计： 上海源规建筑结构设计事务所（普通合伙）
机电单位： 焦学渊、曹阳、陆仁杰、沈建
施工单位： 常州市海格装饰工程有限公司
木结构施工单位： 上海思卡福建筑工程有限公司
摄影： 苏圣亮、吴清山

Cotton Lab Urban Lounge

Type: Building Renovation
Location: Changzhou, Jiangsu Province
Date: 2017/2018
Size: 6,300 m^2
Client: Jiangsu Cotton Lab Clothing Co., Ltd.
Design team: Zhuang Shen, Tang Yu, Zhu Jie, Wang Di, Deng Jian, Ye Yang, You Shiyu, Zhang Haochen, Li Jiahong, Xu Ye(Intern), Zhou Weinan(Intern)
Structural: Shanghai Wildness Structural Des. Firm Inc. (General Partnership)
Mechanical & Electrical Consultant: Jiao Xueyuan, Cao Yang, Lu Renjie, Shen Jian
Constructor: Changzhou Haige Decoration Engineering Co., Ltd.
Wood Structure Constructor: Shanghai SKF Builder
Photographers: Su Shengliang, Wu Qingshan

智慧欧莱托育园

类型： 建筑改造
地点： 上海市宝山区
时间： 2017/2018
建筑面积： 1 467 m^2
业主： 上海鸥翼教育科技有限公司
设计团队： 庄慎、朱捷、张灏宸、周炜楠（实习生）
结构机电： 上海都市建筑结构有限公司
施工单位： 上海林伟建筑工程有限公司
摄影： 吴清山

Wisdom Olion Nursery

Type: Building Renovation
Location: Baoshan District, Shanghai
Date: 2017/2018
Size: 1,467m^2
Client: Shanghai Seagull Wing Education Technology Co., Ltd.
Design Team: Zhuang Shen, Zhu Jie, Zhang Haochen, Zhou Weinan (Intern)
Structural & Engineer: Shanghai Urban Architecture Design LCC.
Constructor: Shanghai Linwei Construction Engineering Co., Ltd.
Photographer: Wu Qingshan

宝山陈化成纪念馆移建改造

类型： 建筑改造
地点： 上海市宝山区临江公园内
时间： 2014/2015
建筑面积： 198 m^2
业主： 上海宝山区文化广播影视管理局
建筑团队： 庄慎、任皓、方昱、田丹妮、杨毓琼、姚文轩、蔡宣皓（实习）、王轶（实习）
结构设计： 上海源规建筑结构设计事务所（普通合伙）
施工单位： 南通华胜建设工程有限公司
摄影： 唐煜

Removal Renovation of Chen Huacheng Memorial

Type: Building Renovation
Location: Baoshan District, Shanghai
Date: 2014/2015
Size: 198 m^2
Client: Shanghai Baoshan Municipal Administration of Culture, Radio and Television
Design Team: Zhuang Shen, Ren Hao, Fang Yu, Tian Danni, Yang Yuqiong, Yao Wenxuan, Cai Xuanhao (Intern), Wang Yi (Intern)
Structural: Shanghai Wildness Structural Des. Firm Inc. (General partnership)
Constructor: Nantong Huasheng Construction Engineering Co., Ltd.
Photographer: Tang Yu

舟山青龙山公园入口改造

类型： 文化
地点： 浙江省舟山市
时间： 2016/2019
建筑面积： 1 080 m^2
业主： 舟山市普陀城市投资集团有限公司
建筑团队： 庄慎、朱捷、邱梅、侯薇、叶阳、褚婉茹、李立德
结构机电： 同济大学建筑设计研究院（集团）有限公司
施工单位： 浙江劲华建筑有限公司
幕墙施工： 上海欣泰建筑装饰设计有限公司
摄影： 陈平楠

Qinglong Mountain Park Entrance Renovation

Type: Cultural
Location: Zhoushan, Zhejiang Province
Date: 2016/2019
Size: 1,080 m^2
Client: Zhoushan Putuo City Construction Investment Co., Ltd.
Design Team: Zhuang Shen, Zhu Jie, Qiu Mie, Hou Mei, Ye Yang, Chu Wanru, Li Lide
Structural & Engineer: Tongji Architectural Design (Group) Co., Ltd.
Constructor: Zhejiang Jinhua Construction Co., Ltd.
Facade Construction: Shanghai Xintai Architectural Decoration Design Co., Ltd.
Photographer: Chen Pingnan

永嘉路口袋广场

类型： 城市更新
地点： 上海市徐汇区
设计 / 建成： 2017/2019
建筑面积： 767 m^2
业主： 上海市徐汇区建设和交通委员会
设计团队： 庄慎、朱捷、李立德、丁心慧、尹济东、陈弘邦、卢穗娟（实习）、邱鑫（实习）
结构设计： 上海源规建筑结构设计事务所（普通合伙）
施工单位： 上海徐房建筑实业公司
摄影： 吴清山

城市内的工作室

长顺路工作室（2009—2014）

地点： 上海市长宁区
规模： 400 m^2
设计团队： 庄慎、任皓、窦义年

桂平路工作室（2014—2015）

地点： 上海市徐汇区
规模： 150 m^2
设计团队： 庄慎、任皓、唐煜、朱捷

龙漕路工作室（2015—2016）

地点： 上海市徐汇区
规模： 320 m^2
设计团队： 庄慎、任皓、唐煜、陈向鹏、吴奇韬

虹口工作室（2016—2017）

地点： 上海市虹口区
规模： 330 m^2
设计团队： 庄慎、任皓、唐煜、朱捷、陈龙

徐家汇工作室（2017—2018）

地点： 上海市徐汇区
规模： 670 m^2
设计团队： 庄慎、任皓、唐煜、朱捷、刘双雁

桂林路小白楼（2018 至今）

地点： 上海市徐汇区
规模： 400 m^2
设计团队： 庄慎、唐煜、朱捷、褚婉茹、张珏

那行文化（2016-2019）

地址： 上海市长宁区
规模： 187 m^2
设计团队： 阿科米星建筑设计事务所、冶是建筑工作室

Pocket Plaza, Yongjia Road

Type: Urban Renewal
Location: Xuhui District, Shanghai
Date: 2017/2019
Size: 767 m^2
Client: Shanghai Xuhui District Construction and Transportation Committee
Design Team: Zhuang Shen, Zhu Jie, Li Lide, Ding Xinhui, Yin Jidong, Chen Hongbang, Lu Shuijuan (Intern), Qiu Xin (Intern)
Structural: Shanghai Wildness Structural Des. Firm Inc. (General Partnership)
Constructor: Shanghai Xufang Construction Industry Co., Ltd.
Photographer: Wu Qingshan

Work Within the City

The Changshun Road Office (2009–2014)

Location: Changning District, Shanghai
Size: 400 m^2
Design Team: Zhuang Shen, Ren Hao, Dou Yinian

The Guiping Road Office (2014–2015)

Location: Xuhui District, Shanghai
Size: 150 m^2
Design Team: Zhuang Shen, Ren Hao, Tang Yu, Zhu Jie

The Longcao Road Office (2015–2016)

Location: Xuhui District, Shanghai
Size: 320 m^2
Design Team: Zhuang Shen, Ren Hao, Tang Yu, Zhu Jie, Chen Xiangpeng, Wu Qitao

The Hongkou Office (2016–2017)

Location: Hongkou District, Shanghai
Size: 330 m^2
Design Team: Zhuang Shen, Ren Hao, Tang Yu, Zhu Jie, Chen Long

The Xujiahui Office (2017–2018)

Location: Xuhui District, Shanghai
Size: 670 m^2
Design Team: Zhuang Shen, Ren Hao, Tang Yu, Zhu Jie, Liu Shuangyan

Little White House (2018–)

Location: Xuhui District, Shanghai
Size: 400 m^2
Design Team: Zhuang Shen, Tang Yu, Zhu Jie, Chu Wanru, Zhang Jue

NEXTMIXING (2016–2019)

Location: Changning District, Shanghai
Size: 187 m^2
Design Team: Atelier Archmixing, YeArch Studio

事务所简介 About Archmixing

阿科米星建筑设计事务所成立于 2009 年 7 月，由庄慎、任皓创建，唐煜、朱捷加入合伙，华霞虹担任学术顾问。

"阿科米星"是 Archmixing 的谐音，意为"混合建筑"，源于成立之初，我们认为："设计需要消除从专业领域到社会价值的各种狭隘的界限，不放弃去尝试各种可能性。"我们的设计无论内容还是范围都不拘一格：从城市设计到小型建筑，从专业建造到民间建造，我们的工作兴趣和实践策略丰富而灵活。

在国内实践二十余年，阿科米星的主持建筑师们秉持知行合一的精神，均拥有丰富的专业实践与研究经验。我们越来越认识到，快速城市化和全面市场化是我们实践的语境，纷繁复杂的中国城市（城乡）现状是我们设计的出发点，也是自己身处其间，无法忽视的日常环境。在这样的实践环境中试图发现一些建筑学的新经验，一直是我们工作的动力。这其中，越来越吸引我们的正是那看上去问题无穷，却又生机勃勃，仿佛蕴藏着巨大力量的日常城市与建筑。

在这样的实践研究里，我们将视野更多地投放到建筑的使用端，发觉了"改变"这个既有建筑学的盲点。我们越来越明确地认识到，"什么改变了既有的城市与建筑""什么会改变未来的建筑"这两个问题会有效地引导我们的设计研究与实践。

Atelier Archmixing was founded by Zhuang Shen and Ren Hao in July 2009. Joined partners are Tang Yu and Zhu Jie. Hua Xiahong serves as an academic advisor.

The studio's Chinese name A Ke Mi Xing echoes the pronunciation of "Archmixing", which means Mixing Architecture, a critical position as well as a strategy we hold from the very beginning. We think it necessary to move beyond our narrow area of expertise to a wider range of unlimited boundaries by seeking every opportunity and employing strategies with no prior notions. The projects we have completed feature in various scales and functions, covering the range from urban design to artistic installation, from professional manufacture to informal locale craftsmanship. We boast in extensive design interests as well as flexible practicing strategies.

Through more than two decades' practices, Atelier Archmixing's principal architects have accumulated plenty of practicing and researching experiences. We see China's rapid urbanization and extensive commercialization as our design context, China's controversial urban and rural status quo, which is also the everyday environment we live in and impossible to ignore, as our departure point. Is it possible to draw new strength for architectural discipline from these circumstances? We especially see opportunities in everyday urban life and built environment, which is problematic but also energetic and vibrant.

Under such a context, we focus more on the user's end of architecture and discover one blind spot in the mainstream Architecture, Change. The following two concerns, "what changes the existing cities and buildings" and "what will change the future architecture" will effectively guide our design research and practice.

阿科米星合伙人庄慎（左2）、任皓（左）、唐煜（右）、朱捷（右2）

Atelier Archmixing's partners: Zhuang Shen (left 2), Ren Hao (left), Tang Yu (right), Zhu Jie (right 2)

阿科米星员工名单 2009–2019

邱 梅	刘 霞	褚 激	王 迪	陈平楠
李立德	张灏宸	游诗雨	吴启晶	赵 颖
邓 健	龚晨曦	郭 炜	黎家泓	徐 帆
唐润恩	杜清娴	张 珏	蒋玲玲	何春瑶
叶 梓	王子潇	马子舒	崔 舟	

陈 龙	姚文轩	蒋卓希	田丹妮	彭 旭
叶 阳	李 敏	顾天恩	方 昱	王 侃
褚婉茹	董智伟	丁心慧	龚佳佩	朱 峰
刘双雁	孙 荃	尹济东	韩 琦	陈弘邦
陈 默	陈卓玮	张 弢	陈向鹏	王骁彬
杨毓琼	李佳颖	唐玄武	张艺露	潘得力
李 聪	黄红宇	黄莉敏	刘羽云	朱成浩
崔惠子	马丽娟	窦义年	黄 东	杨舒婷
范佳峰	侯 薇	孙闻良	陆津烁	杨云樵
张馨培	吴奇韬	杨跃龙	邹红飞	周 静
王志萍	韩术美	王 波	张 楠	王世亮
阿荣高娃	谢安倩	于瑞莹	沈莉莉	甘 云

Archmixing Staff 2009–2019

Qiu Mei	Liu Xia	Chu Ji	Wang Di	Chen Pingnan
Li Lide	Zhang Haochen	You Shiyu	Wu Qijing	Zhao Ying
Deng Jian	Gong Chenxi	Guo Wei	Li Jiahong	Xu Fan
Tang Run'en	Du Qingxian	Zhang Jue	Jiang Lingling	He Chunyao
Ye Zi	Wang Zixiao	Ma Zishu	Cui Zhou	

Chen Long	Yao Wenxuan	Jiang Zhuoxi	Tian Danni	Peng Xu
Ye Yang	Li Min	Gu Tian'en	Fang Yu	Wang Kan
Chu Wanru	Dong Zhiwei	Ding Xinhui	Gong Jiapei	Zhu Feng
Liu Shuangyan	Sun Quan	Yin Jidong	Han Qi	Chen Hongbang
Chen Mo	Chen Zhuowei	Zhang Tao	Chen Xiangpeng	Wang Xiaobin
Yang Yuqiong	Li Jiaying	Tang Xuanwu	Zhang Yilu	Pan Deli
Li Cong	Huang Hongyu	Huang Limin	Liu Yuyun	Zhu Chenghao
Cui Huizi	Ma Lijuan	Dou Yinian	Huang Dong	Yang Shuting
Fan Jiafeng	Hou Wei	Sun Wenliang	Lu Jinshuo	Yang Yunqiao
Zhang Xinpei	Wu Qitao	Yang Yuelong	Zou Hongfei	Zhou Jing
Wang Zhiping	Han Shumei	Wang Bo	Zhang Nan	Wang Shiliang
Arong Gaowa	Xie Anqian	Yu Ruiying	Shen Lili	Gan Yun

阿科米星全体合照2020
Group photo 2020

获奖

2020——IFI卓越设计奖购物空间类银奖［“一条”线下店（莘庄店）］
2019——Archdaily 2019 年度最佳改造建筑大奖（棉仓城市客厅）
2019——AIA 上海卓越设计奖室内设计奖（棉仓城市客厅）
2018——WA 中国建筑奖之 WA 设计实验奖优胜奖（棉仓城市客厅）
2018——《家居廊》最佳年度商用零售空间奖（棉仓城市客厅）
2018——DFA 亚洲最具影响力设计奖银奖（棉仓城市客厅）
2018——入围“Dezeen Awards 2018”零售类室内设计前 5 名（悦阅书店）
2017——上海市建筑学会第七届建筑创作奖之公共建筑优秀类（张江集电港剑腾三期）
2016——自然建造 · 第四届中国建筑传媒奖青年探索奖（庄慎）
2016——上海建筑学会首届科技进步奖三等奖（衡山坊 890 弄 8 号楼外立面）
2015——上海市建筑学会第六届建筑创作奖之佳作奖（衡山坊 890 弄 8 号楼外立面）
2015——第六届最佳设计酒店大奖“最佳新酒店”（富春俱舍）
2012——中国建筑传媒奖居住建筑特别奖提名（黎里）
2012——WA 中国建筑佳作奖（上海文化信息产业园一期 B4/B5 地块）
2010——英国皇家特许建造学会“施工管理杰出成就奖”（嘉定新城规划展示馆）

Major Awards

2020——The silver medal of Commerce Award of IFI. (Office Retail Space of Yitiao. TV-Xinzhuang Store)
2019——The 2nd place Refurbishment in Architecture Award 2019, hosted by Archdaily (Cotton Lab Urban Lounge)
2019——Interior Architecture Honor Award of AIA Shanghai | Beijing Design (Cotton Lab Urban Lounge)
2018——The 1st prize of Design Experiment Award, issued by *World Architecture China*. (Cotton Lab Urban Lounge)
2018——Best Commercial Retail Space Award, issued by *ELLE DECORATION*. (Cotton Lab Urban Lounge)
2018——The silver medal of DFA Design of Asia Awards. (Cotton Lab Urban Lounge)
2017——The 7th Architectural Creation Award for Excellent Public Building from ASSC. (Zhangjiang IC Harbour Phase III China Fortune Wisdom Mark)
2016——The Young Explorer Award of 4th China Architecture Media Award (Zhuang Shen)
2016——The 3rd prize in the Technological Progress Award of ASSC. (Facade Renovation of Building 8, Lane 890, Hengshan Road)
2015——Architectural Creation Award of ASSC. (Facade Renovation of Building 8, Lane 890, Hengshan Road)
2015——Best New Hotel of the Year of the 6th Best Design Hotel. (Fuchun Kosa)
2012——Residential Special Award Nominee, China Architecture Media Award. (LILI)
2012——Best Chinese Architecture Award, issued by *World Architecture China*. (B4/B5 Blocks of Shanghai Culture & Information Industrial Park, Phase I)
2010——Outstanding Achievement in Construction Management Award, issued by CIOB. (Jiading New City Exhibition Hall)

展览

2019——“内的宣言”，那行文化，上海
2019——“未知城市”中国当代建筑装置影像展，深圳
2018——“自由空间”威尼斯建筑双年展，意大利
2017——上海城市空间艺术季：“连接 thisCONNECTION”，上海
2017——“虹口 1617”，阿科米星虹口工作室 / 那行文化，上海
2017——“当代中国建筑师”克拉科夫建筑双年展，波兰
2016——“走向批判的实用主义：当代中国建筑展”，哈佛大学，美国
2016——“中澳设计研究对话展”，墨尔本，澳大利亚
2015——上海城市空间艺术季：“城市更新”，上海
2015——建筑中国 1000 展 (2000–2015)，北京 798 艺术工厂
2015——上海双年展——“城市车间”主题展，上海
2014——“奔牛上海”大型城市公共艺术展，上海
2014——“剩余价值”环保艺术展，上海
2013——西岸建筑与当代艺术双年展，上海
2013——“蜃景——当代中国博物馆建筑的十二种呈现”，上海
2012——首届中国设计大展“时代 · 创造”，深圳
2012——“从研究到设计”米兰三年展，意大利
2011——“公共外延——上海青浦与嘉定的当代建筑实践”深港双年展，深圳 / 香港
2011——“物我之境：田园 / 城市 / 建筑”成都双年展，四川

Exhibitions

2019——“Manifestos on Interiority”, NEXTMIXING, Shanghai
2019——“Unknown City” China Contemporary Architecture and Image Exhibition, Shenzhen
2018——“Freespace”, Venice Architectural Biennale, Italy
2017——“thisCONNECTION”, Shanghai Urban Space Art Season (SUSAS), Shanghai
2017——“Hongkou 1617”, Atelier Archmixing Hongkou Office & NEXTMIXING, Shanghai
2017——“Contemporary Chinese Architecture”, Krakow Architecture Biennale, Poland
2016——“From ‘Experimental Architecture’ to ‘Critical Pragmatism’: Practice of Contemporary Architects in China”, Harvard University, USA
2016——“Sino-Australia Design Research Dialogue & Academic Conference”, Melbourne, Australia
2015——“Urban Renewal”, SUSAS, Shanghai
2015——Architecture China 1000(2000-2015), Beijing.
2015——“Urban Workshop”, 10th Shanghai Biennale, Shanghai
2014——“Cow Parade”, a massive public art exhibition, Shanghai
2014——“Residual Value” Environmental Art Exhibition, Shanghai
2013——“West Bund: A Biennial of Architecture and Contemporary Art”, Shanghai
2013——“Spectacle: 12 Presentations of Contemporary Museum Architecture in China”, Shanghai
2012——“Innovation in New Era” China Design Exhibition, Shenzhen
2012——“From Research to Design”, La Triennal Di Milano, Italy
2011——“Edge of Public: Contemporary Architecture in Shanghai's Qingpu and Jiading New Town”, Hong Kong & Shenzhen Bi-City Biennale of Urbanism/Architecture, Shenzhen/Hong kong
2011——“Holistic Realm: Landscape/Architecture/Urbanism”, Chengdu Biennale, Sichuan

专利

2018——家居功能模块化的住宅（ZL 2018 2 0155441.2）
2016——发光砖和外墙装饰立面（ZL 2016 2 0861769.7）
2016——支腿及家具（ZL 2016 2 1253881.9）
2016——支架（外观设计，ZL 2016 3 0552139.7）
2013——一种成像展示装置（ZL 2013 2 0237210.3）

Patents

2018——Residence with modularized functional components (ZL 2018 2 0155441.2)
2016——Luminous brick and decorative facade (ZL 2016 2 0861769.7)
2016——Table stand and furniture (ZL 2016 2 1253881.9)
2016——Furniture stand (Formal Design, ZL 2016 3 0552139.7)
2013——Image Display Device (ZL 2013 2 0237210.3)

阿科米星论文目录 Archmixing Publication List 2009–2019

阿科米星论著

Publications by Archmixing

1———庄慎．上海嘉定新城规划展示馆 [J]．时代建筑，2009 (6): 138–145．

Zhuang Shen. Urban Planning Exhibition Hall of Jiading New Town, Shanghai [J]. *Time + Architecture*, 2009 (6): 138–145.

2———华霞虹．当代中国的消费梦想和建筑狂欢 [J]．时代建筑，2010 (1): 124–128．

Hua Xiahong. Consumption Dreams and Architectural Carnival in China [J]. *Time + Architecture*, 2010 (1): 124–128.

3———华霞虹．“悬挂的庭院”——上海文化信息产业园 B4/B5 地块的设计策略 [J]．时代建筑，2011 (3): 106–113．

Hua Xiahong. Hanging Courtyard—The Design Strategy for Blocks B4/B5 of Shanghai Culture & Communication Industry District [J]. *Time + Architecture*, 2011 (3): 106–113.

4———庄慎，任皓．个人与大众之间：阿科米星建筑设计事务所办公楼 [J]．城市环境设计，2011 (4): 238–241．

Zhuang Shen, Ren Hao. Between Individual and Public: The Workplace of Atelier Archmixing [J]. *Urban Environment Design*, 2011 (4): 238–241.

5———庄慎，华霞虹．嘉定秋霞圃西侧地块五轮设计方案 [J]．时代建筑，2012 (1): 48–51．

Zhuang Shen, Hua Xiahong. 5 Proposals for the Block West to Qiuxia Garden, Jiading District [J]. *Time + Architecture*, 2012 (1): 48–51.

6———唐煜．上海嘉定新城紫气东来景观配套建筑设计概况 [J]．时代建筑，2012 (1): 58–61．

Tang Yu. Supplemental Buildings for Jiading New Town Ziqidonglai Park, Shanghai [J]. *Time + Architecture*, 2012 (1): 58–61.

7———华霞虹，庄慎．加建 [J]．新建筑，2012 (2): 66–71．

Hua Xiahong, Zhuang Shen. Addition [J]. *New architecture*, 2012 (2): 66–71.

8———庄慎，华霞虹．选择在个人与大众之间 [J]．建筑师，2012 (6): 43–51．

Zhuang Shen, Hua Xiahong. Standing Between the Individual and Public [J]. *The Architect*, 2012 (6): 43–51.

9———王方戟，庄慎．关于黎里“黎里”及陆巷“双栖斋”的对话 [J]．时代建筑，2012 (6): 77–81．

Wang Fangji, Zhuang Shen. A Conversation on LILI in LILI and Twin Trees Pavilion in Luxiang [J]. *Time + Architecture*, 2012 (6): 77–81.

10———庄慎，任皓．上海文化信息产业园 B4/B5 地块 [J]．城市环境设计，2012 (6): 56–61．

Zhuang Shen, Ren Hao. B4/B5 Block of Culture and Information Industrial Park, Shanghai [J]. *Urban Environment Design*, 2012 (6): 56–61.

11———庄慎，任皓．客厅 [J]．城市环境设计，2012 (6): 196–197．

Zhuang Shen, Ren Hao. Living Room [J]. *Urban Environment Design*, 2012 (6): 196–197.

12———田丹妮．双栖斋与黎里的建造 [J]．城市环境设计，2012 (10): 228．

Tian Danni. The Construction of Twin Trees Pavilion and LILI [J]. *Urban Environment Design*, 2012 (10): 228.

13———庄慎，任皓，唐煜，等．双栖斋 [J]．城市环境设计，2012 (10): 229．

Zhuang Shen, Ren Hao, Tang Yu, et al. Twin Trees Pavilion [J]. *Urban Environment Design*, 2012 (10): 229.

14———庄慎，任皓，唐煜，等．黎里 [J]．城市环境设计，2012 (10): 230–231．

Zhuang Shen,, Ren Hao, Tang Yu, et al. LILI[J]. *Urban Environment Design*, 2012 (10): 230–231.

15———赵磊，董功，庄慎．松紧之间：赵磊、董功、庄慎三人谈 [J]．时代建筑，2013 (4): 40–43．

Zhao Lei, Dong Gong, Zhuang Shen. Between Relaxation and Strictness: A Conversation among ZHAO Lei, DONG Gong and ZHUANG Shen [J]. *Time + Architecture*, 2013 (4): 40–43.

16———庄慎．混合设计——一种城市实践的立场与策略 [J]．城市环境设计，2013 (3+4): 104–105．

Zhuang Shen. Mixing Design: An Urban-oriented Practical Strategy [J]. *Urban Environment Design*, 2013 (3+4): 104–105.

17———庄慎，华霞虹．改变即日常：阿科米星的实践综述 [J]．建筑师，2014 (2): 131–137．

Zhuang Shen, Hua Xiahong. Change is More: Atelier Archmixing's Practice [J]. *The Architect*, 2014 (2): 131–137.

18———王方戟，庄慎．关于建筑形式的对话——一个围绕阿科米星建筑事务所最新实践的讨论 [J]．建筑师，2014 (2): 120–131．

Wang Fangji, Zhuang Shen. A Dialogue on Architectural Form: Reviewing Atelier Archmixing's Latest Designs [J]. *The Architect*, 2014 (2): 120–131.

19———唐煜．搭棚记——阿科米星的莫干山蚕种场改造 [J]．城市环境设计，2014 (4): 190–195．

Tang Yu. Build a Shed—Renovation of Moganshan Silkworm Egg Production Farm by Archmixing [J]. *Urban Environment Design*, 2014 (4): 190–195.

20———庄慎，王侃．上海衡山路 890 号 8 号楼立面改造 [J]．时代建筑，2014 (4): 128–131．

Zhuang Shen, Wang Kan. Facade Renovation for Building 8, 890 Hengshan Road, Shanghai [J]. *Time + Architecture*, 2014 (4): 128–131.

21———华霞虹，庄慎．上海嘉定新城双丁路幼儿园设计 [J]．建筑学报，2014 (1): 66–73．

Hua Xiahong, Zhuang Shen. Shuangding Road Kindergarten, Jiading New City [J]. *Architectural Journal*, 2014 (1): 66–73.

22———庄慎，华霞虹．非识别体系的一种高度：杰弗里·巴瓦的建筑世界 [J]．建筑学报，2014 (11): 27–35．

Zhuang Shen, Hua Xiahong. An Altitude of Unrecognizable System: The Architectural World of Geoffrey Bawa [J]. *Architectural Journal*, 2014 (11): 27–35.

23———庄慎，华霞虹．日常·改变·非识别体系 [J]．新建筑，2014 (06): 16–19．

Zhuang Shen, Hua Xiahong. Everyday, Change, and the Unrecognizable System [J]. *New architecture*, 2014 (06): 16–19.

24———庄慎．莫干山庾村蚕种场 [J]．世界建筑，2015 (2): 84–87+132．

Zhuang Shen. Mogan Mountain Silkworm Hatchery, Huzhou, Zhejiang, China, 2013[J]. *World Architectureure*, 2015 (2): 84–87+132.

25———庄慎，徐好好．对话庄慎：改变与反思 [J]．新建筑，2015 (3): 24–29．

Zhuang Shen, Xu Haohao. Dialogue with Architect Zhuang Shen: Changes and Reflections [J]. *New architecture*, 2015 (3): 24–29.

26———鲁安东，庄慎．被栖居的实验室——庄慎谈杰弗里·巴瓦工作室 [J], 世界建筑，2015 (4): 43–45．

Lu Andong, Zhuang Shen. A Dwelled Laboratory: Conversation with ZHUANG Shen on Geoffrey Bawa's Workspaces [J]. *World Architectureure*, 2015 (4): 43–45.

27———张斌，冯路，庄慎，等．从认识到表达 [J]．时代建筑，2015 (5): 82–87．

Zhang Bin, Feng Lu, Zhuang Shen, et al. From Awareness to Expression [J]. *Time + Architecture*, 2015 (5): 82–87.

28———庄慎，华霞虹．城市内的工作室 [J]．时代建筑，2015 (5): 100–107．

Zhuang Shen, Hua Xiahong. Work within the City [J]. *Time + Architecture*, 2015 (5): 100–107.

29———庄慎，华霞虹．空间冗余 [J]．时代建筑，2015 (5): 108–111．

Zhuang Shen, Hua Xiahong. Spatial Redundancy [J]. *Time + Architecture*, 2015 (5): 108–111.

30———庄慎，周渐佳，李丹锋．打开城市的另一种方式：关于上海当代城市研究的一次工作对谈 [J]．时代建筑，2015 (5): 112–116．

Zhuang Shen, Zhou Jianjia, Li Danfeng. An Alternative to Open a City: A Work Session on Urban Research towards Contemporary Shanghai [J]. *Time + Architecture*, 2015 (5): 112–116.

31———庄慎．象征与虚无 [J]．城市环境设计，2015 (5): 179．

Zhuang Shen. Symbolicalness and Immateriality [J]. *Urban Environment Design*, 2015 (5): 179.

32———庄慎，华霞虹．平常的开始，平常的结果——宝山陈化成纪念馆移建改造 [J]．建筑学报，2015 (12): 48–53．

Zhuang Shen, Hua Xiahong. From the Ordinary to the Ordinary: Replacement and Renovation of Baoshan Chen Huacheng Memorial [J]. *Architectural Journal*, 2015 (12): 48–53.

33———庄慎，华霞虹．应用：作为行动和认知 [J]．新建筑，2016 (2): 10–16．

Zhuang Shen, Hua Xiahong. Application as Action and Cognition [J]. *New Architecture*, 2016 (2): 10–16.

34———庄慎，唐煜，王骁彬．体现上海性格的可变肌肤——衡山坊 890 弄 8 号楼外立面改造 [J]．设计家，2016 (2): 52–54．

Zhuang Shen, Tang Yu, Wang Xiaobin. Architectural Facade Renovation, Building 8, Hengshanfang [J]. *Designer & Designing*, 2016 (2): 52–54.

35———庄慎，任皓，唐煜，等．把工作室搬来搬去好不好 [J]．设计家，2016 (3): 124–125．

Zhuang Shen, Ren Hao, Tang Yu, et al. Is It OK that the Office Move too Often [J]. *Designer & Designing*, 2016 (3): 124–125.

36———华霞虹，庄慎．改妆记 [J]．时代建筑，2016 (4): 24–28．

Hua Xiahong, Zhuang Shen. On Facade Renovation [J]. *Time + Architecture*, 2016 (4): 24–28.

37———刘宇扬，庄慎．张江华鑫慧天地 [J]．时代建筑，2016 (4): 68–73．

Liu Yuyang, Zhuang Shen. Huaxin Wisdom Mark, Zhangjiang [J]. *Time + Archi-*

tecture, 2016 (4): 68–73.

38——庄慎．YoungBird 室内空间改造．世界建筑，2016 (4): 98–103．
Zhuang Shen. YoungBird Interior Renovation [J]. *World Architecture*, 2016 (4): 98–103.

39——庄慎．陈化成纪念馆移建改造 [J]．世界建筑，2016 (5): 114–115．
Zhuang Shen. Removal Renovation of Chen Huacheng Memorial Hall, Shanghai, China, 2015 [J]. *World Architecture*, 2016 (5): 114–115.

40——阿科米星．衡山坊 8 号楼立面改造 [J]. 建筑与都市（日文版），2016 (3): 152–155.
Atelier Archmixing. Facade Renovation for Building No. 8 Hengshanfang [J]. *A+U* (Architecture and Urbanism), 2016 (3): 152–155.

41——庄慎，华霞虹．空间冗余与非识别体系 [J]．建筑师，2016 (6): 21–26．
Zhuang Shen, Hua Xiahong. Spatial Redundancy and Unrecognizable System [J]. *The Architect*, 2016 (6): 21–26.

42——庄慎．大进深户型与交错的互跃户型 [J]．时代建筑，2016 (6): 60–65．
Zhuang Shen. Deep-depth Unit and Cross-compound Unit [J]. *Time + Architecture*, 2016 (6): 60–65.

43——庄慎，华霞虹．向日常世界学习 [J]. 建筑研究季刊（英文版），2017，21(3): 222–233
Zhuang Shen, Hua Xiahong. Learning from the Everyday World [J]. *Architectural Research Quarterly*, 2017, 21 (3): 222–233.

44——庄慎．火星救援与二手宇宙 [J]．时代建筑，2017 (5): 7–11．
Zhuang Shen. Rescuing from the Mars and Used Universe [J]. *Time + Architecture*, 2017 (5): 7–11.

45——华霞虹．走向城市建筑学的可能——"虹口 1617 展览暨城市研究"研讨会评述 [J]．建筑学报，2017 (9): 103–109．
Hua Xiahong. Possibilities towards an Urban Architecture: Notes on the Seminar of 'Hongkou 1617 Exhibition and Urban Study' [J]. *Architectural Journal*, 2017 (9): 103–109.

46——庄慎．黎里黎里，江苏，中国 [J]．世界建筑，2017 (9): 98–99+129．
Zhuang Shen. LILI, Jiangsu, China, 2012 [J]. *World Architecture*, 2017 (9): 98–99+129.

47——朱捷，庄慎．敦煌莫高窟窟门的一次公益设计和改造实验 [J]．世界建筑，2017 (12): 22–25．
Zhu Jie, Zhuang Shen. A Volunteer Design and Renovation Experiment for the Gate of Mogao Grottoes in Dunhuang [J]. *World Architecture*, 2017 (12): 22–25.

48——庄慎，华霞虹．逆向还原：城市"空间冗余"的日常逻辑——对话庄慎与华霞虹．城市中国，2017 (10): 48–59．
Zhuang Shen, Hua Xiahong. Reverse Restoration: Daily Logic of Urban Space Evolution/ Interview with ZHUANG Shen and HUA Xiahong [J]. *Urban Wisdom Advancing with China*, 2017 (10): 48–59.

49——华霞虹．一扇门打开两个书店 [J]．时代建筑，2018 (1): 128–133．
Hua Xianghong. One Door Opens to Two Bookstores [J]. *Time + Architecture*, 2018 (1): 128–133.

50——庄慎．看不见的改变：论使用端空间技术化的可能 [J]．时代建筑，2018 (3): 32–35．
Zhuang Shen. Invisible Change: Technologizing Space at the User's End [J]. *Time + Architecture*, 2018 (3): 32–35.

51——庄慎．整体的回响：来自内部的反向思维 [J]．时代建筑，2018 (4): 20–23．
Zhuang Shen. The Echo of the Unity: Reverse Thinking from the Interior [J]. *Time + Architecture*, 2018 (4): 20–23.

52——庄慎，王迪，邓健．建筑结构设备一体化的"屋中屋"[J]．时代建筑，2018 (4): 96–103．
Zhuang Shen, Wang Di, Deng Jian. House within House, the Integration of Architecture, Structure and Equipment [J]. *Time + Architecture*, 2018 (4): 96–103.

53——庄慎，唐煜，华霞虹．柔性倍增——基于人工智能、移动机器人、智能立库的未来居住空间概念研究 [J]．华中建筑，2018 (3): 61–66．
Zhuang Shen, Tang Yu, Hua Xiahong. Flexible Multiplication: Spatial Concept on Future Living Integrating Technology of AI, VAG and Intellectual Elevator-shaft Garage [J]. *Huazhong Architecture*, 2018 (3): 61–66.

54——华霞虹，庄慎．日常·改变·非识别体系 [J]. 建筑设计（英文版），2018 (6): 52–57.
Hua Xiahong, Zhuang Shen. Everyday, Change and the Unrecognisable System [J]. *Architectural Design*, 2018 (6): 52–57.

55——庄慎，华霞虹．棉仓城市客厅 [J]．建筑学报，2018 (7): 42–51．
Zhuang Shen, Hua Xiahong. Cotton Lab Lounge [J]. *Architectural Journal* 2018 (7):42–51

56——陈平楠，庄慎．常州棉仓城市客厅 [J]．城市建筑，2018 (10): 78–83．
Chen Pingnan, Zhuang Shen. Changzhou Cotton Urban Lounge, China [J]. *Urbanism and Architecture*, 2018 (10): 78–83.

57——庄慎，任皓，唐煜，等．常州棉仓城市客厅 [J]．世界建筑，2019 (1): 53–57．
Zhuang Shen, Ren Hao, Tang Yu, et al. Changzhou Cotton Lab Urban Lounge, Jiangsu, China, 2018 [J]. *World Architecture*, 2019 (1): 53–57.

58——庄慎．阿科米星工作室 [J]. 建筑中国（英文版），2019 夏：116–219.
Zhuang Shen. Atelier Archmixing [J]. *Architecture China*, 2019 (summer): 116–129.

刊载阿科米星作品与思想的论著

Publications on Archmixing

59——彼得·罗，夏怡．时代降临：当代中国建筑师第三浪潮的涌现 [J]. 亚洲建筑（英文版），2013 (1): 8–13.
Peter Rowe, Har Ye Kan. Coming of Age: The Emergence of "Third Stream" Architects in Comtemporary China [J]. *Asia Architecture*, 2013 (1): 8–13.

60——上海文化信息产业园 B4/B5 地块 [M] // 支文军，戴春，徐洁．中国当代建筑 (2008–2012)．上海：同济大学出版社，2013：130–139．
B4/B5 Blocks of Culture and Information Industrial Park, Shanghai[M]// Zhi Wenjun, Dai Chun, Xu Jie. *The contemporary Chinese Architecture* (2008–2012). Shanghai: Tongji University Press, 2013: 130–139.

61——庄清湄，等．庄慎：建筑师应该是理想主义者 [J]．外滩画报，第 550 期：44–45．
Zhuang Qingmei, etc.. Zhuang Shen: The Architect Should Be An Idealist [J]. *THE BUND Magazine*, No.550: 44–45.

62——黄正骊．庄慎：混合建筑学 [J]．城市中国，第 60 期：100–107．
Huang Zhengli. A Mixed Architecture: Interview with ZHUANG Shen [J]. *Urban Wisdom Advancing with China*, Vol.60: 100–107.

63——庄慎：走向内心愈深，走向世界愈广 [J]．*Pro. Design*，2014 (2): 84–99．
Zhuang Shen: Walk Deeper in Heart, Walk Wider in World [J]. *Pro Design*, 2014 (2): 84–99.

64——蒋茵苜．庄慎：改变的日常，建筑学的另一种可能 [J]．生活，2014 (12): 64–67．
Jiang Yinmu. ZHUANG Shen, Change is More: Another Possibility for Architecture' Focuses on Atelier Archmixing's Design Philosophy [J]. *Life Magazine*, 2014 (12): 64–67.

65——《建筑七人对谈集》编委会．建筑七人对谈集 [M]. 上海：同济大学出版社，2015.
Book Editorial Board. *Dialogues between Seven Architects*[M]. Shanghai: Tongji University Press, 2015.

66——刘匪思．庄慎：建筑师需要有严肃的思考精神 [J]．室内设计师，2015 (4): 102–109．
Liu Feisi. ZHUANG Shen: Architect Needs Critical Thinking [J]. *Interior Designer*, 2015 (4): 102–109.

67——短暂的建筑——庄慎访谈录 [J]．室内设计与装修，2015 (11): 108．
An Interview with ZHUANG Shen [J]. *Interior Design and Construction*, 2015 (11): 108.

68——同济大学建筑与城市规划学院．同济八骏：中生代的建筑实践 [M] 上海：同济大学出版社，2017.
College of Architecture and Urban Planning, Tongji University. *Architectural Practice of Middle-Aged Generation from Tongji* [M] Shanghai: Tongji University Press, 2017.

69——鲁安东．棉仓城市客厅：一个内部性的宣言 [J]．建筑学报，2018 (7): 52–55．
Lu Andong. Cotton Lab Lounge: A Manifesto of Interiority [J]. *Architectural Journal*, 2018 (7): 52–55.

70——王方戟，王梓童．图纸引导体验——上海宝山贝贝佳欧莱幼儿园设计 [J]. 建筑学报，2018 (11): 90–94.
Wang Fangji, Wang Zitong. Perceptions Guided by Drawings: A Case Study of Baoshan Beibeijia Olion Kindergarten [J]. *Architectural Journal*, 2018 (11): 90–94.

阿科米星微信公众号文章

Archmixing Wechat

71——庄慎．用扩展与对称的方式设计一个小茶室 [EB/OL]．(2016–01–14) [2019–09–09]．
Zhuang Shen. Applying Expansion and Symmetrical Strategy to Design a Small Teahouse[EB/OL]. (2016–01–14)[2019–09–09]. https://mp.weixin.qq.com/s/xe1bLDeHeR9_0c5fGCbVwQ.

72——庄慎．大眼睛与长鼻子 [EB/OL]．(2016–01–18)[2019–09–09]．
Zhuang Shen. Big Eye and Long Nose[EB/OL]. (2016–01–18)[2019–09–09].
https://mp.weixin.qq.com/sqL90TdoHIRRHq3GfwBd7QA.

73——庄慎．"带着鲑鱼去旅行"[EB/OL]．(2016–06–08)[2019–09–09]．
Zhuang Shen. "Il secondo Diario Minimo"[EB/OL]. (2016–06–08)[2019–09–09].
https://mp.weixin.qq.com/s/wmhvsG_OSLAdoh-u_-izlA.

74——庄慎．两种调整 [EB/OL]．(2016–07–21)[2019–09–09]．
Zhuang Shen. Two Adjustments[EB/OL]. (2016–07–21)[2019–09–09].
https://mp.weixin.qq.com/s/oAvq_QUzlo2-NA0qDT6_bQ.

75——庄慎．做户型与抠平面 [EB/OL]．(2016–08–22)[2019–09–09]．
Zhuang Shen. Crafting Plan and Apartment Unit[EB/OL]. (2016–08–22)[2019–09–09].
https://mp.weixin.qq.com/s/5rRagwhFHpqtGaeM4Tt17g.

76——庄慎．把工作室搬来搬去好不好 [EB/OL]．(2016–11–03)[2019–09–09]．
Zhuang Shen. Is It OK to Move Office too Often[EB/OL]. (2016–11–03)[2019–09–09].
https://mp.weixin.qq.com/s/BMIDAIFDIINRCHW9FV37DA.

77——庄慎．局部 [EB/OL]．(2017–04–18)[2019–09–09]．

Zhuang Shen. Parts[EB/OL]. (2017-04-18)[2019-09-09].
https://mp.weixin.qq.com/s/NNr9MMNJNg2G8Uy6AePDeQ.

78———庄慎．内部 [EB/OL]．(2017-05-02)[2019-09-09]．
Zhuang Shen. Interior[EB/OL]. (2017-05-02)[2019-09-09].
https://mp.weixin.qq.com/s/eHtdyylOpssSkCh0dgbsAA.

79———庄慎．有关冗余的笔记 1：poché [EB/OL]．(2017-05-18)[2019-09-09]．
Zhuang Shen. Note on Redundancy1: poché[EB/OL]. (2017-05-18)[2019-09-09].
https://mp.weixin.qq.com/s/5YiAkN5iCVrx-wdxt6Z CaA.

80———庄慎．有关冗余的笔记 2：老白的实验室 [EB/OL]．(2017-07-31)[2019-09-09]．
Zhuang Shen. Note on Redundancy 2: Water White's Laboratory [EB/OL]. (2017-07-31)[2019-09-09].
https://mp.weixin.qq.com/s/exgBZZl4jMpQIj2PiHPELQ.

81———游诗雨．志达书店改造记录 [EB/OL]．(2018-05-10)[2019-09-09]．
You Shiyu. Renovation Record on Zhida Bookstore[EB/OL]. (2018-05-10)[2019-09-09].
https://mp.weixin.qq.com/s/Mx7loNFRpSK5NieLz6RzFg.

82———张灏宸．“舞台建筑”两则 [EB/OL]．(2018-03-21)[2019-09-09]．
Zhang Haochen. Two Articles on "Stage Architecture"[EB/OL]. (2018-03-21)[2019-09-09].
https://mp.weixin.qq.com/s/vkbe0NS5uVrXswvEiWmnoA.

83———张灏宸．欧莱幼儿园改造记 [EB/OL]．(2018-11-29)[2019-09-09]．
Zhang Haochen. Renovation Record on Olion Kindergarten[EB/OL]. (2018-11-29)[2019-09-09].
https://mp.weixin.qq.com/s/cdOr49m5mqeHwTBCNxedEg.

84———庄慎，张灏宸．实用是一种专业常识——阿科米星的两个幼儿园设计 [EB/OL]．(2019-01-10)
Zhuang Shen, Zhang Haochen. Be Practical is a Professional Knowledge: Two Kindergarten Designs by Atelier Archmixing[EB/OL]. (2019-01-10)
https://mp.weixin.qq.com/s?__biz=MzA3NjUxNTQxOQ==&mid=2703430828&idx=1&sn=fe882135f3778996fcba73d432f36e94&chksm=bbb7a5988cc02c8ee52aba6bf8ae869469f787bed8d0141653e28d071f22788e9942df5eec89&token=454719635&lang=zh_CN#rd

阿科米星作品目录 Archmixing Project List 2009–2019

1 **昆明文明街串联庭院设计**
类型：旧城更新
地点：云南 昆明
时间：2006
规模：189 000 m²
状态：方案设计
合作设计：大舍建筑设计事务所 *

Promenade Courtyards, Renewal of Kunming Wenming Street
Type: Old City Renewal
Location: Kunming, Yunnan Province
Date: 2006
Size: 189,000 m²
Status: Schematic Design
Collaborator: Atelier Deshaus*

2 **嘉定新城规划展示馆**
类型：办公、展示
地点：上海 嘉定
时间：2007–2009
规模：6 700 m²
状态：建成
合作设计：大舍建筑设计事务所 *

Urban Planning Exhibition Hall of Jiading New Town
Type: Office & Exhibition
Location: Jiading District, Shanghai
Date: 2007–2009
Size: 6,700 m²
Status: Completed
Collaborator: Atelier Deshaus*

3 **上海文化信息产业园一期 B4/B5 地块**
类型：办公
地点：上海 嘉定
时间：2008–2010
规模：24 900 m²
状态：建成
合作设计：大舍建筑设计事务所 *

B4/B5 Blocks of Shanghai Culture & Information Industrial Park, Phase I
Type: Office
Location: Jiading District, Shanghai
Date: 2008–2010
Size: 24,900 m²
Status: Completed
Collaborator: Atelier Deshaus*

4 **嘉定博物馆新馆**
类型：文化
地点：上海 嘉定
时间：2008–2011
规模：9 600 m²
状态：建成
合作设计：大舍建筑设计事务所 *

Jiading New Mesuem
Type: Cultural
Location: Jiading District, Shanghai
Date: 2008–2011
Size: 9,600 m²
Status: Completed
Collaborator: Atelier Deshaus *

5 **一院一世界**
类型：商业
地点：四川 成都
时间：2010–2012
规模：3 400 m²
状态：建成

A Court, A World
Type: Commercial
Location: Chengdu, Sichuan Province
Date: 2010–2012
Size: 3,400 m²
Status: Completed

6 **嘉定新城双丁路公立幼儿园**
类型：教育
地点：上海 嘉定
时间：2010–2013
规模：6 100 m²
状态：建成

Shuangding Road Public Kindergarten
Type: Educational
Location: Jiading District, Shanghai
Date: 2010–2013
Size: 6,100 m²
Status: Completed

7 **诸暨剧院**
类型：文化
地点：浙江 诸暨
时间：2011–2017
规模：32 000 m²
状态：建成
合作设计：同济大学建筑设计研究院（集团）有限公司

Zhuji New Theatre
Type: Cultural
Location: Zhuji, Zhejiang Province
Date: 2011–2017
Size: 32,000 m²
Status: Completed
Collaborator: TJAD Group

8 **SVA 综合办公楼**
类型：建筑改造
地点：上海 徐汇
时间：2011–2014
规模：8 539 m²
状态：建成

SVA Office Complex
Type: Building Renovation
Location: Xuhui District, Shanghai
Date: 2011–2014
Size: 8,539 m²
Status: Completed

9 **诸暨规划展示馆和科技馆**
类型：文化
地点：浙江 诸暨
时间：2011–2016
规模：46 000 m²
状态：建成
合作设计：同济大学建筑设计研究院（集团）有限公司

Zhuji Urban Planning Exhibition Center & Science and Technology Museum
Type: Cultural
Location: Zhuji, Zhejiang Province
Date: 2011–2016
Size: 46,000 m²
Status: Completed
Collaborator: TJAD Group

10 **黎里**
类型：建筑改造
地点：江苏 黎里
时间：2011–2012
规模：200 m²
状态：建成

LILI
Type: Building Renovation
Location: Lili, Jiangsu Province
Date: 2011–2012
Size: 200 m²
Status: Completed

11 **富春俱舍书院**
类型：文化
地点：浙江 杭州
时间：2011–2013
规模：300 m²
状态：建成

Fuchun Kosa Academy
Type: Cultural
Location: Hangzhou, Zhejiang Province
Date: 2011–2013
Size: 300 m²
Status: Completed

12 **富春俱舍走马楼**
类型：建筑改造
地点：浙江 杭州
时间：2011–2014
规模：464 m²
状态：建成

Fuchun Kosa Zoumalou
Type: Renovation
Location: Hangzhou, Zhejiang Province
Date: 2011–2014
Size: 464 m²
Status: Completed

13 **昆山文化艺术中心景观配套建筑**
类型：服务配套
地点：江苏 昆山
时间：2011–2013
规模：1 261 m²
状态：建成

Kunshan Culture and Art Center Landscape Supporting Buildings
Type: Service Support
Location: Kunshan, Jiangsu Province
Date: 2011–2013
Size: 1,261 m²
Status: Completed

14 **中国福利会嘉定新城幼儿园新建工程**
类型：教育
地点：上海 嘉定
时间：2011–2015
规模：13 073 m²
状态：建成

CWI Kindergarten in Jiading New City
Type: Educational
Location: Jiading District, Shanghai
Date: 2011–2015
Size: 13,073 m²
Status: Completed

15 **张江集电港三期 华鑫慧天地**
类型：办公
地点：上海 浦东
时间：2011–2016
规模：128 760 m²
状态：建成
合作设计：刘宇扬建筑事务所

Zhangjiang IC Harbor Phase III China Fortune Wisdom Mark
Type: Office
Location: Pudong District, Shanghai
Date: 2011–2016
Size: 128,760 m²
Status: Completed
Collaborator: Atelier Liu Yuyang Architects

16 张江集电港一期 华鑫智天地
类型：办公
地点：上海 浦东
时间：2012–2017
规模：54 700 m²
状态：建成
合作设计：刘宇扬建筑事务所

Zhangjiang IC Harbor Phase I
China Fortune Smart Mark
Type: Office
Location: Pudong District, Shanghai
Date: 2012–2017
Size: 54,700 m²
Status: Completed
Collaborator: Atelier Liu Yuyang Architects

17 双栖斋
类型：建筑改造
地点：江苏 苏州
时间：2012
规模：56 m²
状态：建成

Twin Trees Pavilion
Type: Building Renovation
Location: Suzhou, Jiangsu Province
Date: 2012
Size: 56 m²
Status: Completed

18 嘉定区国际汽车城核心区 05B-01 地块初级中学新建工程（安亭中学）
类型：教育
地点：上海 嘉定
时间：2012–2017
规模：15 000 m²
状态：建成

05B-01 Block Middle School,
Jiading International Automobile City
Type: Educational
Location: Jiading District, Shanghai
Date: 2012–2017
Size: 15,000 m²
Status: Completed

19 上下居——芦墟运河边上的院子
类型：建筑改造
地点：江苏 芦墟
时间：2012
规模：50 m²
状态：建成

The Ladder House – Courtyard by the Luxu Canal
Type: Building Renovation
Location: Luxu, Jiangsu Province
Date: 2012
Size: 50 m²
Status: Completed

20 衡山路 890 弄（衡山和集）8 号楼外立面改造
类型：立面改造
地点：上海 徐汇
时间：2012–2014
规模：230 m²
状态：建成

Facade Renovation for No.8 Building,
Lane 890, Hengshan Road
Type: Facade Renovation
Location: Xuhui District, Shanghai
Date: 2012–2014
Size: 230 m²
Status: Completed

21 斜土社区活动中心立面改造
类型：立面改造
地点：上海 徐汇
时间：2013–2014
规模：6 750 m²
状态：建成

Facade Renovation for Xietu Community Center
Type: Facade Renovation
Location: Xuhui District, Shanghai
Date: 2013–2014
Size: 6,750 m²
Status: Completed

22 宝山区奇石陈列馆
类型：文化
地点：上海 宝山
时间：2012–2016
规模：3 600 m²
状态：施工中

Stone Gallery in Baoshan District
Type: Cultural
Location: Baoshan District, Shanghai
Date: 2012–2016
Size: 3,600 m²
Status: Under Construction

23 安龙森林公园东部码头小镇商业建筑
类型：商业
地点：浙江 杭州
时间：2013
规模：2 900 m²
状态：方案设计

Commercial Building at East Dock Town
in Anlong Forest Park
Type: Commercial
Location: Hangzhou, Zhejiang Province
Date: 2013
Size: 2,900 m²
Status: Schematic Design

24 莫干山庾村文化市集蚕种场改造
类型：建筑改造
地点：浙江 湖州
时间：2012–2013
规模：3 000 m²
状态：建成

Silkworm Hatchery Renovation, Yucun Culture Market, Mogan Mountain
Type: Building Renovation
Location: Huzhou, Zhejiang Province
Date: 2012–2013
Size: 3,000 m²
Status: Completed

25 徐汇区龙华街道敬老院立面改造
类型：立面改造
地点：上海 徐汇
时间：2013–2018
规模：10 500 m²
状态：建成

Facade Renovation for Longhua Street Elder Care Center
Type: Facade Renovation
Location: Xuhui District, Shanghai
Date: 2013–2018
Size: 10,500 m²
Status: Completed

26 宝山陈化成纪念馆移建改造
类型：建筑改造
地点：上海 宝山
时间：2014–2015
规模：198 m²
状态：建成

Removal Renovation of Chen Huacheng Memorial
Type: Building Renovation
Location: Baoshan District, Shanghai
Date: 2014–2015
Size: 198 m²
Status: Completed

27 舟山东门车站及周边改造项目一期工程
类型：交通、商业
地点：浙江 舟山
时间：2013–2018
规模：70 329 m²
状态：建成

Renovation of Zhoushan East Gate Station and Surroundings, Phase I
Type: Transportation & Commercial
Location: Zhoushan, Zhejiang Province
Date: 2013–2018
Size: 70 329 m²
Status: Completed

28 陕西南路复兴中路沿街立面改造
类型：立面改造
地点：上海 徐汇
时间：2013
规模：3 800 m²
状态：方案设计

Facade Renovation at the intersection of Shaanxi S. Rd and Fuxing M. Rd
Type: Facade Renovation
Location: Xuhui District, Shanghai
Date: 2013
Size: 3,800 m²
Status: Schematic Design

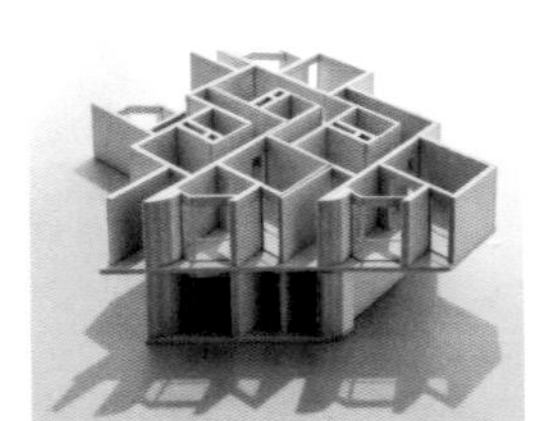

29 南京下关区永宁街地块
类型：商业、办公、住宅
地点：江苏 南京
时间：2013
规模：370 000 m²
状态：方案设计

Yongning Street Plot, Xiaguan District, Nanjing
Type: Commercial & Office & Residence
Location: Jiangsu, Nanjing Province
Date: 2013
Size: 370,000 m²
Status: Schematic Design

30 敦煌莫高窟窟门优化设计
类型：产品设计
地点：甘肃敦煌
时间：2013–2017
状态：建成

Design and Renovation for the Gate of Mogao Grottoes in Dunhuang
Type: Product Design
Location: Dunhuang, Gansu Province
Date: 2013–2017
Status: Completed

31 沪宁高速公路昆山高新区互通项目
类型：交通
地点：江苏 昆山
时间：2014–2018
规模：4 200 m²
状态：建成

Shanghai-Nanjing Expressway Toll Station in Kunshan New & Hi-tech Industrial Development Zone
Type: Transportation
Location: Kunshan, Jiangsu Province
Date: 2014–2018
Size: 4,200 m²
Status: Completed

32 前滩休闲公园 4、5 号建筑

类型：服务配套
地点：上海 浦东
时间：2014
规模：900 m^2
状态：方案设计

No.4 & No.5 Buildings in Qiantan Park

Type: Service Support
Location: Pudong District, Shanghai
Date: 2014
Size: 900 m^2
Status: Schematic Design

33 港城广场

类型：酒店、办公、展览、住宅
地点：上海 浦东
时间：2014–
规模：554 000 m^2
状态：施工中
合作设计：致正建筑工作室

Harbour City Plaza

Type: Hotel & Office & Exhibition & Residence
Location: Pudong District, Shanghai
Date: 2014–
Size: 554,000 m^2
Status: Under Construction
Collaborator: Atelier Z+

34 港城广场展示中心

类型：展示、接待
地点：上海 浦东
时间：2015
规模：700 m^2
状态：建成

Exhibition Center of Harbour City Plaza

Type: Exhibition & Reception
Location: Pudong District, Shanghai
Date: 2015
Size: 700 m^2
Status: Completed

35 彰吴村制衣厂改造

类型：建筑改造
地点：浙江 湖州
时间：2014
规模：2 900 m^2
状态：方案设计

Renovation of Garment Factory in Zhangwu Village

Type: Building Renovation
Location: Huzhou, Zhejiang Provinvce
Date: 2014
Size: 2,900 m^2
Status: Schematic Design

36 桦墅乡村工作室

类型：建筑改造
地点：江苏 南京
时间：2014–2015
规模：252 m^2
状态：建成

Huashu Rural Studio

Type: Building Renovation
Location: Nanjing, Jiangsu Provinvce
Date: 2014–2015
Size: 252 m^2
Status: Completed

37 YoungBird 室内空间改造

类型：室内改造
地点：上海 杨浦
时间：2015
规模：236 m^2
状态：建成

Interior Renovation of YoungBird Office

Type: Interior Renovation
Location: Yangpu District, Shanghai
Date: 2015
Size: 236 m^2
Status: Completed

38 上海嘉北郊野公园北游客中心

类型：服务配套
地点：上海 嘉定
时间：2015–2017
规模：1 598 m^2
状态：建成

North Visitor Center of Shanghai Jiabei Country Park

Type: Service Support
Location: Jiading District, Shanghai
Date: 2015–2017
Size: 1,598 m^2
Status: Completed

39 爱运动的小伙伴

类型：艺术装置
地点：上海 徐汇
时间：2015–2016
状态：建成

Sports Lovers

Type: Art Installation
Location: Xuhui District, Shanghai
Date: 2015–2016
Status: Completed

40 新天地临时读书空间
类型：临时装置
地点：上海 黄浦
时间：2016
规模：143 m^2
状态：建成

Temporary Reading Pavilion in Xintiandi
Type: Temporary Installation
Location: Huangpu District, Shanghai
Date: 2016
Size: 143 m^2
Status: Completed

41 田林路 192 号装饰工程
类型：室内、立面改造
地点：上海 徐汇
时间：2016–2017
规模：2 895 m^2
状态：建成

Renovation of No.192 on Tianlin Road
Type: Interior & Facade Renovation
Location: Xuhui District, Shanghai
Date: 2016–2017
Size: 2,895 m^2
Status: Completed

42 三岔港森林公园单体建筑
类型：服务配套
地点：上海 浦东
时间：2016
规模：1 150 m^2
状态：方案设计

Entrance Building of Sanchagang Forest Park
Type: Service Support
Location: Pudong District, Shanghai
Date: 2016
Size: 1,150 m^2
Status: Schematic Design

43 徐汇养老基地项目
类型：医疗养老
地点：上海 徐汇
时间：2016
规模：43 500 m^2
状态：初步设计

Xuhui Elder Care Center
Type: Healthcare
Location: Xuhui District, Shanghai
Date: 2016
Size: 43,500 m^2
Status: Development Design

44 成都麓湖生态城 D6D7 地块小学幼儿园
类型：教育
地点：四川 成都
时间：2016
规模：31 500 m^2
状态：方案设计
合作设计：致正建筑工作室

Primary School & Kindergarten on Plots D6D7 in Chengdu Luhu Ecological City
Type: Educational
Location: Chengdu, Sichuan Province
Date: 2016
Size: 31,500 m^2
Status: Schematic Design
Collaborator: Atelier Z+

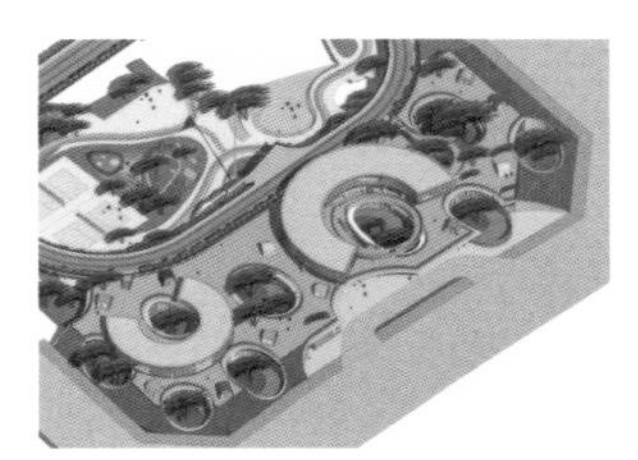

45 申威达厂房改造
类型：建筑改造
地点：上海 徐汇
时间：2016–2019
规模：11 721 m^2
状态：建成

Renovation of Shenweida Workshop Space
Type: Building Renovation
Location: Xuhui District, Shanghai
Date: 2016–2019
Size: 11,721 m^2
Status: Completed

46 东岸云桥：三林北港桥及三林塘港桥
类型：基础设施
地点：上海 浦东
时间：2016–2017
状态：建成
合作设计：冶是建筑工作室

East Bund Footbridge: North Sanlin Bridge and Sanlintang Bridge
Type: Infrastructure
Location: Pudong District, Shanghai
Date: 2016–2017
Status: Completed
Collaborator: YeArch Studio

47 悦阅书店
类型：室内改造
地点：上海 杨浦
时间：2016–2017
规模：94 m^2
状态：建成

Yueyue Bookstore
Type: Interior Renovation
Location: Yangpu District, Shanghai
Date: 2016–2017
Size: 94 m^2
Status: Completed

48 成都麓湖水镇民宿岛 - 酒店 B
类型：酒店
地点：四川 成都
时间：2016
规模：3 377 m^2
状态：初步设计

Homestay Island Hotel B, Luhu, Chengdu
Type: Hotel
Location: Chengdu, Sichuan Province
Date: 2016
Size: 3,377 m^2
Status: Development Design

49 成都麓湖水镇民宿岛 - 酒店 F
类型：酒店
地点：四川 成都
时间：2016
规模：3 050 m^2
状态：初步设计

Homestay Island Hotel F, Luhu, Chengdu
Type: Hotel
Location: Chengdu, Sichuan Province
Date: 2016
Size: 3,050 m^2
Status: Development Design

50 舟山青龙山公园入口改造
类型：文化
地点：浙江 舟山
时间：2016–2019
规模：1 080 m^2
状态：建成

Qinglong Mountain Park Entrance Renovation
Type: Cultural
Location: Zhoushan, Zhejiang Province
Date: 2016–2019
Size: 1,080 m^2
Status: Completed

51 棉仓城市客厅
类型：建筑改造
地点：江苏 常州
时间：2017–2018
规模：6 300 m^2
状态：建成

Cotton Lab Urban Lounge
Type: Building Renovation
Location: Changzhou, Jiangsu Province
Date: 2017–2018
Size: 6,300 m^2
Status: Completed

52 宝山贝贝佳欧莱幼儿园
类型：建筑改造
地点：上海 宝山
时间：2017
规模：5 400 m^2
状态：建成

Baoshan Beibeijia Olion Kindergarten
Type: Building Renovation
Location: Baoshan District, Shanghai
Date: 2017
Size: 5,400 m^2
Status: Completed

53 上海南汇新城中法学院
类型：教育
地点：上海 浦东
时间：2017–
规模：29 000 m^2
状态：施工中

Sino-French Institute, Nanhui City, Shanghai
Type: Educational
Location: Pudong District, Shanghai
Date: 2017–
Size: 29,000 m^2
Status: Under Construction

54 永嘉路口袋广场
类型：城市更新
地点：上海 徐汇
时间：2017–2019
规模：767 m^2
状态：建成

Pocket Plaza, Yongjia Road
Type: Urban Renewal
Location: Xuhui District, Shanghai
Date: 2017–2019
Size: 767 m^2
Status: Completed

55 上海棋院实验小学
类型：教育
地点：上海 静安
时间：2017–
规模：20 095 m^2
状态：施工中

Shanghai Chess Experimental Primary School
Type: Educational
Location: Jing'an District, Shanghai
Date: 2017–
Size: 20,095 m^2
Status: Under Construction

56 徐汇区南部职业学校
类型：教育
地点：上海 徐汇
时间：2017–
规模：92 387 m²
状态：初步设计

Vocational School in the South of Xuhui District
Type: Education
Location: Xuhui District, Shanghai
Date: 2017–
Size: 92,387 m²
Status: Development Design

57 复兴西路街道邻里汇
类型：公共建筑
地点：上海 徐汇
时间：2017
规模：341 m²
状态：方案设计

Community Centre on W. Fuxing Road
Type: Public Building
Location: Xuhui District, Shanghai
Date: 2017
Size: 341 m²
Status: Schematic Design

58 徐汇三江路售楼处改造
类型：建筑改造
地点：上海 徐汇
时间：2017
规模：1 120 m²
状态：方案设计

Renovation of Sales Office in Sanjiang Road
Type: Building Renovation
Location: Xuhui District, Shanghai
Date: 2017
Size: 1,120 m²
Status: Schematic Design

59 智慧欧莱托育园
类型：建筑改造
地点：上海 宝山
时间：2017–2018
规模：1 467 m²
状态：建成

Wisdom Olion Nursery
Type: Building Renovation
Location: Baoshan District, Shanghai
Date: 2017–2018
Size: 1,467 m²
Status: Completed

60 “一条”线下店（莘庄店）
类型：室内改造
地点：上海 闵行
时间：2018
规模：685 m²
状态：建成
合作设计：蘑菇云设计工作室

Offline Retail Space of Yitiao. TV (Xinzhuang Store)
Type: Interior Renovation
Location: Minhang District, Shanghai
Date: 2018
Size: 685 m²
Status: Completed
Collaborator: Supercloud

61 阿那亚金山岭艺术中心
类型：商业
地点：河北 承德
时间：2018–
规模：875 m²
状态：施工中

Art Center of Aranya, Jinshanling
Type: Commerical
Location: Chengde, Hebei Province
Date: 2018–
Size: 875 m²
Status: Under Construction

62 普陀白沙岛沙头交通码头改扩建
类型：交通
地点：浙江 舟山
时间：2018–
规模：285 m²
状态：施工中

Renovation of Shatou Wharf, Baisha Island, Putuo
Type: Transportation
Location: Zhoushan, Zhejiang Province
Date: 2018–
Size: 285 m²
Status: Under Construction

63 石龙路临时公寓
类型：住宅
地点：上海 徐汇
时间：2018–
规模：17 000 m²
状态：施工中

Temporary Apartments in Shilong Road
Type: Residence
Location: Xuhui District, Shanghai
Date: 2018–
Size: 17,000 m²
Status: Under Construction

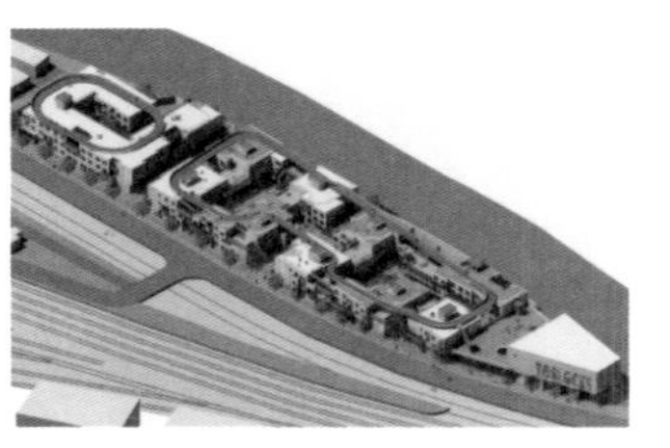

64 宝山绿心幼儿园
类型：教育
地点：上海 宝山
时间：2018–
规模：6 923 m^2
状态：施工中

Greenheart Kindergarten in Baoshan District
Type: Educational
Location: Baoshan District, Shanghai
Date: 2018–
Size: 6,923 m^2
Status: Under Construction

65 昆山实验小学西侧地下车库改扩建
类型：教育、基础设施
地点：江苏 昆山
时间：2018–
规模：36 600 m^2
状态：施工图设计

Expansion and Renovation of Kunshan Experimental Primary School
Type: Education & Infrastructure
Location: Kunshan, Jiangsu Province
Date: 2018–
Size: 36,600 m^2
Status: Construction Design

66 昆山高新区阳科园学校
类型：教育
地点：江苏 昆山
时间：2017–
规模：77 188 m^2
状态：初步设计

Yangkeyuan School in Kunshan New & Hi-tech Industrial Development Zone
Type: Educational
Location: Kunshan, Jiangsu Province
Date: 2017–
Size: 77,188 m^2
Status: Development Design

67 虹桥好望角商业广场
类型：立面改造
地点：上海 青浦
时间：2019–
规模：12 359 m^2
状态：施工中

Hongqiao Cape Commercial Plaza
Type: Facade Renovation
Location: Qingpu District, Shanghai
Date: 2019–
Size: 12,359 m^2
Status: Under Construction

68 贝尔厂房改造
类型：立面改造
地点：上海 静安
时间：2019–
规模：15 188 m^2
状态：施工中

Renovation of Bell Factory
Type: Facade Renovation
Location: Jing'an District, Shanghai
Date: 2019–
Size: 15,188 m^2
Status: Under Construction

69 天等路项目装饰工程
类型：建筑改造
地点：上海 徐汇
时间：2019–
规模：11 252 m^2
状态：施工中

Tiandeng Road Decoration Project
Type: Building Renovation
Location: Xuhui District, Shanghai
Date: 2019–
Size: 11,252 m^2
Status: Under Construction

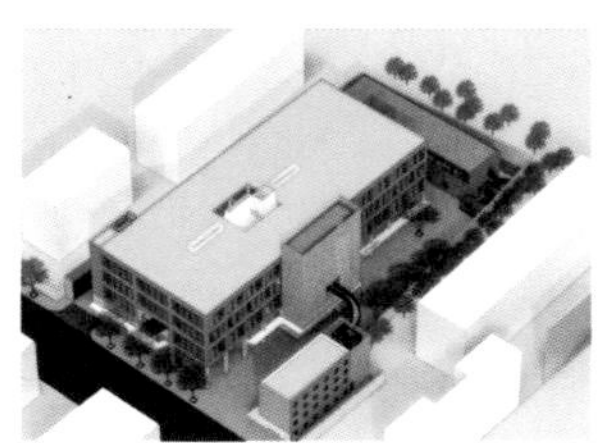

* 庄慎主创

* Created by Zhuang Shen